Classical Artinian Rings and Related Topics

$$R = \begin{pmatrix} Q & \cdots & Q \\ & \cdots & \\ Q & \cdots & Q \end{pmatrix}_{\sigma,c,n}$$

$$0 \longrightarrow X \xrightarrow{\ i\ } B = B_1 \oplus B_2$$
$$f \downarrow \qquad\quad h_1 \downarrow \quad \uparrow h_2$$
$$A \qquad\quad = A_1 \oplus A_2$$
$$\uparrow$$
$$0$$

$$0$$
$$\uparrow$$
$$A \qquad\quad = A_1 \oplus A_2$$
$$f \downarrow \qquad\quad h_1 \downarrow \quad \uparrow h_2$$
$$0 \longleftarrow X \xleftarrow{\ g\ } B = B_1 \oplus B_2$$

$$\begin{pmatrix} Q & Q\alpha_{12} & Q\alpha_{13} & Q\alpha_{14} \\ Q\beta_{21} & Q & Qc^2\alpha_{23} & Qc\alpha_{24} \\ Qc\beta_{31} & Qc\beta_{32} & Q & Q\alpha_{34} \\ Qc^2\beta_{41} & Qc\beta_{42} & Q\beta_{43} & Q \end{pmatrix} \Big/ \begin{pmatrix} 0 & 0 & Qc^2\alpha_{13} & Qc^2\alpha_{14} \\ 0 & 0 & Qc^3\alpha_{23} & Qc^2\alpha_{24} \\ Qc^2\beta_{31} & Qc^2\beta_{32} & 0 & 0 \\ Qc^3\beta_{41} & Qc^2\beta_{42} & 0 & 0 \end{pmatrix}$$

$$\begin{pmatrix} Q^* & Qc^2 & Qc^2 & Qc\alpha_{12} \\ Qc^2 & Q^* & Qc^2 & Qc\alpha_{12} \\ Qc^2 & Qc^2 & Q & Q\alpha_{12} \\ Qc\beta_{21} & Qc\beta_{21} & Q\beta_{21} & Q \end{pmatrix} \Big/ \begin{pmatrix} 0 & Qc^3 & Qc^3 & Qc^2\alpha_{12} \\ Qc^3 & 0 & Qc^3 & Qc^2\alpha_{12} \\ Qc^3 & Qc^3 & 0 & 0 \\ Qc^2\beta_{21} & Qc^2\beta_{21} & 0 & 0 \end{pmatrix}$$

Classical Artinian Rings and Related Topics

$$R = \begin{pmatrix} Q & \cdots & Q \\ \vdots & \ddots & \vdots \\ Q & \cdots & Q \end{pmatrix}_{\sigma,c,n}$$

$$
\begin{array}{ccc}
0 \longrightarrow X & \overset{i}{\longrightarrow} & B = B_1 \oplus B_2 \\
f \downarrow & h_1 \downarrow \quad \uparrow h_2 \\
A & = A_1 \oplus A_2 \\
& \uparrow \\
& 0
\end{array}
$$

$$
\begin{array}{ccc}
0 \\
\uparrow \\
A & = A_1 \oplus A_2 \\
f \downarrow & h_1 \downarrow \quad \uparrow h_2 \\
0 \longleftarrow X & \overset{g}{\longleftarrow} & B = B_1 \oplus B_2
\end{array}
$$

$$
\begin{pmatrix}
Q & Q\alpha_{12} & Qc\alpha_{13} & Qc\alpha_{14} \\
Q\beta_{21} & Q & Qc^2\alpha_{23} & Qc\alpha_{24} \\
Qc\beta_{31} & Qc\beta_{32} & Q & Q\alpha_{34} \\
Qc^2\beta_{41} & Qc\beta_{42} & Q\beta_{43} & Q
\end{pmatrix}
\Big/
\begin{pmatrix}
0 & 0 & Qc^2\alpha_{13} & Qc^2\alpha_{14} \\
0 & 0 & Qc^3\alpha_{23} & Qc^2\alpha_{24} \\
Qc^2\beta_{31} & Qc^2\beta_{32} & 0 & 0 \\
Qc^3\beta_{41} & Qc^2\beta_{42} & 0 & 0
\end{pmatrix}
$$

$$
\begin{pmatrix}
Q^* & Qc^2 & Qc^2 & Qc\alpha_{12} \\
Qc^2 & Q^* & Qc^2 & Qc\alpha_{12} \\
Qc^2 & Qc^2 & Q & Q\alpha_{12} \\
Qc\beta_{21} & Qc\beta_{21} & Q\beta_{21} & Q
\end{pmatrix}
\Big/
\begin{pmatrix}
0 & Qc^3 & Qc^3 & Qc^2\alpha_{12} \\
Qc^3 & 0 & Qc^3 & Qc^2\alpha_{12} \\
Qc^3 & Qc^3 & 0 & 0 \\
Qc^2\beta_{21} & Qc^2\beta_{21} & 0 & 0
\end{pmatrix}
$$

Yoshitomo Baba
Osaka Kyoiku University, Japan

Kiyoichi Oshiro
Yamaguchi University, Japan

World Scientific

NEW JERSEY · LONDON · SINGAPORE · BEIJING · SHANGHAI · HONG KONG · TAIPEI · CHENNAI

Published by

World Scientific Publishing Co. Pte. Ltd.

5 Toh Tuck Link, Singapore 596224

USA office: 27 Warren Street, Suite 401-402, Hackensack, NJ 07601

UK office: 57 Shelton Street, Covent Garden, London WC2H 9HE

Library of Congress Cataloging-in-Publication Data
Baba, Yoshitomo.
  Classical artinian rings and related topics / Yoshitomo Baba & Kiyoichi Oshiro.
    p. cm.
  Includes bibliographical references and index.
  ISBN-13: 978-981-4287-24-1 (hardcover : alk. paper)
  ISBN-10: 981-4287-24-5 (hardcover : alk. paper)
  1. Artin rings. 2. Commutative rings. 3. Rings (Algebra). I. Oshiro, Kiyoichi, 1943–
II. Title.
  QA251.3.B325 2009
  512'.4--dc22

                                                                          2009018193

British Library Cataloguing-in-Publication Data
A catalogue record for this book is available from the British Library.

Printed in Singapore.

Dedicated to Professor Manabu Harada

Preface

One century or more has passed since the study of noncommutative ring theory began with the pioneering work of its forefathers Wedderburn, Frobenius, Noether, Artin, von Neumann and many others. Since its birth, the theory has developed extensively, both widely and deeply. A survey of its history reveals the following key topics: the representation of finite groups, Frobenius algebras, quasi-Frobenius rings, von Neumann regular rings, homological algebra, category theory, Morita equivalence and duality, the Krull-Remak-Schmidt-Azumaya Theorem, classical quotient rings, maximal quotient rings, torsion theory, the representation of quivers, etc. In spite of its wide range, ring theory is roughly divided into four branches; namely noetherian rings, artinian rings, von Neumann regular rings and module theory with Wedderburn-Artin's structure theorem on semisimple rings located at the crossroads.

One of roots of the theory of artinian rings is the historical article [51] of Frobenius in 1903. In this paper, Frobenius studied finite dimensional algebras over a field whose left right regular representations are equivalent. In the late 1930s and early 1940s, these "Frobenius algebras" were reviewed and studied in Brauer-Nesbitt [24], Nesbitt [138], Nakayama-Nesbitt [136] and Nakayama [132], [133]. In particular, Nakayama developed these algebras by giving ring-theoretic characterizations and introduced Frobenius rings and quasi-Frobenius rings (QF-rings) in the eliminated forms from the dependency of the operation of the field. Thereafter QF-rings were continuously studied and became one of the central subjects in the study of ring theory.

Let us briefly retrace the progress of QF-rings. By the method of homological algebra, QF-rings have been studied by many peaple. In 1951,

Ikeda [82] (cf. [83]) characterized *QF*-rings as the self-injective artinian rings, and in 1958, Morita published his famous paper [122] (cf. [123]), in which a *QF*-ring *R* was characterized as an artinian ring with the property that $\mathrm{Hom}_R(-, R)$ defines a duality between the category of finitely generated left *R*-modules and that of finitely generated right *R*-modules. Prior to these works, in 1948, Thrall generalized *QF*-rings by introducing *QF-1* rings, *QF-2* rings and *QF-3* rings in [174] (cf. Nesbitt-Thrall [139]). *QF*-rings are important in the study of representations of finite groups since each group algebra of a finite group over a field is *QF*. We emphasis that *QF*-rings are artinian rings which have an abundance of properties and a deep structure, but there are still several outstanding unsolved problems on these rings such as the Nakayama conjecture and the Faith conjecture, to name just a few.

In the same [133], Nakayama also introduced another artinian rings which are called generalized uniserial rings. These rings are just artinian serial rings, and are also called Nakayama rings in honor of Nakayama and in our book we use this latter terminology.

Comparing Nakayama rings with *QF*-rings, we are able to get a deep insight into the structure of Nakayama rings. In [134] (cf. Eisenbud-Griffith [43]), Nakayama showed that every module over a Nakayama ring can be expressed as a direct sum of uniserial modules. This fact shows that Nakayama rings are the most typical rings of finite representation type.

Nakayama rings appear in the study of non-commutative noetherian rings. Indeed, the Eisenbud-Griffith-Robson Theorem ([43], [44]) states that every proper factor ring of a hereditary noetherian prime ring is a Nakayama ring, therefore this classical theorem turns the spotlight on Nakayama rings. Kupisch made a major contribution in the study of Nakayama rings. He introduced what is now called the Kupisch series and, at the same time, was the first to discover skew matrix rings. As we will see in this book, skew matrix rings over local Nakayama rings are the essence of Nakayama rings.

In the current of the study of *QF*-rings, there are two features: one is to give characterizations of these rings and the other is to generalize these rings. However in this book we study these rings from different viewpoints by exhibiting several topics.

In the early 1980s, Harada found two new classes of artinian rings which contain *QF*-rings and Nakayama rings. These rings are introduced as mutually dual notions. Oshiro [147] studied these new ring classes, naming them Harada rings (*H*-rings) and co-Harada rings (*co-H-rings*) and showed the unexpected result that left H-rings and right co-H-rings coincide.

The main objective of this book is to present the structure of left H-rings and to apply this to artinian rings, including QF-rings and Nakayama rings, giving a new perspective on these classical artinian rings. In particular, several fundamental results on QF-rings and Nakayama rings are described, and a classification on Nakayama rings is exhibited. We hope that, through this book, the readers can come to appreciate left H-rings and gain a better understanding of QF-rings and Nakayama rings.

For the study of direct sums of indecomposable modules, Azumaya's contribution is inestimable. In the original Krull-Remak-Schmidt Theorem, the indecomposable modules were assumed to have both the ascending and descending chain conditions. However, in [10], Azumaya considered completely indecomposable modules, that is, modules with local endomorphism rings, replacing the chain conditions, and improved the Krull-Remak-Schmidt Theorem to what is now referred to as the Krull-Remak-Schmidt-Azumaya Theorem, a very useful tool in the study of module theory. The final version of the Krull-Remak-Schmidt-Azumaya Theorem was completed by Harada (see [64]) using the factor categories induced from completely indecomposable modules. It can be stated as follows: If M is a direct sum of completely indecomposable modules $\{M_\alpha\}_{\alpha \in I}$, then M satisfies the (finite) exchange property if and only if $\{M_\alpha\}_{\alpha \in I}$ is locally semi-T-nilpotent in the sense of Harada-Kanbara [71].

As one of the applications of his work on the Krull-Remak-Schmidt-Azumaya Theorem, in the late 1970s Harada introduced extending and lifting properties in certain classes of modules. Although these properties were initially rather specialized, these lead to their introduction in a more general setting in [147], [148] as extending modules and lifting modules. However, it should be noted that, in the early days of ring theory, the extending property explicitly appeared in Utumi's work [176]; in particular, in today's terminology, Utumi showed that a von Neumann regular ring R is right continuous if and only if the right R-module R is extending. Furthermore, he introduced the notion of a continuous ring by using the extending property. On the other hand, the lifting property implicitly appeared in Bass's work [21], where, in essence, it was proved that a ring R is semiperfect if and only if the right R-module R is lifting.

Generalizing the concept of continuous rings to module theory, continuous modules and quasi-continuous modules were introduced by Jeremy [87] in 1974 by using the extending property. This concept and its generalizations remained relatively dormant until Harada's work on lifting and extending properties appeared. Indeed, from the early 1980s, many ring theorists began to get interested in the study of the extending and lifting properties and, during the last 30 years, this field has developed at an incredible rate. This situation is witnessed in the following books on this field: M. Harada, *Factor Categories with Applications to Direct Decomposition of Modules* (Marcel Dekker, 1983); Mohamed-Müller, *Continuous Modules and Discrete Modules* (Cambridge Univ. Press, 1990); Dung-Huynh-Smith-Wisbauer *Extending Modules* (Pitman, 1994); and Clark-Lomp-Vanaja-Wisbauer, *Lifting modules* (Birkhäuser, 2007).

Although Harada introduced his classes of H-rings and co-H-rings prior to his study of lifting and extending properties, it seems to be unclear whether he was conscious of the relationships between these properties and rings. However, in [147], [148], module-theoretic characterizations of left H-rings and Nakayama rings were given using the properties. Indeed, the results obtained in these papers lead us to claim that the extending and lifting properties really have their origin in the theory of artinian rings and so, in this sense, these concepts have been studied extensively.

In Chapter 1, we provide a background sketch of fundamental concepts and well-known facts on artinian rings and related materials. In particular, since materials in this book is based on semiperfect rings, we give known facts on these rings by using the lifting property of modules. Basic QF-rings, Nakayama permutations and Nakayama automorphisms as rings are also considered here for later use.

In Chapter 2, we present Fuller's Theorem which provides a criterion for the injectivity of an indecomposable projective module over one-sided artinian rings. We present an improved version of Fuller's Theorem for semiprimary rings by Baba and Oshiro [20], and introduce a more general presentation due to Baba [11].

In Chapter 3, we introduce one-sided H-rings and one-sided co-H-rings and give module-theoretic characterizations of these rings, and show that left H-rings and right co-H-rings coincide. Among the several characterizations of left H-rings, we show that they are precisely the rings R for which the class of projective right R-modules is closed under taking essential extensions.

In Chapter 4, fundamental structure theorems of H-rings are explored. We establish the matrix representation of left H-rings. It is shown that for a given basic indecomposable left H-ring R, there exists a QF-subring $F(R)$ of R, from which R can be constructed as an upper staircase factor ring of a block extension of $F(R)$. Because of this feature, we say that $F(R)$ is the *frame* QF-subring of R.

In Chapter 5, we study self-duality of left H-rings. For a basic indecomposable QF-ring F, it is shown that every basic indecomposable left H-ring R with $F(R) \cong F$ is self-dual if and only if F has a Nakayama automorphism. In Koike [98], it is noted that an example in Kraemer [103] provides a basic indecomposable QF-ring without a Nakayama automorphism. The difference between QF-rings with or without Nakayama automorphisms affects self-duality of their factor rings and block extensions.

In Chapter 6, we introduce skew matrix rings and develop fundamental properties of these rings. As mentioned earlier, these rings were introduced by Kupisch [105] and independently by Oshiro [149] through the study of Nakayama rings. In this book, we use the definition of skew matrix rings given in the latter paper. We will illustrate the usefulness of skew matrix rings for the study on QF-rings and Nakayama rings.

In Chapter 7, we develop a classification of Nakayama rings as an application of left H-rings. We study basic indecomposable Nakayama rings by analyzing certain types of Nakayama permutations of their frame QF-subrings. It is shown that a basic indecomposable Nakayama rings is represented as an upper staircase factor ring of a block extension of the frame QF-subring and, in particular, every basic indecomposable Nakayama ring whose frame QF-subring is not weakly symmetric can be directly represented as an upper staircase factor ring of a skew matrix ring over a local Nakayama ring. As an application of left H-rings, we confirm the self-duality of Nakayama rings by applying the fact that skew matrix rings over local Nakayama rings have Nakayama automorphisms.

In Chapter 8, we give several module-theoretic characterizations of Nakayama rings by using extending and lifting properties. These characterizations provide relationships between QF-rings, Nakayama rings and one-sided H-rings.

In Chapter 9, we study Nakayama algebras over an algebraically closed field. It is shown that such algebras are represented as factor rings of skew matrix rings over one variable polynomial rings over algebraically closed fields. This result inspires us to study Nakayama group algebras over such fields. We summarize several results on these group algebras in relationships to skew matrix rings.

In Chapter 10, we introduce the Faith conjecture, and as a by-product of the study of this conjecture, we give a new way of constructing local QF-rings with Jacobson radical cubed zero. This construction produces many local QF-rings which are not finite dimensional algebras.

We end the text with several questions on QF-rings and Nakayama rings.

Though we have tried to make this book as self-contained as possible, for basic notions and well-known facts on non-commutative ring theory (such as semiprimary rings, semiperfect rings, quasi-Frobenius rings, and Morita duality etc.), the reader is referred to the books Faith [46], [47], Anderson-Fuller [5], Harada [64], Lam [107], and Wisbauer [182], and, in addition, to the books Dung-Huynh-Smith-Wisbauer [41] for extending modules, Puninski [162] for serial rings, Xue [184] for Morita duality, and Nicholson-Yousif [142] and Tachikawa [172] for QF-rings. In particular, for the background of Frobenius algebras and QF-rings, the reader is referred to Faith [49], Lam [107], Nakayama-Azumaya [137], Nicholson-Yousif [142], and Yamagata [189].

We are indebted to a large number of people for this overall effort possible. In particular we are thankful to John Clark, Jiro Kado, Kazutoshi Koike as well as Akihide Hanaki, Shigeo Koshitani, Isao Kikumasa, and Hiroshi Yoshimura for much of their help. Thanks also to Derya Keskin, Yosuke Kuratomi, Cosmin Roman and Masahiko Uhara, and Kota Yamaura for their comments which improved the overall presentation.

Contents

List of Symbols

R : an associative ring with $1 \neq 0$.

M_R : a unitary right module over a ring R.

$_RM$: a unitary left module over a ring R.

$Pi(R)$: a complete set of orthogonal primitive idempotents of a semiperfect ring R.

$Z(M)$: the singular submodule of a module M.

id_X or, simply, id if no confusion arises : the identity map of X.

$J(M)$: the Jacobson radical of a module M.

$J_i(M)$: the i-th radical of a module M.

$J(R)$ or, simply, J : the Jacobson radical of a ring R.

$S(M)$: the socle of a module M.

$S_k(M)$: the k-th socle $\{\, x \in M \mid xJ^k = 0 \text{ (resp. } J^k x = 0) \,\}$ of a right (resp. left) R-module M.

$T(M)$: the top $M/J(M)$ of M.

$E(M)$: an injective hull of a module M.

$L(M)$: the Loewy length of a module M.

$|\,M\,|$: the composition length of a module M.

$N \subseteq_e M$: N is an essential submodule of a module M.

$N \subseteq_c M$: N is a co-essential submodule of a module M.

$N \ll M$: N is a small (superfluous) submodule of a module M.

$N \lesssim M$: N is isomorphic to a submodule of M.

$(a)_L : S \to aS$: the left multiplication map by a, where S is a subset of a ring R and $a \in R$.

$(a)_R : S \to Sa$: the right multiplication map by a, where S is a subset of a ring R and $a \in R$.

$r_S(T)$: the right annihilator of T in S.

$l_S(T)$: the left annihilator of T in S.

$\#S$: the cardinal of a set S.

$\mathbb{N}$: the set of positive integers.

$\mathbb{N}_0$: the set of non-negative integers.

$M^{(I)}$: a direct sum of $\#I$ copies of a module M.

M^I : a direct product of $\#I$ copies of a module M.

$R^{(\mathbb{N})}$: a direct sum of countable copies of a ring R.

Mod-R (resp. *R-Mod*) : the category of all right (resp. left) R-modules.

FMod-R (resp. *R-FMod*) : the category of all finitely generated right (resp. left) R-modules.

$(Q)_{\sigma,c,n}$: the skew matrix ring over Q with respect to (σ, c, n).

$\{\alpha_{ij}\}$, $\{\,\alpha_{ij} \mid i < j\,\} \cup \{\,\beta_{ij} \mid i < j\,\}$: the skew matrix units.

$\langle\, a\,\rangle_{ij}$: the matrix with (i, j)-entry 0 and other entries 0.

$\langle\, X\,\rangle_{ij}$: the set $\{\langle\, x\,\rangle_{ij} \mid x \in X\}$.

$[i]$: the least positive residue modulo m.

$F(R)$: the frame QF-subring of R.

$\dim(V_D)$: the dimension of a right vector space V_D over the division ring D.

$A \subsetneqq B$: A is isomorphic to a submodule of B.

Chapter 1

Preliminaries

1.1 Background Sketch

In this section, for the reader's convenience, we list some basic definitions, notation and well-known facts that will be needed in this book. For details, the reader is referred to standard books and recent books, in particular, to Anderson-Fuller [5], Faith [47], Harada [64], Lam [107], Mohamed-Mueller [115], Nicholson-Yousif [141], Dung-Huynh-Smith-Wisbauer [41], Clark-Lomp-Vanaja-Wisbauer [31] and Xue [184].

Throughout this book, all rings R considered are associative rings with identity and all R-modules are unital. Departing from convention, when we consider subrings of a ring, we do not assume that they have the same identity. All R-homomorphisms between R-modules are written on the opposite side of scalars. The notation M_R (resp. $_RM$) is used to stress that M is a right (resp. left) R-module. For a module M, we use $E(M)$ to denote its injective hull. An idempotent e of R is called *primitive* if eR_R is an indecomposable module or, equivalently, if $_RRe$ is an indecomposable module. A set $\{e_1, \ldots, e_n\}$ of orthogonal primitive idempotents of R with $1 = e_1 + \cdots + e_n$ is called a *complete set of orthogonal primitive idempotents* of R. We put $Pi(R) = \{e_1, \ldots . e_n\}$.

Let M be a right R-module and N a submodule of M. Then N is called an *essential* submodule of M (or M is an *essential* extension of N) if $N \cap X \neq 0$ for any non-zero submodule X of M and we denote

1

this by $N \subseteq_e M$. We use $Z(M)$ to denote the *singular* submodule of M, i.e., $Z(M) = \{\, m \in M \mid r_R(m) \subseteq_e R_R \,\}$, where $r_R(m)$ means the right annihilator of m in R. Dually, a submodule N of M is called a *small* submodule (or *superfluous* submodule) of M, abbreviated $N \ll M$, if, given a submodule $L \subseteq M$, $N + L = M$ implies $L = M$.

For an R-module M, the (*Jacobson*) *radical* $J(M)$ of M is defined as the intersection of all maximal submodules of M, i.e., $J(M) = \cap\{\, K \le M \mid K$ is a maximal submodule of $M \,\}$. If M has no maximal submodule, we define $J(M) = M$.

For a ring R, we say that $J(R_R)$ $(= J(_R R))$ is the *Jacobson radical* of R and denote it by $J(R)$ or, simply, by J if no confusion arises.

For a module M_R, by transfinite induction, *the upper Loewy series*

$$J_0(M) \supseteq J_1(M) \supseteq J_2(M) \supseteq \cdots$$

of M is defined as follows:

$J_0(M) = M$, $J_1(M) = J(M)$,
$J_\alpha(M) = J(J_{\alpha-1}(M))$ if α is a successor ordinal, and
$J_\alpha(M) = \cap_{\beta < \alpha} J_\beta(M)$ if α is a limit ordinal.

Here $J_\alpha(M)$ is called the α-*th radical* of M.

We record some fundamental properties of $J(M)$ (see, for example, [5]).

Proposition 1.1.1. *Let M be a right R-module. Then*

$$J(M) = \sum\{\, L \mid L \ll M \,\}.$$

The following basic result is due to Bass [21].

Theorem 1.1.2. *Let R be a ring and P a non-zero projective right R-module. Then the following hold:*

(1) $J(P) = PJ$.

(2) $J(P) \ne P$, *i.e., P has a maximal submodule.*

Theorem 1.1.3. *Let M be a finitely generated right R-module. Then the following hold:*

(1) $MJ \subseteq J(M)$.

(2) $J(M) \ll M$.

Remark 1.1.4. Theorem 1.1.3 (2) is well-known as *Nakayama's lemma*. We can easily prove this using Theorem 1.1.2 (1) and the fact that every submodule of M is contained in a maximal submodule of M. Nakayama's lemma is important for its use in various situations in the ring and module theory.

Throughout this book, $\mathbb{N}$ denotes the set of positive integers while $\mathbb{N}_0$ denotes the set of non-negative integers.

A ring R is called a *right noetherian (artinian)* ring if R satisfies the ascending (descending) chain condition (abbreviated $ACC\,(DCC)$) for right ideals. A *left noetherian (artinian)* ring is similarly defined. A left and right noetherian (artinian) ring is called a *noetherian (artinian)* ring. A ring R is called a *semiprimary* ring if R/J is semisimple and J is nilpotent, i.e., there exists $k \in \mathbb{N}$ with $J^k = 0$.

The following theorem records some fundamental facts of ring theory.

Theorem 1.1.5. *Let R be a right artinian ring. Then the following hold:*

(1) *R is a right noetherian ring.*

(2) *J is nilpotent.*

(3) *R/J is semisimple.*

(4) *There exists a complete set $\{e_i\}_{i=1}^n$ of orthogonal primitive idempotents of R (so that $1 = e_1 + \cdots + e_n$ and $R = e_1R \oplus \cdots \oplus e_nR$) and each e_iJ is a maximal submodule of e_iR.*

(5) *Every complete set of orthogonal primitive idempotents of $\overline{R} = R/J$ lifts to a complete set of orthogonal primitive idempotents of R.*

Given a right R-module M, the *socle* of M is defined as the sum of all the simple submodules of M_R and is denoted by $S(M)$. When M does not have simple submodules, we put $S(M) = 0$. By transfinite induction, *the lower Loewy series*

$$S_0(M) \subseteq S_1(M) \subseteq S_2(M) \subseteq \cdots$$

of M is defined as follows:

$S_0(M) = 0$, $S_1(M) = S(M)$,

$S_\alpha(M)/S_{\alpha-1}(M) = S(M/S_{\alpha-1}(M))$ if α is a successor ordinal, and

$S_\alpha(M) = \cup_{\beta<\alpha} S_\beta(M)$ if α is a limit ordinal.

Here $S_\alpha(M)$ is called the α-*th socle* of M.

As a dual to Proposition 1.1.1, we have the following proposition (see, for example, [5, 9.7. Proposition]).

Proposition 1.1.6. *Let M be a right R-module. Then*

$$S(M) = \bigcap \{\, L \mid L \subseteq_e M \,\}.$$

Let R be a ring. We will denote the category of all right (left) R-modules by *Mod-R* (*R-Mod*), and use *FMod-R* (*R-FMod*) to denote the category of all finitely generated right (left) R-modules. A right R-module M is called a *generator* in *Mod-R* if, for any right R-module X, there exists a direct sum $\sum M$ of copies of M and an epimorphism from $\sum M$ to X, equivalently, R is isomorphic to a direct summand of a finite direct sum of copies of M. Dually, a right R-module M is called a *cogenerator* in *Mod-R* if any right R-module X can be embedded in a direct product of copies of M. It is well-known that a right R-module M is a cogenerator in *Mod-R* if and only if, for any simple right R-module S, M contains an injective module which is isomorphic to the injective hull of S.

Let R and S be rings. Then *Mod-R* and *Mod-S* are called *equivalent* if there exist additive covariant functors $F : Mod\text{-}R \to Mod\text{-}S$ and $G : Mod\text{-}S \to Mod\text{-}R$ such that GF and FG are isomorphic to the identity functors of *Mod-R* and *Mod-S*, respectively. In this case, we also say that the rings R and S are *Morita equivalent*.

Given $n \in \mathbb{N}$, we let $(R)_n$ denote the ring of $n \times n$-matrices over the ring R. Given any R-module M, we let $\operatorname{End}(M)$ denote the ring of endomorphisms of M.

Theorem 1.1.7. (Morita equivalence) *For two rings R and S, the following are equivalent:*

(1) *Mod-R is equivalent to Mod-S.*

(2) *R-Mod is equivalent to S-Mod.*

(3) *There exists a finitely generated projective generator P_R such that $\operatorname{End}(P_R) \cong S$.*

(4) *There exist $n \in \mathbb{N}$ and an idempotent e of $T = (R)_n$ such that $TeT = T$ and $eTe \cong S$.*

Let P_R be a projective module as given in Theorem 1.1.7 (3) above. Then, setting $Q = \operatorname{Hom}_R({}_SP_R, R)$, the functors $\operatorname{Hom}_R({}_SP_R, -)$ and $\operatorname{Hom}_S({}_RQ_S, -)$ realize the equivalence between *Mod-R* and *Mod-S*.

A property $(\mathcal{P})$ in the class of all rings is called *Morita invariant* if, whenever R has the property $(\mathcal{P})$ and S is Morita equivalent to R, then S has the property $(\mathcal{P})$ as well.

Let R and S be rings. Let $\mathcal{C}$ and $\mathcal{D}$ be full subcategories of *Mod-R* and *S-Mod*, respectively. If there exist additive contravariant functors $F : \mathcal{C} \to \mathcal{D}$ and $G : \mathcal{D} \to \mathcal{C}$ such that GF and FG are isomorphic to the identity functors of $\mathcal{C}$ and $\mathcal{D}$, respectively, then we say that the pair (F, G) *is a duality* between $\mathcal{C}$ and $\mathcal{D}$. Further we say that a duality (F, G) between $\mathcal{C}$ and $\mathcal{D}$ is a *Morita duality* if $R_R \in \mathcal{C}$ and ${}_SS \in \mathcal{D}$ and $\mathcal{C}$ and $\mathcal{D}$ are closed under submodules and factor modules.

Consider a bimodule ${}_SE_R$. For a module $M = M_R$, we put $M^* = \operatorname{Hom}_R(M, E)$. Then M^* is a left S-module. For a module $N = {}_SN$, we also put $N^* = \operatorname{Hom}_S(N, E)$. Then N^* is a right R-module. For a right R-module M, the map $\sigma_M : M \to M^{**}$ given by

$$\sigma_M(m)(f) = f(m) \qquad (m \in M, \ f \in M^*)$$

is a homomorphism, which we call an *evaluation map*. When $M \cong M^{**}$ under σ_M, M is called an *E-reflexive* right R-module. Similarly, beginning with a left S-module N, we may define N^{**} and the evaluation map σ_N and, when $N \cong N^{**}$ under σ_N, N is called an *E-reflexive* left S-module.

We let $R[E]_R$ and $_S R[E]$ denote the family of all E-reflexive right R-modules and that of all E-reflexive left S-modules, respectively. Then the pair of the functors $F(-) = \mathrm{Hom}_R(-, E)$ and $G(-) = \mathrm{Hom}_S(-, E)$ induces a duality between $R[E]_R$ and $_S R[E]$.

For a given bimodule $_S E_R$, there exist canonical ring homomorphisms $S \to \mathrm{End}(E_R)$ and $R \to \mathrm{End}(_S E)$. If these are isomorphisms, $_S E_R$ is called a *faithfully balanced* bimodule.

The next two theorems give the essence of Morita duality.

Theorem 1.1.8. (Morita duality I) *Let $\mathcal{C}$ and $\mathcal{D}$ be full subcategories of Mod-R and S-Mod, respectively, and let (F, G) be a Morita duality between $\mathcal{C}$ and $\mathcal{D}$. Then there exists a bimodule $_S E_R$ such that*

(1) *$_S E \cong F(R_R)$ and $E_R \cong G(_S S)$,*

(2) *there exist natural isomorphisms $F(-) \cong \mathrm{Hom}_R(-, E)$ and $G(-) \cong \mathrm{Hom}_S(-, E)$ and*

(3) *all modules M_R in $\mathcal{C}$ and all modules $_S N$ in $\mathcal{D}$ are E-reflexive.*

As the above result shows, every Morita duality can be realized as functors $\mathrm{Hom}_R(-, E)$ and $\mathrm{Hom}_S(-, E)$ for a suitable bimodule $_S E_R$. We say that a bimodule $_S E_R$ *defines a Morita duality* in case the pair of the functors $\mathrm{Hom}_R(-, E)$ and $\mathrm{Hom}_S(-, E)$ induces a Morita duality.

Theorem 1.1.9. (Morita duality II) *The following are equivalent for a bimodule $_S E_R$:*

(1) *$_S E_R$ defines a Morita duality.*

(2) *Every factor module of R_R, $_S S$, E_R and $_S E$ are E-reflexive.*

(3) *$_S E_R$ is a faithfully balanced bimodule and E_R and $_S E$ are injective cogenerators.*

The following theorem is known as the Morita-Azumaya Theorem:

Theorem 1.1.10. (Azumaya [9], Morita [122]) *Let R be a right artinian ring and S a ring. For a bimodule $_SE_R$, the following are equivalent:*

(1) $_SE_R$ *defines a Morita duality between S-FMod and FMod-R.*

(2) E_R *is a finitely generated injective cogenerator and the canonical ring homomorphism $S \to$ End (E_R) is an isomorphism.*

When this is so, the following hold:

(1) S *is left artinian, FMod-R = $R[E]_R$ and S-FMod = $_SR[E]$.*

(2) *Every indecomposable injective right R-module and every indecomposable injective left S-module are finitely generated.*

From this theorem we know that, for a right artinian ring R, if $E = E((R/J)_R)$ is finitely generated, then $_SE_R$ defines a Morita duality between *S-FMod* and *FMod-R*, where $S = \text{End}(E_R)$. We note that this important fact is also shown in Tachikawa [172].

Let R be a right or left noetherian ring. We say that R is *self-dual* or has a *self-duality* if there is a duality between *FMod-R* and *R-FMod*.

For an R-module M, the top $M/J(M)$ of M is denoted by $T(M)$. The right (resp. left) annihilator of a subset T of M_R (resp. $_RM$) in a subset S of R is denoted by $r_S(T)$ (resp. $l_S(T)$).

A ring R is called a *quasi-Frobenius* ring (abbreviated as *QF-ring*) if R is a left and right artinian and left and right self-injective ring. *QF*-rings were introduced by Nakayama [133], as artinian rings with the condition (5) in the theorem below (see Section 1.3).

The following theorem records well-known characterizations of *QF*-rings.

Theorem 1.1.11. *The following are equivalent for a ring R:*

(1) R *is QF.*

(2)   *R is left or right artinian and left or right self-injective.*

(3)   *R is left or right noetherian and left or right self-injective.*

(4)   *R is a left or right noetherian ring and $_R R_R$ defines a self-duality between R-FMod and FMod-R.*

(5)   *R is a left or right artinian ring and, for any primitive idempotent e of R, there exists a primitive idempotent f of R such that $(eR_R; _R Rf)$ is an i-pair, i.e., $S(eR_R) \cong T(fR_R)$ and $S(_R Rf) \cong T(_R Re)$.*

(6)   *R is a left or right noetherian ring with double annihilator conditions, i.e.,*

$$l_R(r_R(I)) = I \text{ for any left ideal } I \text{ of } R \text{ and}$$
$$r_R(l_R(K)) = K \text{ for any right ideal } K \text{ of } R.$$

(7)   *Every projective left (or right) R-module is injective.*

(8)   *Every injective left (or right) R-module is projective.*

A right R-module M is said to be *uniserial* if the family of all submodules of M is linearly ordered by inclusion. A right (resp. left) artinian ring is called a *right* (resp. *left*) *Nakayama* ring if, for any primitive idempotent e of R, eR_R (resp. $_R Re$) is uniserial. The ring R is called a *Nakayama ring* if it is a right and left Nakayama ring.

Theorem 1.1.12. ([133], cf. [43]) *Let R be a Nakayama ring. Then every right R-module can be expressed as a direct sum of uniserial modules.*

For a given right R-module M, there exists an injective module $E(M)$ containing M as an essential submodule. The existence of $E(M)$ is known as the Eckmann-Schopf Theorem ([42]), and $E(M)$ is called an *injective hull* of M.

For a module M and an indexed set A, we let $M^{(A)}$ denote the direct sum of $\sharp A$ copies of M. An injective module M is said to be $\sum$-*injective* if $M^{(A)}$ is injective for every indexed set A.

The following fundamental theorem is due to Faith.

Theorem 1.1.13. ([47, 20.3 A and 20.6 A] or [5, Theorem 25.1]) *The following are equivalent for an injective module M:*

   (1) *M is $\sum$-injective.*

   (2) *ACC holds on $\{\, r_R(X) \mid X$ is a subset of $M \,\}$.*

   (3) *$M^{(\mathbb{N})}$ is injective.*

We say that a module M has *the exchange property* if, for every R-module L containing M and for submodules N and $\{L_i\}_{i \in I}$ of L, decompositions

$$L = M \oplus N = \oplus_{i \in I} L_i$$

imply the existence of a submodule K_i of L_i for every $i \in I$ satisfying

$$L = M \oplus (\oplus_{i \in I} K_i).$$

We note that every quasi-injective R-module has the exchange property (see, for instance, Fuchs [52], Warfield [180]). M is said to have the *finite exchange property* if this condition is satisfied whenever the above indexed set I is finite.

Let M be a right R-module with a decomposition $M = \oplus_{i \in I} A_i$. We say that the decomposition $M = \oplus_{i \in I} A_i$ is *exchangeable* (or *complements direct summands*) if, for any direct summand N of M, there exists a submodule A_i' of A_i for every $i \in I$ such that $M = N \oplus (\oplus_{i \in I} A_i')$.

Theorem 1.1.14. ([179]) *Let A be an indecomposable module and $M = A_1 \oplus A_2$ with $A_i \cong A$ for $i = 1, 2$. Then the following are equivalent:*

   (1) *A satisfies the (finite) exchange property.*

   (2) *A satisfies the exchange property for $A \oplus A$, i.e., for any direct summand N of M with $N \cong A$, either $M = N \oplus A_1$ or $M = N \oplus A_2$.*

   (3) *$M = A_1 \oplus A_2$ is exchangeable.*

   (4) *End(A) is a local ring.*

An infinite family $\{M_\alpha\}_{\alpha \in I}$ of modules is called *locally semi-transfinitely nilpotent* (abbreviated lsTn) if, for any subfamily $\{M_{\alpha_i}\}_{i \in \mathbb{N}}$ of $\{M_\alpha\}_{\alpha \in I}$ with distinct α_i, any family $\{f_i : M_{\alpha_i} \to M_{\alpha_{i+1}}\}_{i \in \mathbb{N}}$ of non-isomorphisms and any $x \in M_{\alpha_1}$, there exists $n \in \mathbb{N}$ (depending x) such that $f_n \cdots f_2 f_1(x) = 0$.

Theorem 1.1.15. ([64], cf. [191]) *Let $M_R = \oplus_{i \in I} M_i$ be an indecomposable decomposition, where $\mathrm{End}(M_i)$ is a local ring for all $i \in I$. Then the following are equivalent:*

    (1) *M has the (finite) exchange property.*

    (2) *$M_R = \oplus_{i \in I} M_i$ is exchangeable.*

    (3) *(If I is an infinite set) the family $\{M_i\}_{i \in I}$ is lsTn.*

Given $M_R \supseteq B_R \supseteq A_R$, we may interpret the dual of "$A \subseteq_e B$" in M to mean "B/A is a small submodule of M/A" or, equivalently, "for any submodule X of M with $A \subseteq X$, $M = X + B$ implies $X = M$". If this dual condition holds, A is called a *co-essential* submodule of B (in M) and this is denoted by $A \subseteq_c B$ *in* M.

An R-module M is called *uniform* (resp. *hollow*) if any non-zero (resp. proper) submodule of M is an essential (resp. a small) submodule of M.

Let A be a submodule of M_R. We say that M satisfies the *extending property for A* if A can be essentially extended to a direct summand of M. In case any uniform submodule of M can be essentially extended to a direct summand of M, we say that M satisfies the *extending property for uniform submodules*. We say that M is an *extending* module or a *CS*-module if M satisfies the extending property for all of its submodules.

Note that, for an R-module M_R and a submodule A of M, we can find a submodule B of M such that $A \subseteq_e B$ and there exists no proper essential extension of B in M. A submodule B which has no proper essential extension in M is called a *closed* submodule of M. So we may say that M_R is an extending module if any closed submodule of M is a direct summand. (The alternative CS name comes from this "closed submodule is a

<u>s</u>ummand" property.)

Dually, given an R-module M and a submodule A of M, we say that M has the *lifting property for M/A* or A is *co-essentially lifted* to a direct summand of M if there exists a decomposition $M = A^* \oplus A^{**}$ satisfying $A^* \subseteq A$ and $A/A^* \ll M/A^*$, i.e., $A \cap A^{**} \ll A^{**}$.

This leads to the following dual notions of "extending property for uniform modules" and "extending module". If M satisfies the lifting property for all of its hollow factor modules, we say that M satisfies the *lifting property for hollow factor modules*. And if it satisfies the lifting property for all its factor modules, we say that M is a *lifting* module. Furthermore we say that M satisfies the *lifting property for simple factor modules* if it satisfies the lifting property for all of its simple factor modules.

The following result for the extending case follows from the fact that, for any $A \subseteq B \subseteq C$ such that A is a closed submodule of B and B is a closed submodule of C, A is a closed submodule of C (see Gooderal [55], or Oshiro [145]). The result for the lifting case follows quickly from the definition.

Proposition 1.1.16. *Any direct summand of an extending (resp. a lifting) module is an extending (resp. a lifting) module.*

Let M and N be R-modules. Then M is said to be *N-injective* if, for any submodule X of N and any homomorphism $\varphi : X \to M$, there exists $\tilde{\varphi} \in \mathrm{Hom}_R(N, M)$ such that the restriction map $\tilde{\varphi}|_X$ coincides with φ. If in this definition we only consider homomorphisms φ with simple images, M is then said to be *N-simple-injective*.

Dually, M is said to be *N-projective* if, for any submodule X of N and any homomorphism $\varphi : M \to N/X$, there exists $\tilde{\varphi} \in \mathrm{Hom}_R(M, N)$ such that $\pi_X \tilde{\varphi} = \varphi$, where $\pi_X : N \to N/X$ is the natural epimorphism.

A module M is then called *quasi-injective* (resp. *quasi-projective*) if M is M-injective (resp. M-projective).

Let A and B be right R-modules. A is said to be *generalized B-injective* (or *B-ojective*) if, for any submodule X of B and any homomorphism $f : X \to A$, there exist decompositions $A = A_1 \oplus A_2$, $B = B_1 \oplus B_2$, a homomorphism $h_1 : B_1 \to A_1$, and a monomorphism $h_2 : A_2 \to B_2$ satisfying

(a) $X \subseteq B_1 \oplus h_2(A_2)$, and

(b)   for any $x \in X$ with $x = x_1 + x_2$, where $x_i \in B_i$ for $i = 1, 2$, we have $f(x) = h_1(x_1) + h_2^{-1}(x_2)$.

The following diagram illustrates this definition.

$$
\begin{array}{ccc}
0 \longrightarrow X \xrightarrow{\;i\;} B & \qquad & 0 \longrightarrow X \xrightarrow{\;i\;} B = B_1 \oplus B_2 \\
f \downarrow & & f \downarrow \qquad h_1 \downarrow \quad \uparrow h_2 \\
A & & A \qquad = A_1 \oplus A_2 \\
& & \uparrow \\
& & 0
\end{array}
$$

A right R-module A is said to be *essentially B-injective* if, for any submodule X of B, any homomorphism $f : X \to A$ with $\operatorname{Ker} f \subseteq_e X$ can be extended to a homomorphism $B \to A$.

Remark 1.1.17. It is easy to see that if A is generalized B-injective, then A is essentially B-injective (cf. [57]). From this, it follows that, for uniform right R-modules A and B, the following are equivalent:

(1) A is generalized B-injective.

(2)   For any submodule X of B and any homomorphism $f : X \to A$, if f is not monomorphic, then f extends to $B \to A$, and if f is monomorphic, then either f extends to $B \to A$ or f^{-1} extends to $A \to B$.

The following theorem is one of the fundamental results on extending modules.

Theorem 1.1.18. ([57], [116])   *Let $A_1, \ldots, A_n$ be extending right R-*

modules and set $M = A_1 \oplus \cdots \oplus A_n$. *Then the following are equivalent:*

(1) M *is an extending module and the decomposition* $M = A_1 \oplus \cdots \oplus A_n$ *is exchangeable.*

(2) A_i *is generalized* $\oplus_{j \neq i} A_j$*-injective for any* $i \in \{1, \cdots, n\}$.

(3) $\oplus_{j \neq i} A_j$ *is generalized* A_i*-injective for any* $i \in \{1, \cdots, n\}$.

In particular, when each A_i *is uniform, the following are also equivalent to these conditions:*

(4) *The direct sum* $A_i \oplus A_j$ *is extending and exchangeable for any* $i \neq j$.

(5) A_i *and* A_j *are matually relative generalized injective modules for any* $i \neq j$.

For $M = A_1 \oplus \cdots \oplus A_n$ in this theorem, we note that, if A_i and A_j are mutually relative injective for any $i \neq j$, then M is an extending module (cf. [115], [75]).

We now define the dual notion of generalized B-injectivity. Given modules A and B, we say that A is *generalized B-projective* (or *dual B-ojective*) if, for any module X, any homomorphism $f : A \to X$ and any epimorphism $g : B \to X$, there exist decompositions $A = A_1 \oplus A_2$, $B = B_1 \oplus B_2$, a homomorphism $h_1 : A_1 \to B_1$, and an epimorphism $h_2 : B_2 \to A_2$ such that $gh_1 = f|_{A_1}$ and $fh_2 = g|_{B_2}$. The following diagram shows the decompositions and maps involved.

$$
\begin{array}{ccc}
 & 0 & \\
 & \uparrow & \\
A & = A_1 \oplus A_2 & \\
f \downarrow \quad & h_1 \downarrow \quad \uparrow h_2 & \\
0 \longleftarrow X \xleftarrow{\ g\ } B & = B_1 \oplus B_2 &
\end{array}
$$

A right R-module A is said to be *small B-projective* if, for any module X, any homomorphism $f : A \to X$ with $\operatorname{Im} f \ll X$ and any epimorphism $g : B \to X$, there exists a homomorphism $h : A \to B$ satisfying $gh = f$.

Remark 1.1.19. (cf. [106]) We can easily see that if A is a generalized B-projective module and B is lifting, then A is small B-projective. From this fact, we note that, when A and B are hollow modules, the following are equivalent:

(1) A is generalized B-projective.

(2) For any X, any homomorphism $f : A \to X$ and any epimorphism $g : B \to X$, if f is not epimorphic, there exists a homomorphism $h : A \to B$ satisfying $gh = f$, and if f is epimorphic, then there exists an epimorphism $h : A \to B$ or $B \to A$ satisfying $gh = f$ or $fh = g$, respectively.

Theorem 1.1.20. ([106], [117], cf. [31]) *Let* $A_1, \ldots, A_n$ *be lifting right R-modules and set* $M = A_1 \oplus \cdots \oplus A_n$. *Then the following are equivalent:*

(1) M *is a lifting module and the decomposition* $M = A_1 \oplus \cdots \oplus A_n$ *is exchangeable.*

(2) A_i' *and* T *are mutually generalized projective for any summand* A_i' *of* A_i *and any summand* T *of* $\oplus_{j \neq i} A_j$ *for any* $i \in \{1, \cdots, n\}$.
In particular, when each A_i *is a hollow module, the following are equivalent to these conditions:*

(3) *The direct sum* $A_i \oplus A_j$ *is an exchangeable and lifting module for any* $i \neq j$.

(4) A_i *and* A_j *are mutually relative generalized projective modules for any* $i \neq j$.

For $M = A_1 \oplus \cdots \oplus A_n$ in this theorem, we note that, if A_i and A_j are mutually relative projective for any $i \neq j$, then M is a lifting module and the direct sum $A_i \oplus A_j$ is exchangeable ([106], cf. [92]).

For a right R-module M, we consider the following conditions:

(C_1) M is an extending module.

(C_2) If a submodule X of M is isomorphic to a direct summand of M, then X is a direct summand.

(C_3) If M_1 and M_2 are direct summands of M with $M_1 \cap M_2 = 0$, then $M_1 \oplus M_2$ is a direct summand of M.

(D_1) M is a lifting module.

(D_2) If X is a submodule of M such that M/X is isomorphic to a direct summand of M, then X is a direct summand of M.

(D_3) If M_1 and M_2 are direct summands of M with $M = M_1 + M_2$, then $M_1 \cap M_2$ is a direct summand of M.

M is called a *continuous* module if (C_1) and (C_2) hold, and is called a *quasi-continuous module* if (C_1) and (C_3) hold. Dually M is called a *discrete* module (or *semiperfect* module) if (D_1) and (D_2) hold, and is called a *quasi-discrete* (or *quasi-semiperfect*) module if (D_1) and (D_3) hold.

Remark 1.1.21. We have the following implications for these module properties:

(1) injective $\Rightarrow$ quasi-injective $\Rightarrow$ continuous $\Rightarrow$ quasi-continuous $\Rightarrow$ extending

(2) projective $\Rightarrow$ quasi-projective $\nRightarrow$ discrete $\Rightarrow$ quasi-discrete $\Rightarrow$ lifting

(3) (quasi-)projective lifting $\Rightarrow$ discrete.

It follows from the last of these and Theorem 1.1.20 that any finite direct sum of projective lifting modules is lifting.

The following results give fundamental facts on (quasi-)discrete modules.

Theorem 1.1.22. ([146], cf. [115]) *If M is a quasi-discrete module, then M can be expressed as a direct sum $\oplus_{i \in I} M_i$ of hollow modules M_i. In particular, if M is discrete, then each $\mathrm{End}(M_i)$ is a local ring and $M = \oplus_{i \in I} M_i$ is exchangeble, and hence the family $\{M_i\}_{i \in I}$ is lsTn. Consequently, M has* *the exchange property by Theorem 1.1.15.*

Theorem 1.1.23. ([146], cf. [115]) *Let M be a discrete right R-module.*

(1) *If $M = \sum_{i \in I} M_i$, M_i is a summand of M and $M_i \cap (\sum_{j \in I - \{i\}} M_j) \ll M$ for any $i \in I$, then $M = \sum_{i \in I} M_i$ is a direct sum.*

(2) *If $M = \sum_{i \in I} M_i$ is an irredundant sum of indecomposable submodules M_i, then the sum $M = \sum_{i \in I} M_i$ is a direct sum.*

Let M be a right R-module and N a submodule of M. We say that N is a *fully invariant* submodule of M if N is a right R- left $\mathrm{End}(M_R)$-subbimodule of M. Let M and P be right R-modules. An epimorphism $\varphi : P \to M$ is called *small* (or *superfluous*) if $\mathrm{Ker}\,\varphi \ll P$. A pair (P, φ) is called a *projective cover* of M if P is projective and there exists a small epimorphism $\varphi : P \to M$. In this case we usually just say that $\varphi : P \to M$ is a projective cover or, more simply, that P is a projective cover of M.

To finish this section, we record several properties of quasi-injective modules and quasi-projective modules.

Theorem 1.1.24. *Let M be a right R-module.*

(1) *If M is a quasi-injective module, then it is a fully invariant submodule of $E(M)$.*

(2) *If M is a quasi-injective module with injective hull $E(M)$, then any direct sum decomposition $E(M) = E_1 \oplus \cdots \oplus E_n$ induces $M = (M \cap E_1) \oplus \cdots \oplus (M \cap E_n)$.*

(3) *If M is a quasi-projective module with projective cover $\varphi : P \to M$, $\mathrm{Ker}\,\varphi$ is a fully invariant submodule of P; consequently, any endomorphism of P induces an endomorphism of M.*

(4) *If M is a quasi-projective module with projective cover $\varphi : P \to M$, then any decomposition $P = P_1 \oplus \cdots \oplus P_n$ induces a decomposition $M = \varphi(P_1) \oplus \cdots \oplus \varphi(P_n)$.*

(5) If $\varphi : P \to M$ is an epimorphism with P projective and M has a projective cover, then there exists a decomposition $P = P_1 \oplus P_2$ such that $\varphi(P_2) = 0$ and the restriction map $\varphi|_{P_1} : P_1 \to M$ is a projective cover.

We note that (1), (3), (5) are easily verified, and (2), (4) can be proved using (1), (3), respectively.

1.2 Semiperfect Rings and Perfect Rings

Semiperfect rings and perfect rings were introduced by Bass [21] in 1960. Since our book is based on these rings, for the sake of the reader's convenience, we shall recall fundamental facts on these rings.

In [21], as a dual to the notion of an injective hull, Bass introduced a projective cover of a module. In general projective covers need not exist. A ring R is called *semiperfect* (resp. *right perfect*) if every finitely generated right module (resp. every right R-module) has a projective cover.

Let $M = M_1 \oplus M_2$ and let $\varphi : M_1 \to M_2$ be an R-homomorphism. Let $\langle M_1 \xrightarrow{\varphi} M_2 \rangle$ denote the submodule of M given by $\{\, m_1 - \varphi(m_1) \mid m_1 \in M_1 \,\}$. This is called the *graph* of φ. Note that $M = M_1 \oplus M_2 = \langle M_1 \xrightarrow{\varphi} M_2 \rangle \oplus M_2$.

Proposition 1.2.1. *Let R be a ring such that every maximal right ideal is a direct summand of R_R. Then R is a semisimple ring.*

Proof. Assume that $S(R_R) \subsetneqq R_R$. Then there exists a maximal right ideal I_R of R such that $S(R_R) \subseteq I_R$. By hypothesis, there exists a decomposition $R_R = I \oplus X$. Then X is a simple submodule of R_R and hence $X \subseteq S(R_R) \subseteq I$, a contradiction. Hence $R = S(R_R)$.

$\square$

Proposition 1.2.2. *Let e be an idempotent of a ring R. For any $s \in eRe$, $s \in J(eRe) = eJe$ if and only if $sR \ll eR$.*

Proof. ($\Rightarrow$). Let $s \in J(eRe) = eJe$. Then $sR \subseteq eJ$, so $sR \ll R$. Hence $sR \ll eR$.

($\Leftarrow$). Assume $sR \ll eR$. Then $sR \ll R$ and hence $sR \subseteq J$. Hence $esRe \subseteq eJe$. Thus $s = ese \in eJe$. $\square$

For a subset S of a ring R and $a \in R$, the left multiplication map $: S \to aS$ defined by $s \mapsto as$ is denoted by $(a)_L$. (Similarly, the right multiplication map $: S \to Sa$ defined by $s \mapsto sa$ is denoted by $(a)_R$.)

Proposition 1.2.3. *For an idempotent e of a ring R, the following are equivalent:*

(1) *eRe is a local ring.*

(2) *eJ is the unique maximal submodule of eR_R.*

(3) *Je is the unique maximal submodule of $_RRe$.*

Proof. It suffices to show (1) $\Leftrightarrow$ (2).

(1) $\Rightarrow$ (2). Let K be a proper submodule of eR_R and let $eR = K + L$. Then $eR/K \cong L/(L \cap K)$. Consider the canonical epimorphism $f : L \to L/(L\cap K)$. Since eR_R is projective, there exists a homomorphism $\rho : eR \to L$ such that $\operatorname{Im}\rho + \operatorname{Ker} f = L$. Since $\rho \in \operatorname{End}(eR_R)$, ρ is realized by a left multiplication $(s)_L$ for some $s \in eRe$. Since $\operatorname{Ker} f \neq L$, $\operatorname{Im}\rho$ is not small in L. By Proposition 1.2.2, $s \notin J(eRe) = eJe$. Then, since eRe is a local ring, s is unit and hence $\operatorname{Im}\rho = L = eR$. Hence $K \ll eR$, and $K \subseteq eJ$.

(2) $\Rightarrow$ (1). Since eJ is the unique maximal submodule of eR_R, we see that the set of all non-unit elements is a two-sided ideal of $\operatorname{Hom}_R(eR, eR) \cong eRe$, whence eRe is a local ring. $\square$

An idempotent e of R is called a *local* idempotent if eRe is a local ring.

Proposition 1.2.4. *Let R be a ring such that R_R is a lifting module. Then the following hold:*

(1) *$\overline{R} = R/J$ is a semisimple ring.*

(2) *If e is a primitive idempotent of R, then eJ is the unique maximal submodule of eR_R, i.e., eRe is a local ring.*

(3) *Every complete set of orthogonal (primitive) idempotents of $\overline{R} = R/J$ lifts to a complete set of orthogonal (primitive) idempotents of R.*

Proof. (1). Though this follows from Theorem 1.1.22, we give a direct proof. Let A be a submodule of R_R with $A \supseteq J$. We put $\overline{A} = A/J$ and $\overline{R} = R/J$. We may show that $\overline{A} \oplus^< \overline{R}$. Since R_R is lifting, there exists a decomposition $R_R = A^* \oplus A^{**}$ such that $A^* \subseteq A$ and $A \cap A^{**} \ll R$. Hence $\overline{A} = \overline{A^*} \oplus^< \overline{R}$, as required.

(2). Consider a proper submodule $K_R \subsetneq eR$. Since eR_R is an indecomposable lifting module, we have $K \ll eR$. Hence $K \subseteq eJ$. Therefore eJ is the unique maximal submodule of eR_R.

(3). Let $\overline{R} = \overline{g_1 R} \oplus \cdots \oplus \overline{g_n R}$, where $\{\overline{g_1}, \ldots, \overline{g_n}\}$ is a complete set of orthogonal idempotents of $\overline{R}$. Let $\varphi : R \to \overline{R}$ be the canonical epimorphism. Since R_R is lifting, there exists a decomposition $R_R = A_i \oplus A_i^*$ such that $A_i \subseteq \varphi^{-1}(\overline{g_i R})$ and $\varphi^{-1}(\overline{g_i R}) \cap A_i^* \ll A_i^*$ for $i = 1, 2, \ldots, n$. Then $R = A_1 + \cdots + A_n + \mathrm{Ker}\,\varphi$. Since $\mathrm{Ker}\,\varphi \ll R_R$, this implies that $R = A_1 + \cdots + A_n$. Moreover, since $A_i \cap (\sum_{j \neq i} A_j) \ll R_R$, we see that $R = A_1 \oplus \cdots \oplus A_n$ by Theorem 1.1.22. Then $\overline{A_i} = \overline{g_i R}$ for $i = 1, \ldots, n$. Now take a complete set $\{e_1, \ldots, e_n\}$ of orthogonal idempotents of R such that $A_i = e_i R$ for $i = 1, 2, \ldots, n$. Since $\varphi(1) = \overline{1} = \overline{e_1} + \cdots + \overline{e_n} = \overline{g_1} + \cdots + \overline{g_n}$, we see that $\overline{e_i} = \overline{g_i}$ for $i = 1, \ldots, n$. $\qquad\square$

Proposition 1.2.5. *If M_R has a projective cover, then, for any epimorphism $\varphi : P \to M$, where P_R is a projective module, there exists a decomposition $P = P_1 \oplus P_2$ such that $P_1 \subseteq \mathrm{Ker}\,\varphi$ and $\varphi|_{P_2} : P_2 \to M$ is a projective cover of M.*

Proof. Let $f : Q \to M$ be a projective cover. Then we have a homomorphism $h : P \to Q$ satisfying $fh = \varphi$. Noting that $\mathrm{Ker}\,f \ll Q$, we see that h is epimorphic. Since Q is projective, h splits, i.e., there exists an R-homomorphism $g : Q \to P$ such that $hg = 1_Q$, and hence $P = \operatorname{Im} g \oplus \operatorname{Ker} h$. Put $P_2 = \operatorname{Im} g$ and $P_1 = \operatorname{Ker} h$. Then $P_1 \subseteq \operatorname{Ker} \varphi$ since $\operatorname{Ker} h \subseteq \operatorname{Ker} fh = \operatorname{Ker} \varphi$. Since $P_2 \cong Q$ by $h|_{P_2}$ and $fh|_{P_2} = \varphi|_{P_2}$, we see that $\varphi|_{P_2} : P_2 \to M$ is a projective cover of M. $\qquad\square$

Proposition 1.2.6. *Let P be a projective module. Then the following are equivalent:*

    (1) *Every factor module of P has a projective cover.*

    (2) *P is a lifting module.*

    Proof. (1) $\Rightarrow$ (2). Let A be a submodule of P and let $\varphi : P \to P/A$ be the canonical epimorphism. Then, by Proposition 1.2.5, there exists a decomposition $P = P_1 \oplus P_2$ such that $P_1 \subseteq A$ and $\varphi|_{P_2} : P_2 \to P/A$ is a projective cover. Since $A \cap P_2 \subseteq \operatorname{Ker} \varphi|_{P_2} \ll P_2$, P satisfies the lifting property for A.

    (2) $\Rightarrow$ (1). We must show that, for any submodule A of P, P/A has a projective cover. Since P is a lifting module, we have a decomposition $P = P_1 \oplus P_2$ such that $P_1 \subseteq A$ and $A \cap P_2 \ll P_2$. Then $\varphi|_{P_2} : P_2 \to P/A$ is epimorphic and $\operatorname{Ker} \varphi|_{P_2} = A \cap P_2$. Thus $\varphi|_{P_2} : P_2 \to P/A$ is a projective cover of P/A. $\qquad\square$

Using similar arguments, we can show the following two propositions.

Proposition 1.2.7. *Let R be a ring. The following are equivalent:*

    (1) *Every cyclic right R-module has a projective cover.*

    (2) *R_R is a lifting module.*

Proposition 1.2.8. *Let R be a ring. The following are equivalent:*

    (1) *Every simple right R-module has a projective cover.*

    (2) *R_R satisfies the lifting property for simple factor modules.*

Proposition 1.2.9. *If $P_1, \ldots, P_n$ are projective lifting R-modules, then $P = P_1 \oplus \cdots \oplus P_n$ is a lifting module.*

Proof. Since any projective lifting module is a discrete module, the statement follows from Theorem 1.1.20. □

Proposition 1.2.10. *Let R be a ring such that R/J is semisimple and every idempotent of $\overline{R} = R/J$ lift modulo J. Then R satisfies the lifting property for simple factor modules.*

Proof. Let M be a maximal right ideal. By the assumption, we can take an idempotent e of R such that $\overline{eR} = \overline{M}$. Then $\overline{(1 - e)R}$ is simple and $eR + J = M + J = M$. Hence $M = eR \oplus (M \cap (1 - e)R) \subseteq eR \oplus (1 - e)J$. Since $\overline{(1 - e)R}$ is simple, $(1 - e)J$ is the unique maximal submodule of $(1 - e)R_R$. Hence $M \cap (1 - e)R \ll (1 - e)R$. □

Proposition 1.2.11. *Let R be a ring and A a right ideal of R. If $R/(A + J)_R$ has a projective cover, then so does R/A_R.*

Proof. Consider the canonical epimorphisms: $R \xrightarrow{f} R/A \xrightarrow{g} R/(A+J)$. Then, by Proposition 1.2.5, we can take an idempotent $e \in R$ for which $gf|_{eR} : eR \to R/(A + J)$ is a projective cover. Then $\mathrm{Ker}(gf|_{eR}) \ll eR$. Since $R = eR + A + J$, we have $R = eR + A$. Hence $f|_{eR} : eR \to R/A$ is epimorphic. Since $\mathrm{Ker}(f|_{eR}) \subseteq \mathrm{Ker}(gf|_{eR}) \ll eR$, $f|_{eR} : eR \to R/A$ is a projective cover. □

Proposition 1.2.12. *Let R be a ring such that R_R satisfies the lifting property for simple factor modules. Then R_R is a lifting module. In other words, if every simple R-module has a projective cover, then every cyclic R-module has a projective cover. (Hence R is a lifting module by Proposition 1.2.7.)*

Proof. Let M be a maximal right ideal of R. By assumption, we have a decomposition $R = X \oplus Y$ such that $X \subseteq M$ and $M \cap Y \ll Y$. Hence $(M + J)/J = (X + J)/J$ and $R/J = (X + J)/J \oplus (Y + J)/J$. Thus every maximal right ideal of R/J is a direct summand, and hence R/J is semisimple by Proposition 1.2.1. Let $A_R \subseteq R_R$. We show that R/A has a projective cover. By Proposition 1.2.11, we may assume that $J \subseteq A$. Noting that $(R/J)/(A/J) \cong R/A$, we can show that R/A is expressible as a direct sum of simple submodules. Since any simple module has a projective cover, R/A has a projective cover. $\square$

Theorem 1.2.13. *Let R be a ring. Then the following are equivalent:*

(1) *R is semiperfect.*

(2) *R/J is semisimple and idempotents of R/J lift modulo J.*

(3) *R/J is semisimple and every complete set of orthogonal (primitive) idempotents of R/J lifts to a complete set of orthogonal (primitive) idempotents of R.*

(4) *R is expressed as $R = e_1 R \oplus \cdots \oplus e_n R$, where $\{e_i\}_{i=1}^n$ is a complete set of orthogonal primitive idempotents of R and each e_i is a local idempotent.*

(5) *R is expressed as $R = e_1 R \oplus \cdots \oplus e_n R$, where $\{e_i\}_{i=1}^n$ is a complete set of orthogonal primitive idempotents of R and each $e_i J$ is the unique maximal submodule of $e_i R_R$.*

(6) *Every cyclic right R-module has a projective cover.*

(7) *Every simple right R-module has a projective cover.*

(8) *Every finitely generated projective right R-module is a lifting module.*

(9) *R_R is a lifting module.*

(10) *R_R satisfies the lifting property for simple factor modules.*

Proof. $(1) \Rightarrow (6) \Rightarrow (7)$ are obvious. $(3) \Rightarrow (4)$ is obvious. $(1) \Leftrightarrow (8)$ follows from Proposition 1.2.6. Proposition 1.2.7 gives $(6) \Leftrightarrow (9)$. (7)

$\Leftrightarrow$ (10) follows from Proposition 1.2.8. (8) $\Rightarrow$ (9) $\Rightarrow$ (10) are obvious. Proposition 1.2.12 gives (10) $\Rightarrow$ (9). (9) $\Rightarrow$ (8) follows from Proposition 1.2.9. Proposition 1.2.4 gives (9) $\Rightarrow$ (3), (4). (2) $\Rightarrow$ (10) follows from Proposition 1.2.10. (3) $\Rightarrow$ (2) is obvious. (4) $\Leftrightarrow$ (5) follows from Proposition 1.2.3.

We finish by showing (5) $\Rightarrow$ (9). Since $e_i J$ is the unique maximal submodule of $e_i R_R$, any proper submodule of $e_i R_R$ is small. Hence each $e_i R_R$ is a lifting module and hence R is a lifting module by Proposition 1.2.9.

$\square$

We note that the notion of a semiperfect ring is left-right symmetric by (2) in Theorem 1.2.13.

Let R be a semiperfect ring. By Theorem 1.2.13, R has a complete set of orthogonal primitive idempotents. Henceforth, we let $Pi(R)$ denote a complete set of orthogonal primitive idempotents of R. Let $Pi(R) = \{e_1, ..., e_n\}$. In case $e_i R_R \not\cong e_j R_R$ for any distinct $i, j \in \{1, ..., n\}$, R is called a *basic* semiperfect ring. If R is not basic, we may partition $E = E_1 \cup \cdots \cup E_k$ such that, for any $e \in E_i$ and any $f \in E_j$, $e R_R \cong f R_R$ if and only if $i = j$. If we choose one e_i in E_i for each $i = 1, ..., k$, and set $e = e_1 + \cdots + e_k$, then eR is a finitely generated projective generator in *Mod-R*, and hence R is Morita equivalent to eRe. (eRe is called a *basic* subring of R.) For this reason, we usually restrict our attention to basic semiperfect rings.

Let R be a basic semiperfect ring with $Pi(R) = \{e_1, ..., e_n\}$, and represent R as

$$R = \begin{pmatrix} e_1 R e_1 & e_1 R e_2 & \cdots & \cdots & e_1 R e_n \\ e_2 R e_1 & e_2 R e_2 & \cdots & \cdots & e_2 R e_n \\ & \cdots & \cdots \cdots \\ e_n R e_1 & e_n R e_2 & \cdots & \cdots & e_n R e_n \end{pmatrix}.$$

Then we have

$$J = \begin{pmatrix} e_1 J e_1 & e_1 R e_2 & \cdots & \cdots & e_1 R e_n \\ e_2 R e_1 & e_2 J e_2 & \cdots & \cdots & e_2 R e_n \\ & & \cdots & \cdots \cdots & \cdots \\ e_n R e_1 & e_n R e_2 & \cdots & \cdots & e_n J e_n \end{pmatrix}.$$

Using this representation, we can see that, for any subset $\{f_1, \ldots, f_k\} \subseteq Pi(R)$, fRf is also a basic semiperfect ring, where $f = f_1 + \cdots + f_k$.

Theorem 1.2.14. *Let R be a semiperfect ring. For two complete sets $\{e_i\}_{i=1}^n$ and $\{f_i\}_{i=1}^m$ of orthogonal primitive idempotents of R, the following hold:*

(1) $n = m$.

(2) *There exist a permutation ρ of $\{1, \ldots, n\}$ and isomorphisms g_i : $e_i R_R \cong f_{\rho(i)} R_R$ for any i.*

(3) *There exists an inner automorphism τ of R satisfying $\tau(e_i) = f_{\rho(i)}$ for any i.*

Proof. (1), (2). These follow from the Krull-Remak-Schmidt-Azumaya Theorem.

(3). We take $\{g_i\}_{i=1}^n$ from (2). The direct sum map $g = \oplus_{i=1}^n g_i$ is an automorphism of R_R such that $g(e_i R) = f_{\rho(i)} R$. If $x = g(1)$, then $g(r) = g(1)r = xr$ for any $r \in R$. Since g is an automorphism, $1 = g(a) = xa$ for some $a \in R$. Since $1 = xa$, we see that $aR_R \cong R_R$ and $R = aR \oplus (1 - ax)R$. Then using (1), we see that $(1 - ax)R = 0$, so $1 = ax$ and hence x is a unit of R. Hence $xe_i Rx^{-1} = xe_i R = g(e_i R) = f_{\rho(i)} R$ for any i. Since $1 = \sum_{i=1}^n e_i = x1x^{-1} = \sum_{i=1}^n xe_i x^{-1} = \sum_{i=1}^n f_{\rho(i)}$ and $xe_i x^{-1} \in f_{\rho(i)} R$, we see that $x^{-1} e_i x = f_{\rho(i)}$ for any i. $\qquad\square$

A subset I of a ring R is called *right T-nilpotent* if, for every sequence $a_1, a_2, \ldots$ in I, there exists $n \in \mathbb{N}$ with $a_n a_{n-1} \cdots a_1 = 0$. (Similarly, I is called *left T-nilpotent* if, for any sequence $a_1, a_2, \ldots$ in I, we have $a_1 a_2 \cdots a_n = 0$ for some n.) We note that, if I is left or right T-nilpotent, then it is nil because $a, a, \ldots$ is a sequence in I whenever $a \in I$.

Theorem 1.2.15. (cf. [5]) *Let I be a right ideal of a ring R. Then the following are equivalent:*

(1) *I is right T-nilpotent.*

(2) *$MI \neq M$ for every non-zero right R-module M.*

(3) *$MI \ll M$ for every non-zero right R-module M.*

(4) *$FI \ll F$ for the countably generated free module $F = R^{(\mathbb{N})}$.*

The following theorem due to Bass is one of the fundamental results in ring theory. (For its proof, refer to Anderson-Fuller [5].)

Theorem 1.2.16. ([21]) *The following are equivalent for a ring R:*

(1) *R is right perfect.*

(2) *R/J is semisimple and J is right T-nilpotent.*

(3) *R/J is semisimple and every non-zero right R-module contains a maximal submodule.*

(4) *Every right flat R-module is projective.*

(5) *R satisfies DCC on principal left ideals.*

(6) *R contains no infinite set of orthogonal idempotents and every non-zero left R-module contains a minimal submodule.*

(7) *Every countably generated free right R-module is lifting.*

We now give a further characterization perfect rings using the lifting property.

Theorem 1.2.17. *Let R be a ring. The following are equivalent:*

(1) *R is a right perfect ring.*

(2) *Every projective right R-module is a lifting module.*

(3) *Every quasi-projective right R-module is a lifting module.*

(4) *$R^{(\mathbb{N})}$ is a lifting module.*

Proof. (1) $\Leftrightarrow$ (2). This follows from Proposition 1.2.6.

(2) $\Rightarrow$ (3). Let Q_R be a quasi-projective module and let A be a submodule of Q. Consider the canonical epimorphism $f : Q \to Q/A$. Now choose

an epimorphism $g : P \to Q$, where P_R is a projective module. Then, since P is a lifting module, by Proposition 1.2.5, there is a decomposition $P = P_1 \oplus P_2$ such that $P_1 \subseteq g^{-1}(A)$ and $fg|_{P_2} : P_2 \to Q/A$ is a projective cover. Since Q is quasi-projective, the decomposition $P = P_1 \oplus P_2$ induces a direct decomposition $Q = g(P_1) \oplus g(P_2)$ by Theorem 1.1.24 (4). Thus it follows that $g(P_1) \subseteq A$ and $g(P_2) \cap A \ll g(P_2)$.

(3) $\Rightarrow$ (2) is obvious.

(1) $\Rightarrow$ (4) follows from Theorem 1.2.16.

(4) $\Rightarrow$ (1). By (4), R is a semiperfect ring and R/J is semisimple. Since $R^{(\mathbb{N})}$ is lifting, there exists a decomposition $R^{(\mathbb{N})} = X \oplus Y$ such that $X \subseteq J(R^{(\mathbb{N})})$ and $J(R^{(\mathbb{N})}) \cap Y \ll Y$. Since $J(R^{(\mathbb{N})}) = J(X) \oplus J(Y)$ and $X \subseteq J(R^{(\mathbb{N})})$, we see that $J(X) = X$, which implies $X = 0$ and $R^{(\mathbb{N})}J = J(R^{(\mathbb{N})}) \ll R^{(\mathbb{N})}$. Hence, by Theorem 1.2.15, J is right T-nilpotent. Thus R is a right perfect ring. $\qquad\square$

Theorem 1.2.18. ([113]) *For a projective module P, the following are equivalent:*

(1) *P is a lifting module.*

(2) (a) *$J(P) \ll P$,*

     (b) *$\overline{P} = P/J(P)$ is semisimple, and*

     (c) *every decomposition of $\overline{P}$ lifts to a decomposition of P.*

Proof. Note that, for any indecomposable projective lifting module T, $J(T) = TJ$ is a maximal submodule.

(1) $\Rightarrow$ (2). (a) There exists a direct decomposition $P = P_1 \oplus P_2$ such that $J(P) = P_1 \oplus (J(P) \cap P_2)$ and $J(P) \cap P_2 \ll P$. Then $J(P_1) = P_1$ and hence, by Theorem 1.1.2, $P_1 = 0$. Therefore $P = P_2$ and hence $J(P) \ll P$. (b) By Theorem 1.1.22, P can be expressed as a direct sum $\oplus_{i \in I} P_i$ of cyclic indecomposable projective lifting modules $\{P_i\}_{i \in I}$. Since $J(P_i)$ is the unique maximal submodule of P_i, $\overline{P}$ is semisimple. (c) Consider a decomposition $\overline{P} = \oplus_{i \in I} \overline{A_i}$. Then $A_i = A_i^* \oplus B_i$ for some direct summand

A_i^* of P and some small submodule B_i of P. Since $J(P) \ll P$, it follows that $P = \sum_{i \in I} A_i^*$ and $\overline{P} = \oplus_{i \in I} \overline{A_i^*}$. By Theorem 1.1.23, this implies $P = \oplus_{i \in I} A_i^*$.

(2) $\Rightarrow$ (1). Let A be a submodule of P. Then $\overline{P} = \overline{A} \oplus \overline{B}$ for some submodule B by (b). From (c), this decomposition lifts to a decomposition $P = P_1 \oplus P_2$ with $\overline{P_1} = \overline{A}$ and $\overline{P_2} = \overline{B}$. Let $\pi \colon P = P_1 \oplus P_2 \to P_1$ be the projection. Then it follows from $\overline{P_1} = \overline{A}$ and (a) that $\pi(A) = P_1$. Since P is projective, there exists a direct summand A^* of P such that $A = A^* \oplus (A \cap P_2)$. Thus, by $\overline{P_1} = \overline{A}$ and (a), we see that $A \cap P_2 \ll P_2$. $\square$

Theorem 1.2.19. ([146], cf. [70] or [188]) *Every projective right R-module over a right perfect ring is a discrete module, and hence has the exchange property.*

Proof. Let P_R be a projective module. By Theorem 1.2.17, P is lifting, and hence P is discrete from Remark 1.1.21 and Theorem 1.1.22. $\square$

1.3 Frobenius Algebras, and Nakayama Permutations and Nakayama Automorphisms of QF-Rings

In this section, we recall the definition of Frobenius algebras and Nakayama automorphisms for Frobenius algebras, and we introduce Nakayama automorphisms for Frobenius rings for later use.

Let R be an n-dimensional algebra over a field k; say $R = u_1 k \oplus \cdots \oplus u_n k$. For any $a \in R$, there are $n \times n$ matrices $L(a)$ and $R(a) \in (k)_n$ satisfying

$$a(u_1, \cdots, u_n) = (u_1, \cdots, u_n) R(a), \quad {}^T(u_1, \cdots, u_n) a = L(a)^T (u_1, \cdots, u_n).$$

These $R(a)$ and $L(a)$ are said to be *right* and *left regular representations*, respectively. If there exists a regular $n \times n$ matrix $P \in (k)_n$ satisfying $PL(a) = R(a)P$, then $R(a)$ and $L(a)$ are said to be *equivalent*. Now R

is called a *Frobenius algebra* if, for any $a \in R$, its right and left regular representations are equivalent.

There are several characterizations of Frobenius algebras. We present three versions:

(A) Put $R^* = \mathrm{Hom}_k(R, k)$. Then R^* is an (R, R)-bimodule. Then R is a Frobenius algebra if and only if $R \cong R^*$ as right R-modules. More generally, R is called a *quasi-Frobenius algebra* if R_R and R_R^* have the same distinct representative indecomposable components, that is, for any indecomposable direct summand P_R of R_R, there exists a direct summand T_R of R_R^* such that $P \cong T$ and a similar condition holds for any indecomposable direct summand of R_R^*. In addition, if $R \cong R^*$ as (R, R)-modules, then R is called a *symmetric algebra*. If $S(eR_R) \cong T(eR_R)$ and $S(_RRe) \cong T(_RRe)$ for any primitive idempotent e of R, then R is called a *weakly symmetric algebra*.

(B) R is a Frobenius algebra if and only if the following hold:

For any right ideal A and any left ideal B of R,

$$rl(A) = A, \quad lr(B) = B,$$
$$\dim(A) + \dim(l(A)) = \dim(R),$$
$$\dim(B) + \dim(r(B)) = \dim(R),$$

where $l(X)$ and $r(X)$ denote the left annhilator ideal and right annihilator ideal of X, respectively, and $\dim(X)$ denotes the dimension of X over k.

(C) We arrange $Pi(R)$ as $Pi(R) = \{e_{ij}\}_{i=1,j=1}^{m, \ p(i)}$ with $e_{ij}R_R \cong e_{kl}R_R$ if $i = k$, and $e_{ij}R_R \not\cong e_{kl}R_R$ if $i \le k$. Put $e_i = e_{i1}$. R is a Frobenius algebra if and only if the following hold:

(a) There is a pemutation of π of $\{e_1, e_2, \ldots, e_m\}$:

$$\pi = \begin{pmatrix} e_1 & e_2 & \cdots & e_m \\ e_{\pi(1)} & e_{\pi(2)} & \cdots & e_{\pi(m)} \end{pmatrix}$$

such that $S(e_iR) \cong e_{\pi(i)}R/e_{\pi(i)}J$ holds for $i = 1, \ldots, m$.

(b) $m(i) = m(\pi(i))$ holds for $i = 1, \ldots, m$.

In view of the characterization (C), the field k does not appear. From this point of view, in [133] Nakayama called R a quasi-Frobenius algebra, and introduced quasi-Frobenius rings and *Frobenius rings* as artinian rings with the condition (a) and ones with the conditions (a) and (b), respectively. It is easy to see that a QF-ring R is a Frobenius ring iff $S(R_R)_R \cong (R/J)_R$.

Remark 1.3.1.

(1) The permutation (a) above for a quasi-Frobenius algebra R or for a quasi-Frobenius ring R is called a Nakayama permutation of R. By Theorem 1.2.14, we note that a Nakayama permutation is uniquely determined up to an inner automorphism.

(2) Let R be a basic quasi-Frobenius ring R with $Pi(R) = \{e_1, \ldots, e_m\}$. For each $e_i R$, we take $f_i \in Pi(R)$ such that $f_i R$ is a projective cover of $e_i R$. Then the permutation

$$\begin{pmatrix} e_1 & e_2 & \cdots & e_m \\ e_{\pi(1)} & e_{\pi(2)} & \cdots & e_{\pi(m)} \end{pmatrix}$$

is a Nakayama permutation.

Next we recall Nakayama automorphisms for Frobenius algebras and introduce Nakayama automorphisms as rings for QF-rings. Let R be a Frobenius algebra and let $\varphi : R_R \to R^* = \operatorname{Hom}_k(R, k)_R$ be an isomorphism. Such φ is called a *hyper plane*. We recall the definition of the Nakayama automorphism σ of R. Firstly the map $f : R \times R \to k$ given by $(r, s) \mapsto \varphi(r)(s)$ is a nonsingular associative k-bilinear map, where "non-singular" means that $f(a, A) = 0$ implies $a = 0$, and "associative" means that $f(ab, c) = f(a, bc)$ for any a, b, $c \in R$. For each $a \in R$, there exists a unique b for which $f(a, x) = f(x, b)$ holds for any $x \in R$. Then the map $\sigma : R \to R$, $a \mapsto b$ is a k-algebra automorphism of R and is called a Nakayama automorphism of R. Though this Nakayama automorphism σ depends on the choice of the isomorphism φ, it is known that σ is uniquely determined up to an inner automorphism, that is, for another (similarly defined) Nakayama automorphism σ' with respect to other hyper plane φ', there exists a unit u in R such that $\sigma'(x) = u\sigma(x)u^{-1}$ for any x in R.

Nakayama automorphisms have a remarkable property which effects Nakayama permutations as follows:.

Theorem 1.3.2. ([133]) *Let R be a Frobenius algebra. For any its Nakayama permutation:*

$$\pi = \begin{pmatrix} e_1 & e_2 & \cdots & e_m \\ e_{\pi(1)} & e_{\pi(2)} & \cdots & e_{\pi(m)} \end{pmatrix}$$

there exists a Nakayama automorphism σ which induces this permutation, that is, $\sigma(e_i) = e_{\pi(1)}$ for $i = 1, \ldots, m$.

By adopting the property in this theorem, we introduce *Nakayama automorphisms for a Frobenius ring R* as a ring automorphism σ which induces a Nakayama permutation of R.

Remark 1.3.3.

(1) Nakayama automorphisms for Frobenius algebras are Nakayama automorphisms as QF-rings. However Nakayama automorphisms as QF-rings for Frobenius algebras are not necessarily Nakayama automorphisms as algebras (see Example 1.3.4 below). Therefore we use the term "Nakayama automorphism as a ring" in a broad sense than "Nakayama automorphism as an algebra".

(2) Frobenius algebras have always Nakayama automorphisms as algebras. But, in general QF-rings need not have Nakayama automorphisms as rings as we see in Example 5.3.2 due to Koike [98].

Example 1.3.4. Let k be a field, let p be a non-zero element in K and let $R = k\langle x, y \rangle / (xy - pyx, x^2, y^2)$. Then R is a local *Frobenius algebra*. By direct calculation, we can show that the map $\sigma : R \to R$ defined by

$$a + b\overline{x} + c\overline{y} + d\overline{xy} \;\mapsto\; a + pb\overline{y} + c\overline{x} + pd\overline{yx}$$

for any $a, b, c, d \in K$ is a Nakayama automorphism of R as an algebra with $\sigma(\overline{x}) = p\overline{y}$ and $\sigma(\overline{y}) = \overline{x}$. Since σ is neither the identity map nor an inner automorphism, the identity map is not a Nakayama automorphism of R as an algebra. Since R is a local ring, the identity map is of course a Nakayama automorphism of R as a ring.

We close this section with the following known facts:

(1) We have the following hierarchy:
symmetric algebra $\Rightarrow$ weakly symmetric algebra $\Rightarrow$ Frobenius algebra $\Rightarrow$ quasi-Frobenius algebra.

(2) Basic quasi-Frobenius algebras are Frobenius algebras.

(3) For a finite-dimensional algebra R over a field k, R is a quasi-Frobenius algebra if and only if R is a quasi-Frobenius ring.

(4) Let R be a Frobenius k-algebra and let σ be a Nakayama automorphism of R. Then, for any unit $u \in R$, the map: $R \to R$, $x \mapsto u\sigma(x)u^{-1}$ is also a Nakayama automorphism of R.

(5) A Nakayama automorphism of a symmetric algebra is the identity map of R.

(6) For a finite group G and a field k, the group algebra kG is a symmetric algebra. Therefore its Nakayama automorphisms as an algebra or a ring are the identity map.

For more detailed information on these algebras, the reader is referred to Lam [107], Nagao-Tushima [130] and Yamagata [189].

1.4 Notation in Matrix Representations of Rings

Let R be a ring and $\{e_i\}_{i=1}^n$ a complete set of orthogonal idempotents of R. We may represent R as

$$R = \begin{pmatrix} e_1 R e_1 & \cdots & e_1 R e_n \\ & \cdots & \\ e_n R e_1 & \cdots & e_n R e_n \end{pmatrix}.$$

In this representation, for each $x \in e_i R e_j$, we let $\langle x \rangle_{ij}$ denote the matrix whose (i,j) entry is x and all other entries are 0. Moreover, for any $X \subseteq e_i R e_j$, we let $\langle X \rangle_{ij} = \{ \langle x \rangle_{ij} \mid x \in X \}$. We also use these notations when we consider other generalized matrix rings.

Let R and T be rings with matrix representations:

$$R = \begin{pmatrix} A_{11} & \cdots & A_{1n} \\ & \cdots & \\ A_{n1} & \cdots & A_{nn} \end{pmatrix}, \quad T = \begin{pmatrix} B_{11} & \cdots & B_{1n} \\ & \cdots & \\ B_{n1} & \cdots & B_{nn} \end{pmatrix}.$$

A ring homomorphism

$$\tau = \begin{pmatrix} \tau_{11} & \cdots & \tau_{1n} \\ & \cdots & \\ \tau_{n1} & \cdots & \tau_{nn} \end{pmatrix} : \begin{pmatrix} A_{11} & \cdots & A_{1n} \\ & \cdots & \\ A_{n1} & \cdots & A_{nn} \end{pmatrix} \rightarrow \begin{pmatrix} B_{11} & \cdots & B_{1n} \\ & \cdots & \\ B_{n1} & \cdots & B_{nn} \end{pmatrix}$$

is said to be a *matrix ring homomorphism* when, for each i and j, τ_{ij} is a map of A_{ij} to B_{ij} and $\tau(\langle x \rangle_{ij}) = \langle \tau_{ij}(x) \rangle_{ij}$ for any $x \in A_{ij}$.

Let R be a ring with a matrix representation:

$$R = \begin{pmatrix} Q_1 & A_{12} & \cdots & A_{1s} & \cdots & A_{1n} \\ A_{21} & Q_2 & \ddots & & & A_{2n} \\ \vdots & \ddots & \ddots & \ddots & & \vdots \\ A_{s1} & & \ddots & Q_s & \ddots & A_{sn} \\ \vdots & & & \ddots & \ddots & \vdots \\ A_{n1} & A_{n2} & \cdots & A_{ns} & \cdots & Q_n \end{pmatrix}.$$

Let T be any ring and let $Q_s \in \{Q_1, \ldots, Q_n\}$. If there exists a ring isomorphism $\rho : T \rightarrow Q_s$, then, by replacing Q_s with T in the matrix representation, we can make a new ring R' which is canonically isomorphic to R. We represent R' by

$$R' = \begin{pmatrix} Q_1 & A_{12} & \cdots & \cdots & A_{1s} & \cdots & & \cdots & & A_{1n} \\ A_{21} & Q_2 & \ddots & & & & & & & A_{2n} \\ \vdots & \ddots & \ddots & \ddots & & & & & & \vdots \\ \vdots & & \ddots & Q_{s-1} & \ddots & & & & & \vdots \\ A_{s1} & & & \ddots & T^\rho & \ddots & & & & A_{sn} \\ \vdots & & & & \ddots & Q_{s+1} & \ddots & & & \vdots \\ \vdots & & & & & \ddots & \ddots & & & A_{n-1,n} \\ A_{n1} & A_{n2} & \cdots & \cdots & A_{ns} & \cdots & A_{n,n-1} & & & Q_n \end{pmatrix}$$

and identify R with R'.

COMMENTS

Extending modules and lifting modules are important as generalizations of injective modules and supplemented projective modules, respectively, and these modules have been extensively studied as mentioned in the preface. (A module M is called a supplemented module if, for any submodule X, there exists a submodule Y such that $X \cap Y \ll M$. Right projective modules over right perfect rings are supplemented modules.) Relative generalized injectivity and relative generalized projectivity are introduced in the study of the following open problems: When is the direct sum of extending modules extending? And dually, when is the direct sum of lifting modules lifting? Theorems 1.1.18 and 1.1.20 give partial answers for these problems. More work on semiperfect rings and perfect rings by using lifting modules in Section 1.2 is presented in Oshiro [146]. For Morita duality, the reader is referred to Anderson-Fuller [5], Faith [47], Lam [107], and Xue [184], and, in addition, to Nicholson-Yousif [142], and Tachikawa [172] for QF-rings.

Chapter 2

A Theorem of Fuller

Let R be a left or right artinian ring and let $e \in Pi(R)$. In 1969, Fuller showed the following useful theorem.

Theorem 2.A (Fuller's Theorem [54]). *The following are equivalent:*

(1) eR_R *is injective.*

(2) *There exists* $f \in Pi(R)$ *such that*

$$(2^*) \quad S(eR_R) \cong T(fR_R) \text{ and } S(_RRf) \cong T(_RRe).$$

(3) *There exists* $f \in Pi(R)$ *such that*

$$(3l) \quad l_{eR}(r_{Rf}(I)) = I \text{ for any left } eRe\text{-submoudles } I \text{ of } eR,$$

and

$$(3r) \quad r_{Rf}(l_{eR}(K)) = K \text{ for any right } fRf\text{-submodule } K \text{ of}$$

Rf,

where $r_{Rf}(I) = \{\, a \in Rf \mid Ia = 0 \,\}$ *and* $l_{eR}(K) = \{\, b \in eR \mid bK = 0 \,\}$.

In 1993, Baba and Oshiro extended Fuller's Theorem to semiprimary rings by using the notion of relative simple injectivity due to Harada [66] (see also ([69]).

Let R be a semiperfect ring and $e, f \in Pi(R)$. $(eR; Rf)$ is called an *i-pair* if the condition (2^*) above is satisfied.

Baba-Oshiro's version is the following:

Theorem 2.B (Baba-Oshiro's Theorem [20]). *Let R be a semiprimary ring and let $e, f \in Pi(R)$. Then the following hold:*

(a) eR_R *is injective if and only if the conditions (2) and (3r) of Theorem 2.A are satisfied.*

(b) *If R satisfies the condition (2) of Theorem 2.A and the condition (∗) below, then eR_R is injective.*

(∗) *The lattice $\{r_{Rf}(X) \,|\, X \subseteq eR\}$ satisfies ACC.*

(c) *If $(eR; Rf)$ is an i-pair, where $e, f \in Pi(R)$, then the following are equivalent:*

(c1) Rf_{fRf} *is artinian.*

(c2) $_{eRe}eR$ *is artinian.*

(c3) eR_R *and $_RRf$ are injective.*

(d) *If R_R is R_R-simple-injective, then R_R is injective.*

The purpose of this chapter is to introduce these improved results. We assume that the ring R considered in this chapter is a semiperfect ring and put $J = J(R)$. For a module M_R, we note that the length of the upper Loewy series coincides with that of the Loewy lower series of M_R. Consequently, we may call it simply the *Loewy length* of M and denote it by $L(M)$.

2.1   Improved Versions of Fuller's Theorem

The following lemma is frequently used in this book:

Lemma 2.1.1. *Let $e, f \in Pi(R)$.*

(1) *If S is a simple right R-module, then Sf_{fRf} is simple or 0.*

(2) *If $S(eR_R)$ is an essential simple socle with $S(eR_R)f \neq 0$, then $S(eRf_{fRf}) = S(eR_R)f$ and it is an essential simple right fRf-submodule of eRf.*

(3) *If both $S(eRf_{fRf})$ and $S(_{eRe}eRf)$ are simple and essential in eRf,*

then $S(eRf_{fRf}) = S(_{eRe}eRf)$.

(4) *Assume that R is basic. If $S(eR_R)$ is an essential simple socle and fR_R is a projective cover of $S(eR_R)$, then*

(a) $S(eR_R) = S(eRf_{fRf})$.

Moreover if eR_R is injective, then the following hold:

(b) $S(eR_R)$ *is a simple left ideal.*

(c) $S(_RRf)$ *is an essential simple socle with projective cover* $_RRe$.

(d) $S(_RRf) = S(eR_R) = S(eRf_{fRf}) = S(_{eRe}eRf)$.

(e) $S(eRf_{fRf})_{fRf}$ *and* $_{eRe}S(_{eRe}eRf)$ *are simple.*

(f) $(eR; Rf)$ *is an i-pair.*

(g) *If $X = \mathrm{Hom}_R(fR, S(eR_R))$, then $_{eRe}X$ and X_{fRf} are simple (and so $_{eRe}X \cong S(_{eRe}eRf)$ and $X_{fRf} \cong S(eRf_{fRf})$), canonically).*

Proof. (1). If $0 \neq s \in Sf$, then $s(fRf) = sRf = Sf$. Hence Sf_{fRf} is simple or 0.

(2). Since $S(eR_R)$ is simple and $S(eR_R)f \neq 0$, $S(eR_R)f$ is a simple right fRf-submodule of eRf by (1). Thus $S(eRf_{fRf}) \supseteq S(eR_R)f$. On the other hand, $S(eR_R)f$ is an essential right fRf-submodule of $S(eRf_{fRf})$ since $S(eR_R)$ is an essential right R-submodule of eR. Therefore $S(eRf_{fRf}) = S(eR_R)f$.

(3). It is obvious that $eRe\, S(eRf_{fRf}) \subseteq S(eRf_{fRf})$, i.e., $S(eRf_{fRf})$ is a left eRe-module. Therefore we have $S(eRf_{fRf}) \supseteq S(_{eRe}eRf)$ since $S(_{eRe}eRf)$ is simple and essential in eRf. The reverse inclusion is obtained by a converse argument.

(4). (a) Since fR_R is a projective cover of $S(eR_R)$, $S(eR_R) = S(eR_R)f$. Hence, from (1), $S(eR_R) = S(eR_R)f = S(eR_R)fRf$. Again from (1), $S(eR_R)$ is a simple right fRf-module. This implies that $S(eR_R)$ is simple as a right R-module.

(b) To show that $S(eR_R)$ is a left ideal, suppose that $RS(eR_R) \not\subseteq S(eR_R)$. Then there exist h ($\neq e$) $\in Pi(R)$ and $r \in hR$ such that $rS(eR_R) \not\subseteq S(eR_R)$. Consider the homomorphism $\varphi : eR \to hR$ given by $y \mapsto ry = hrey = hry$. We show that φ is monomorphic. To prove this, suppose that there exists $0 \neq y \in S(eR_R)$ with $ry = 0$. Since $S(eR_R) \subseteq_e eR_R$, we have $s \in R$ with $0 \neq ys \in S(eR_R)$. Then $ysR = S(eR_R)$ and $0 = rysR = rS(eR_R) \not\subseteq S(eR_R)$, a contradiction. Hence φ is monomorphic, and hence isomorphic. Since R is basic, this implies that $e = h$, a contradiction. Accordingly $S(eR_R)$ is a left ideal of R.

Next we show that $_RS(eR_R)$ is simple. It suffices to show that, for non-zero elements $x, y \in S(eR_R)$, there exists $r \in R$ satisfying $rx = y$. The map $\tau : xR \to yR$ given by $xr \mapsto yr$ is an R-homomorphism. To confirm this, suppose that there exists $r' \in R$ such that $xr' = 0$ but $yr' \neq 0$. Since $x, y, yr' \in S(eR_R) = S(eR_R)fRf$, it induces that $xfr'f = 0$ but $yfr'f \neq 0$. Then $fr'f$ is an invertible element of fRf and hence $x = 0$, a contradiction. Hence τ is well defined. Since eR_R is injective, we can extend τ to an isomorphism $\tau^* : eR_R \to eR_R$. Hence we can take an element $r \in eRe$ satisfying $rx = y$ as desired.

(c) Let $0 \neq rf \in Rf$. Take $h \in Pi(R)$ with $hrf \neq 0$. From (4)(b) we see that $S(eR_R) \subseteq S(_RRf)$. Take $0 \neq x \in S(eR_R)(= S(eRf_{fRf}))$. We must show that there exists $r' \in R$ with $r'hrf = x$. We can define an epimorphism $\varphi : fR_R \to S(eR_R)$ by $f \mapsto x$. Consider the left multiplication map $(hrf)_L : fR \to hrfR$. Then we have an epimorphism $\psi : hrfR \to S(eR_R)$ with $\psi(hrf)_L = \varphi$, i.e., $\psi(hrf) = x$. Since eR_R is injective, there exists an extension map $\tilde{\psi} : hR \to eR$ of ψ. Put $r' = \tilde{\psi}(h)$. Then $x = \varphi(f) = \psi(hrf)_L(f) = \psi(hrf) = r'hrf$.

(d) From (b), (c) we see that $S(eR_R) = S(_RRf)$. Hence the statement follows from (a), (c).

(e) This follows from (2), (c).

(f) This follows from the assumption and (c).

(g) We must show that $X = eRe\varphi = \varphi fRf$ holds for any $0 \neq \varphi \in X$. Let φ and ψ be non-zero elements of X. Since fR_R is projective, there exists $\xi \in \mathrm{Hom}_R(fR, fR)$ satisfying $\varphi\xi = \psi$. This shows that $X = \varphi fRf$. On the other hand, φ and ψ induce isomorphisms $\overline{\varphi} : T(fR_R) \to S(eR_R)$ and $\overline{\psi} : T(fR_R) \to S(eR_R)$, respectively. Since eR_R is injective, we can take $\eta \in \mathrm{Hom}_R(eR, eR)$ satisfying $\overline{\psi}\overline{\varphi}^{-1} = \eta\,|_{S(eR_R)}$. Hence $\overline{\psi} = \eta\overline{\varphi}$ and hence $\psi = \eta\varphi$. As a result, $X = eRe\varphi$. $\qquad\square$

Notation:

(1) For $e, g \in Pi(R)$, we define $V\langle n.e, g\rangle$ for $n \geq 2$ as follows: $V\langle 2, e, g\rangle = eJg$ and, for $n \geq 3$, $V\langle n, e, g\rangle = \{eJh_{n-2}Jh_{n-3}Jh_{n-4}\cdots h_2Jh_1Jg \mid h_1, \cdots, h_{n-2} \in Pi(R)\}$.

(2) For an R-module M_R and $x \in M$, $|x| = i$ means $x \in S_i(M) - S_{i-1}(M)$.

Lemma 2.1.2. *Let R be a perfect ring and let $e, f \in Pi(R)$ such that $S(_RRf)$ is simple and $_RRe$ is a projective cover of $S(_RRf)$. Then the following hold for any $i \in \mathbb{N}$:*

(1) *For $a \in S_i(_RRf) - S_{i-1}(_RRf)$, there exist $g \in Pi(R)$ and $\alpha \in V\langle i, e, g\rangle$ such that $0 \neq \alpha ga \in S(_RRf)$.*

(2) *$S_i(_RRf)fJf \subseteq S_{i-1}(_RRf)$.*

Proof. (1). Let $a \in S_i(_RRf) - S_{i-1}(_RRf)$. Then $ga \in S_i(_RRf) - S_{i-1}(_RRf)$ for some $g \in Pi(R)$. We claim that there exists $u_1 \in J$ with $u_1ga \in S_{i-1}(_RRf) - S_{i-2}(_RRf)$. Suppose that $u_1g \notin Jg$, then $Rg = Ru_1g$, and hence $g = xu_1g$ for some $x \in R$, from which $ga = xu_1ga \in S_{i-1}(_RRf)$, a contradiction. Then we can take $h_1 \in Pi(R)$ with $0 \neq h_1u_1ga \in S_{i-1}(_RRf) - S_{i-2}(_RRf)$. Similarly we can take $h_2 \in Pi(R)$ and $u_2 \in J$ for which $h_2u_2h_1u_1ga \in S_{i-2}(_RRf) - S_{i-3}(_RRf)$. Continuing this procedure, we can take $h_1, h_2, \ldots, h_{i-1}$ and $u_1, u_2, \ldots, u_{i-1} \in J$ such that $0 \neq h_{i-1}u_{i-1}h_{i-2}u_{i-2}h_{i-3}\cdots h_1u_1gaf \in S(_RRf)$. Here, since $_RRe$

is a projective cover of $S(_RRf)$, we can take e as h_{i-1}. Thus putting $\alpha = eu_{i-1}h_{i-2}u_{i-2}h_{i-3}\cdots h_1u_1g$, $\alpha \in V\langle i, e, g\rangle$ and $0 \neq \alpha ga \in S(_RRf)$.

(2). When $i = 1$: Suppose that $S_1(_RRf)fJf \neq 0$ and take a non-zero element $afjf \in S_1(_RRf)fJf$ with $j \in J$. Since $S_1(_RRf)$ is simple, we see $S_1(_RRf) = S_1(_RRf)fjf$; whence $S_1(_RRf) = S_1(_RRf)(fjf)^n$ for $n = 1, 2, \cdots$, which contradicts the nilpotency of fjf. When $i \geq 2$: Suppose that $S_i(_RRf)fJf \nsubseteq S_{i-1}(_RRf)$ and take $afjf \in S_i(_RRf)fJf - S_{i-1}(_RRf)$. Then $gafjf \in S_i(_RRf)fJf - S_{i-1}(_RRf)$ for some $g \in Pi(R)$. Using (1) for $gafjf$, we can take $\alpha \in V\langle i, e, g\rangle$ such that $\alpha ga \in S(_RRf)$ and $0 \neq \alpha gafjf$, a contradiction. $\qquad \square$

Let $e, f \in Pi(R)$ such that $(eR; Rf)$ is an i-pair. We say that $(eR; Rf)$ *satisfies DCC for left (resp. right) annihilators* if the descending chain condition holds for $\{\, l_{eR}(X) \mid X_{fRf} \subseteq Rf \,\}$. Similarly the term $(eR; Rf)$ *satisfies ACC for left (or right) annihilators* is defined. Further we say that $(eR; Rf)$ *satisfies the condition* (α) *(or* (β)*)* if the following (α) (or (β)) holds:

(α) $r_{Rf}l_{eR}(X) = X$ for any right fRf-submodule X of Rf_{fRf}.

(β) $l_{eR}r_{Rf}(Y) = Y$ for any left eRe-submodule Y of $_{eRe}eR$.

The following lemma is easily shown.

Lemma 2.1.3. *Let $e, f \in Pi(R)$ with $(eR; Rf)$ an i-pair. Then the following are equivalent:*

(1) *For any $Y_{fRf} \subsetneqq X_{fRf} \subseteq Rf$, there exists $r \in eR$ such that $rY = 0$ but $rX \neq 0$.*

(2) *For any $g \in Pi(R)$ and any $Y_{fRf} \subsetneqq X_{fRf} \subseteq gRf$, there exists $r \in eR$ such that $rY = 0$ but $rX \neq 0$.*

(3) *$r_{Rf}l_{eR}(X) = X$ for any right fRf-submodule X of Rf_{fRf}, i.e., $(eR; Rf)$ satisfies the condition (α).*

(4) *$r_{gRf}l_{eRg}(X) = X$ for any $g \in Pi(R)$ and any right fRf-submodule X of Rf_{fRf}.*

Proposition 2.1.4. *Let R be a semiprimary ring and let $e \in Pi(R)$. If eR_R is simple eR-injective, then $S(eR_R)$ is simple.*

Proof. Let $t \in \mathbb{N}$ with $eJ^{t+1} = 0$ but $eJ^t \neq 0$. Let X_R be a simple submodule of eJ^t and let $S(eR_R) = X_R \oplus Y_R$. Suppose $Y \neq 0$. Let $\pi : S(eR_R) = X \oplus Y \to X$ be the projection. Since eR_R is simple eR-injective, there exists $\tilde{\pi} : eR \to eR$ which is an extension of π. Since $\operatorname{Ker} \tilde{\pi} \neq 0$, we see that $\tilde{\pi}(eR) \subseteq eJ$. It follows that $\tilde{\pi}(eR)J^t = 0$. Therefore $X = \pi(X) \subseteq \tilde{\pi}(eJ^t) = 0$, a contradiction. Hence Y must be 0 and $S(eR_R)$ is simple. $\qquad\square$

Proposition 2.1.5. *Let R be a semiprimary ring and let $e \in Pi(R)$. If eR_R is simple gR-injective for any $g \in Pi(R)$, then eR_R is injective.*

Proof. By Proposition 2.1.4, $S(eR_R)$ is simple. Let I be a non-zero submodule of gR_R and let $\varphi : I_R \to eR_R$ be a homomorphism. We show that φ can be extended to a homomorphism: $gR_R \to eR_R$. We take an integer $t_i \geq 0$ such that $\operatorname{Ker} \varphi \supseteq IJ^{t_1+1}$ but $\operatorname{Ker} \varphi \not\supseteq IJ^{t_1}$. Since $\varphi(IJ^{t_1}) = S(eR_R)$, there exists $\tilde{\varphi}_1 : gR_R \to eR_R$ such that $\tilde{\varphi}_1|_{IJ^{t_1}} = \varphi|_{IJ^{t_1}}$. If $\tilde{\varphi}_1|_I = \varphi$, then $\tilde{\varphi}_1$ is a desired homomorphism. Suppose that $\tilde{\varphi}_1|_I \neq \varphi$. Put $\varphi_2 = \varphi - (\tilde{\varphi}_1|_I) : I \to eR$. Then $\operatorname{Ker} \varphi_2 \supseteq IJ^{t_1}$. Let t_2 be an integer such that $\operatorname{Ker} \varphi_2 \supseteq IJ^{t_2+1}$ but $\operatorname{Ker} \varphi_2 \not\supseteq IJ^{t_2}$. Then $t_2 < t_1$ and $\varphi_2(IJ^{t_2}) = S(eR_R)$. Hence there exists a homomorphism $\tilde{\varphi}_2 : gR \to eR$ which is an extension of $\varphi_2|_{IJ^{t_2}}$. If $\tilde{\varphi}_2|_I = \varphi_2$, then $\tilde{\varphi}_1 + \tilde{\varphi}_2$ is a required homomorphism. In the case when $\tilde{\varphi}_2|_I \neq \varphi_2$, we consider $\varphi_3 = \varphi_2 - (\tilde{\varphi}_2|_I) : I \to eR$ and proceed with the same argument. Since this procedure must finitely terminate, there exists $n \in \mathbb{N}$ such that $\tilde{\varphi}_1 + \tilde{\varphi}_2 + \cdots + \tilde{\varphi}_n : gR \to eR$ is an extension of φ. $\qquad\square$

Proposition 2.1.6. *Let R be a semiprimary ring and let $e, f \in Pi(R)$ such that $(eR; Rf)$ is an i-pair. If $(eR; Rf)$ satisfies the condition (α), then eR_R is simple gR-injective for any $g \in Pi(R)$.*

Proof. Let $g \in Pi(R)$, A a submodule of gR_R and φ a homomorphism from A to $S(eR_R)$. Put $K = \operatorname{Ker} \varphi$. Then $Af \supsetneq Kf$. Hence, by the condition(α), there exists $r \in eRg$ such that $rA = S(eR_R)$ but $rK = 0$. This implies that $\varphi = (r'r)_L$ for some $r' \in eRe$ by Lemma 2.1.1 (1) since $S(_RRf) \cong T(_RRe)$. $\qquad\square$

Proposition 2.1.7. *Let R be a semiprimary ring and let $e, f \in Pi(R)$. If $S(eR_R) \cong T(fR_R)$ and eR_R is a simple gR-injective for any $g \in Pi(R)$, then $S(_RRf)$ is simple, $S(_RRf) = RS(eR_R)f$ and $S(_RRf) \cong T(_RRe)$; Therefore $(eR; Rf)$ is an i-pair.*

Proof. First we show that $RS(eR_R)f$ is a simple left ideal. To show this, we may show that, for non-zero elements $x, y \in RS(eR_R)f$, there exists $r \in R$ with $rx = y$. For such x, y, we claim that an R-homomorphism $\varphi : xR \to yR$ can be defined by $\varphi(x) = y$. Asssume that there exists $r \in R$ such that $xr = 0$ but $yr \neq 0$. Since $y \in RS(eR_R)f$, we have $f' \in Pi(R)$ such that $f'R_R \cong fR_R$ and $yrf' = yfrf' \neq 0$. Then there eixsts $r' \in R$ such that $frf'r'f$ is a unit element in fRf because $frf' \notin J$. Hence $x \neq 0$ induces $xrf'r'f = xfrf'r'f \neq 0$ and $xr \neq 0$, a contradiction. Since eR_R is simple eR-injective, $\varphi = (t)_L$ for some $t \in eRe$. Hence $tx = y$.

Next we show that $RS(eR_R)f$ is an essential submodule of $_RRf$. Let $0 \neq rf \in Rf$, where $r \in R$. Then $0 \neq grf \in Rf$ for some $g \in Pi(R)$. Let $\pi : fR_R \to S(eR_R)$ be an epimorphism and let $(gr)_L : fR \to grfR$. We can take a homomorphism $\varphi : grfR \to S(eR_R)$ with $\varphi(gr)_L = \pi$, i.e., the diagram

$$(gr)_L$$
$$fR \longrightarrow grfR$$
$$\searrow \pi \quad \downarrow \varphi$$
$$S(eR_R)$$

is commutative. Since eR_R is simple gR-injective, φ can be extended to a homomorphism $\tilde{\varphi} : gR \to eR$. Hence there exists $s \in eRg$ with $0 \neq srf \in S(eR_R)$. Thus $RS(eR_R)f$ is an essential submodule of Rf.

Hence $S(_RRf) = RS(eR_R)f \subseteq ReRf$, from which we can easily see that $_RRe$ is a projective cover of $S(_RRf) = RS(eR_R)f$.   □

Theorem 2.1.8. *Let R be a semiprimary ring. For $e \in Pi(R)$, the following are equivalent:*

   (1) *eR_R is injective.*

   (2) *eR_R is simple gR-injective for any $g \in Pi(R)$.*

   (3) *There exists $f \in Pi(R)$ such that $(eR; Rf)$ is an i-pair and satisfies the condition (α).*

Proof. (1) $\Rightarrow$ (2). Trivial.

(2) $\Rightarrow$ (1). This follows from Proposition 2.1.5.

(3) $\Rightarrow$ (2). This follows from Proposition 2.1.6.

(2) $\Rightarrow$ (3). By Propositions 2.1.4 and 2.1.7, there exists $f \in Pi(R)$ such that $(eR; Rf)$ is an i-pair. Let $g \in Pi(R)$ and consider submodules $K_{fRf} \subsetneq A_{fRf} \subseteq gRf$ such that A/K_{fRf} is simple. Since $S(eR_R)f_{fRf}$ is simple by Lemma 2.1.1 (1), there exists a fRf-homomorphism $\varphi : A \to S(eR_R)f$ with $\operatorname{Ker} \varphi = K$. Then we claim that the map $\varphi' : AR \to S(eR_R)$ given by $\sum_{i=1}^{n} a_i r_i \mapsto \sum_{i=1}^{n} \varphi(a_i)r_i$ is well defined. If $\sum_{i=1}^{n} a_i r_i = 0$ but $\sum_{i=1}^{n} \varphi(a_i)r_i \neq 0$, then we can take $s \in R$ such that $0 \neq \sum_{i=1}^{n} \varphi(a_i)r_i s \in S(eR_R)f$. But it follows that $0 \neq \sum_{i=1}^{n} \varphi(a_i)r_i s = \sum_{i=1}^{n} \varphi(a_i)r_i sf = \sum_{i=1}^{n} \varphi(a_i r_i sf) = \varphi((\sum_{i=1}^{n} a_i r_i)sf) = 0$, a contradiction. Now, since eR_R is simple gR-injective, there exists $r \in eRg$ such that $(r)_L = \varphi'$. This implies $rK = 0$ but $rA \neq 0$. Thus we see from Lemma 2.1.3 that $(eR; Rf)$ satisfies the condition (α).   □

Lemma 2.1.9. *Let R be a semiprimary ring and let $e, f \in Pi(R)$ such that $(eR; Rf)$ is an i-pair and satisfies DCC for left annihilators. Then, for any $g \in Pi(R)$ and any right R-submodules K and A of gR_R with $K \subsetneq A$ and $Kf \neq Af$, there exists $r \in eRg$ such that $rA \neq 0$ but $rK = 0$.*

Proof. Since $S(eR_R) \subseteq_e eR_R$ and $S(fR_R) \cong T(fR_R)$, we see that, for any $r \in eR$, $rKf = 0$ implies $rK = 0$. Hence we may show that there exists $r \in eRg$ such that $rAf \neq 0$ but $rKf = 0$. Further we may assume that Kf is a maximal right fRf-submodule of Af. Now $(S_n(_RRf) \cap Af)/(S_{n-1}(_RRf) \cap Af)_{fRf}$ is semisimple by Lemma 2.1.2 (2) because it is canonically isomorphic to a submodule of $S_n(_RRf)/S_{n-1}(_RRf)_{fRf}$. We put $n = max\{\,|x|\,|\,x \in Af\,\}$ with respect to $_RRf$. Then we can take $x_0 \in Af - Kf$ such that $|x_0| = n$. Put $V = V\langle n, e, g \rangle$. If $Kf \subseteq S_{n-1}(_RRf)$, then, clearly, $VKf = 0$ but $Vx_0 \neq 0$. Therefore we may consider the case $Kf \not\subseteq S_{n-1}(_RRf)$. In this case, we put $Q = fRf$ and express Af as

$$Af = x_0Q + x_1Q + \cdots + x_kQ + \sum_{\omega \in \Omega} x_\omega Q + (S_{n-1}(_RRf) \cap Af),$$

where

(1) $n = |x_\alpha|$ for any $\alpha \in \{0, 1, \ldots, k\} \cup \Omega$,

(2) $\{x_0Q, x_1Q, \ldots, x_kQ\} \cup \{x_\omega Q \mid \omega \in \Omega\}$ is independent modulo $(S_{n-1}(_RRf) \cap Af)_{fRf}$,

(3) $K \supseteq x_0J(Q) + \sum_{i=1}^{k} x_iQ + \sum_{\omega \in \Omega} x_\omega Q$, and

(4) $l_{eR}(\{x_1, x_2, \ldots, x_k\}) = l_{eR}(\{x_1, x_2, \ldots, x_k\} \cup \{x_\omega \mid \omega \in \Omega\})$.

First we consider the case $k = 1$ and $Kf \supseteq S_{n-1}(_RRf) \cap Af$. Then

$$Kf = x_0J(Q) + x_1Q + \sum_{\omega \in \Omega} x_\omega Q + (S_{n-1}(_RRf) \cap Af)$$

$$\subsetneq Af = x_0Q + x_1Q + \sum_{\omega \in \Omega} x_\omega Q + (S_{n-1}(_RRf) \cap Af).$$

Thus we may only show that there exists $r \in V$ such that $rx_0 \neq 0$ but $rx_1 = 0$. (By this r, $rAf \neq 0$ but $rKf = 0$.) Suppose that, for any $r \in V$, $rx_1 = 0$ implies $rx_0 = 0$. Since $|x_0| = |x_1| = n$, $_{eRe}Vx_1 \cong {}_{eRe}Vx_0$ by $rx_1 \mapsto rx_0$, where $r \in V$. Now we can take $v \in V$ such that $0 \neq vAf$. Then $vAf = eS(_RRf) = S(_{eRe}eRf) = S(eRf_{fRf})$ which is a simple right fRf-module. Therefore the kernel of the map $(v)_L : Af \to vAf$ is a maximal right fRf-submodule of Af by Lemma 2.1.1 (2), (3). If $x_1 \in \text{Ker}(v)_L$, then $vx_1 = 0$ and hence $vx_0 = 0$. This implies that $vAf = 0$, a contradiction.

Hence $x_1 \notin \mathrm{Ker}(v)_L$, and we have $Af = x_1Q + \mathrm{Ker}(v)_L$. We write $x_0 = x_1q + y$, where $q \in Q$ and $y \in \mathrm{Ker}(v)_L$. Then we note that $vx_1 \neq 0$ but $vy = 0$. On the other hand, since $\{x_0Q, x_1Q\}$ is independent modulo $(S_{n-1}({}_RRf) \cap Af)_{fRf}$, $|y|$ must be n. Hence ${}_{eRe}Vx_1 \cong {}_{eRe}Vy$ by $rx_1 \mapsto ry$, where $r \in V$. This contradicts the fact that $vx_1 \neq 0$ but $vy = 0$.

Next we consider the case $k \geq 2$ and $Kf \supseteq S_{n-1}({}_RRf) \cap Af$. We only have to show that there exists $r \in V$ such that $rx_0 \neq 0$ but $rx_1 = rx_2 = \cdots = rx_k = 0$. Suppose that, for any $r \in V$, $rx_1 = rx_2 = \cdots = rx_k = 0$ implies $rx_0 = 0$. Then we consider the map $\varphi : V(x_1, \ldots, x_k)$ $(\subseteq Vx_1 \times \cdots \times Vx_k) \to Vx_0$ given by $r(x_1, \ldots, x_k) \mapsto rx_0$, where $r \in V$. By applying the argument above for

$$K' := x_0J(Q) + x_1Q + (S_{n-1}({}_RRf) \cap Af)$$

$$\subsetneq A' = x_0Q + x_1Q + (S_{n-1}({}_RRf) \cap Af),$$

we can take $r' \in V$ such that $r'x_0 = 0$ but $r'x_1 \neq 0$. Thus $\mathrm{Ker}\,\varphi \neq 0$. Suppose that $V(x_1, 0, \ldots, 0) + V(0, x_2, 0, \ldots, 0) + \cdots + V(0, \ldots, 0, x_{l-1}, 0, \ldots, 0) \subseteq \mathrm{Ker}\,\varphi$ but $V(0, \ldots, 0, x_l, 0, \ldots, 0) \not\subseteq \mathrm{Ker}\,\varphi$. Then there exists $s \in V$ such that $(0, \ldots, 0, sx_l, sx_{l+1}, \ldots, sx_k) \in \mathrm{Ker}\,\varphi$ and $sx_l \neq 0$. We may assume that $sx_l \neq 0, \ldots, sx_{l+j} \neq 0$ and $sx_{l+j+1} = \cdots = sx_k = 0$. We take $p_{l+i} \in Q - J(Q)$ such that $sx_{l+i}p_{l+i} = sx_l$ for $i = 1, \ldots, j$ and put

$$x_i' = \begin{cases} x_i & \text{for } 1 \leq i \leq l \text{ or } l+j+1 \leq i \leq k, \\ x_ip_i - x_l & \text{for } l+1 \leq i \leq l+j. \end{cases}$$

Then we see that, for any $r \in V$, $r(x_1, \ldots, x_k) = 0$ iff $r(x_1', \ldots, x_k') = 0$, and

$$K = x_0J(Q) + \sum_{i=1}^k x_i'Q + \sum_{\omega \in \Omega} x_\omega Q + (S_{n-1}({}_RRf) \cap Af).$$

And the kernel of the map $\varphi' : V(x_1', \ldots, x_k') \to Vx_0$ given by $r(x_1', \ldots, x_k') \mapsto rx_0$ contains

$$V(x_1', 0, \ldots, 0) + V(0, x_2', 0, \ldots, 0) + \cdots + V(0, \ldots, 0, x_l', 0, \ldots, 0)$$

since $s(0, \ldots, 0, x'_l, 0 \ldots 0) = s(0, \ldots, 0, x'_l, x'_{l+1}, \ldots x'_k) \in \operatorname{Ker} \varphi'$ induces $V(0, \ldots, 0, x'_l, 0, \ldots, 0) \subseteq \operatorname{Ker} \varphi'$. By this argument, we can assume that $V(x'_1, 0, \ldots, 0) + \cdots + V(0, \ldots, 0, x'_k) \subseteq \operatorname{Ker} \varphi'$. But then $Vx_0 = \operatorname{Im} \varphi = 0$, a contradiction.

Finally we consider the case that $Kf \not\supseteq S_{n-1}(_RRf) \cap Af$. Since $Af/Kf_{fRf} \cong S(eR_R)f_{fRf}$, there exists a homomorphism $\psi : Af_{fRf} \to S(eR_R)f_{fRf}$ such that $\operatorname{Ker} \psi = Kf$. Let $l_1 \in \mathbb{N}$ such that $Kf \supseteq S_{l_1-1}(_RRf) \cap Af$ but $Kf \not\supseteq S_{l_1}(_RRf) \cap Af$. Then $\operatorname{Im} \psi|_{S_{l_1}(_RRf) \cap Af} = RS(eR_R)f$. Put $K_1 = \operatorname{Ker}(\psi|_{S_{l_1}(_RRf) \cap Af})$. Since $K_1 \supseteq S_{l_1-1}(_RRf) \cap Af$, we see by above that there exists $r \in V\langle l_1, e, g \rangle$ such that $r(S_{l_1}(_RRf) \cap Af) \neq 0$ but $rK_1 = 0$. Since $eRer(S_{l_1}(_RRf) \cap Af) = \psi(S_{l_1}(_RRf) \cap Af) = RS(eR_R)f$, we can take $r' \in eRe$ such that $\psi|_{S_{l_1}(_RRf) \cap Af} = (r'r)_L$ ($: S_{l_1}(_RRf) \cap Af \to S(eR_R)f$). Put $s_1 = r'r$ and we consider $(s_1)_L : Af_{fRf} \to S(eR_R)f_{fRf}$. If $\psi = (s_1)_L$, then $s_1 Af \neq 0$ but $s_1 Kf = 0$, i.e., s_1 is a desired element in eRg. Suppose that $\psi \neq (s_1)_L$ and put $\psi_1 = \psi - (s_1)_L$. Then $\psi_1(S_{l_1}(_RRf) \cap Af) = 0$. We take $l_2 (> l_1)$ such that $\psi_1(S_{l_2}(_RRf) \cap Af) \neq 0$ but $\psi_1(S_{l_2-1}(_RRf) \cap Af) = 0$. Then, as above, we can take $s_2 \in eRg$ such that $\psi_1|_{S_{l_2}(_RRf) \cap Af} = (s_2)_L|_{S_{l_2}(_RRf) \cap Af}$, where we let $(s_2)_L : Af \to S(eR_R)f$. If $\psi_1 = (s_2)_L$, then $s_1 + s_2$ is a desired element. When $\psi_1 \neq (s_2)_L$, we proceed further. But this procedure must finitely terminate. Therefore we can take a desired element in eRg. $\qquad\square$

Corollary 2.1.10. *Let R be a semiprimary ring and let $e, f \in Pi(R)$. If $(eR; Rf)$ is an i-pair and satisfies DCC for left annihilators, then $(eR; Rf)$ satisfies the condition (α).*

Proof. This follows from Lemmas 2.1.3 and 2.1.9. $\qquad\square$

Remark 2.1.11.

(1) Let $e, f \in Pi(R)$ such that $(eR; Rf)$ is an i-pair. If $_{eRe}eR$ is artinian, then $(eR; Rf)$ satisfies for DCC for left annihilators.

(2)   In Lemma 2.1.9, in view of the proof, the following fact holds without the assumption that $(eR; Rf)$ satisfies for DCC left annihilators: For any $g \in Pi(R)$ and any submodules K and A of gR_R with $K \subsetneqq A$ and $Kf \neq Af$, where K and A are finitely generated, there exists $r \in eRg$ satisfying $rA \neq 0$ but $rK = 0$.

Proposition 2.1.12. *Let R be a semiprimary ring and let $e, f \in Pi(R)$ such that $(eR; Rf)$ is an i-pair. If $(eR; Rf)$ satisfies DCC for left annihilators, then Rf_{fRf} is artinian and $(eR; Rf)$ satisfies the condition (α); whence $(eR; Rf)$ satisfies DCC for right annihilators, $_{eRe}eR$ is also artinian and $(eR; Rf)$ satisfies the condition (β).*

Proof. We may show that Rf_{fRf} is noetherian by Corollary 2.1.10. Assume that there exists an ascending chain $I_1 \subsetneqq I_2 \subsetneqq \cdots$ of submodules of Rf_{fRf}. Then there exists $g \in Pi(R)$ such that $gI_1 \subsetneqq gI_2 \subsetneqq \cdots$. Hence there exists $n \in \mathbb{N}$ such that $l_{eRg}(gI_n) = l_{eRg}(gI_{n+1}) = \cdots$ by the assumption since $l_{eRg}(gI_1) \supseteq l_{eRg}(gI_2) \supseteq \cdots$. However, by Lemma 2.1.9, $l_{eRg}(gI_n) \neq l_{eRg}(gI_{n+1})$, a contradiction. □

Theorem 2.1.13. *Let R be a semiprimary ring and let $e, f \in Pi(R)$ such that $(eR; Rf)$ is an i-pair. Then the following are equivalent:*

(1)   *Rf_{fRf} is artinian.*

(2)   *$_{eRe}eR$ is artinian.*

(3)   *eR_R and $_RRf$ are injective.*

Proof. We may show (2) $\Leftrightarrow$ (3).

(2) $\Rightarrow$ (3).   This follows from Theorem 2.1.8, Remark 2.1.11 and Proposition 2.1.12.

(3) $\Rightarrow$ (2).   Suppose that both eR_R and $_RRf$ are injective. Let $g \in Pi(R)$ and put $X_i = S_i(_RRf) \cap gRf$ for each $i = 0, 1, 2, \ldots$. We may show that X_i is noetherian as a right fRf-module for any i. Assume that

$X_1, X_2, \ldots, X_{k-1}$ are noetherian but X_k is not so. We put $\overline{X}_k = X_k/X_{k-1}$. Since $\overline{X}_i$ is a semisimple right fRf-module by Lemma 2.1.2 (2), we write $\overline{X}_k$ as $\overline{X}_k = \oplus_{\alpha \in \Lambda} \overline{x}_\alpha fRf$, where $x_\alpha \in X_k$ and each $\overline{x}_\alpha fRf_{fRf}$ is simple. For each $\alpha \in \Lambda$, put $M_\alpha = x_\alpha J + \sum_{\beta \in \Lambda - \{\alpha\}} x_\beta fR + X_{k-1}R$. Then M_α is a maximal right R-submodule of $X_k R$ with $X_k R/M_\alpha \cong T(fR_R) \cong S(eR_R)$. Since eR_R is injective, there exists $y_\alpha \in eRg \cap eJ^{k-1}$ such that $0 \neq y_\alpha x_\alpha \in S(eR_R)$ but $y_\alpha M_\alpha = 0$. Then we claim that $\{Ry_\alpha\}_{\alpha \in \Lambda}$ is independent modulo $J^k g$. Let $\xi = r_1 y_{\alpha_1} + \cdots + r_n y_{\alpha_n} \in Ry_{\alpha_1} + \cdots + Ry_{\alpha_n}$, where $r_i \in Re$, $\alpha_i \in \Lambda$ and $\alpha_i \neq \alpha_j$ if $i \neq j$, and suppose that $\xi \in J^k g$. Since $r_i y_{\alpha_i} x_{\alpha_j} = 0$ if $i \neq j$, $0 \neq y_{\alpha_i} x_{\alpha_i}$ ($\in S(eR_R)$) and $\xi x_{\alpha_i} = 0$, we see that $r_i \in J$ for any i. Hence each $r_i y_{\alpha_i}$ lies in $J^k g$ as desired. Take any $\alpha' \in \Lambda$. Since $\{Ry_\alpha\}_{\alpha \in \Lambda}$ is independent modulo $J^k g$, we also see that $\{Ry_{\alpha'}\} \cup \{R(y_{\alpha'} - y_\beta)\}_{\beta \in \Lambda - \{\alpha'\}}$ is independent modulo $J^k g$. Put $T = Ry_{\alpha'} + \sum_{\beta \in \Lambda - \{\alpha'\}} R(y_{\alpha'} - y_\beta) + J^k g$ and $W = Jy_{\alpha'} + \sum_{\beta \in \Lambda - \{\alpha'\}} R(y_{\alpha'} - y_\beta) + J^k g$. Then T/W is a simple left R-module with $T/W \cong T(_R Re) \cong S(_R Rf)$. Since $_R Rf$ is injective, we can take $x \in gRf$ such that $Tx \neq 0$ but $Wx = 0$. Since $y_{\alpha'} x \neq 0$ but $J^k x = 0$, we see that $x \in X_k - X_{k-1}$. We express x as $x = x_{\alpha'_1} r'_1 + \cdots + x_{\alpha'_n} r'_n + z$, where $r'_i \in fRf$, $\alpha'_i \in \Lambda$ and $z \in X_{k-1}$. Since $y_{\alpha'} x \neq 0$ but $Wx = 0$, we see that $0 \neq y_{\alpha'} x = y_\beta x$ for any $\beta \in \Lambda$. As Λ is an infinite set, we can take an element $\beta \in \Lambda - \{\alpha'_1, \ldots, \alpha'_n\}$. Then $0 \neq y_\beta x$ but $y_\beta x = y_\beta x_{\alpha'_1} r'_1 + \cdots + y_\beta x_{\alpha'_n} r'_n = 0$, a contradiction. This completes the proof. $\square$

Theorem 2.1.14. *Let R be a semiprimary ring and $e, f \in Pi(R)$. Assume that $(eR; Rf)$ is an i-pair and $_{eRe}eR$ is artinian. Then the following hold:*

(1) *Both $_{eRe}eRf$ and eRf_{fRf} are injective.*

(2) *There exists a duality between eRe-FMod and FMod-fRf.*

Proof. (1). By Theorem 2.1.13, Rf_{fRf} is also artinian. Let I be a right ideal of fRf and let $\varphi : I \to eRf_{fRf}$ be a homomorphism. Since $S(eR_R) \cong T(fR_R)$, we can see that φ canonically extends to a homomorphism $\varphi' : IR_R \to eR_R$. Because eR_R is injective, there exists $\tilde{\varphi}' : fR \to eR$ which is an extension of φ'. Then the restriction map $\tilde{\varphi}'|_{fRf}$ is an extension of φ. Hence eRf_{fRf} is injective. Similarly we can show that $_{eRe}eRf$ is injective.

(2). Since eRf_{fRf} (resp. $_{eRe}eRf$) is a finitely generated injective cogenerator, we may show by the Azumaya-Morita Theorem 1.1.10 that

$$\cdot\mathrm{End}_{fRf}(eRf_{fRf}) \cong eRe \quad \text{and} \quad \mathrm{End}_{eRe}(_{eRe}eRf) \cong fRf.$$

We define $\varphi : eRe \rightarrow \mathrm{End}_{fRf}(eRf_{fRf})$ by $ere \mapsto (ere)_L$ for $r \in R$. If $0 \neq ere$, then we can take $s \in eRf$ such that $0 \neq eres \in S(eR_R)f = S(eRf_{fRf})$. Hence φ is a monomorphism. Let $\alpha \in \mathrm{End}_{fRf}(eRf_{fRf})$. Since $S(eR_R) \cong T(fR_R)$, α extends to a homomorphism $\alpha' : eR_R \rightarrow eR_R$ defined by $xr \mapsto \alpha(x)r$, where $x \in eRf$ and $r \in R$. Because eR_R is injective, there exists $u \in eRe$ such that $(u)_L = \alpha'$. Hence φ is an epimorphism. Accordingly $eRe \cong \mathrm{End}_{fRf}(eRf_{fRf})$. Similarly $fRf \cong \mathrm{End}_{eRe}(_{eRe}eRf)$.□

Example 2.1.15. Let D and F be division rings with $D \subseteq F$. We put

$$S = \begin{pmatrix} D & D & F \\ 0 & D & F \\ 0 & 0 & D \end{pmatrix}, \qquad T = \begin{pmatrix} 0 & 0 & F \\ 0 & 0 & 0 \\ 0 & 0 & 0 \end{pmatrix}.$$

Then S becomes a ring and T is an ideal of S. We see that S/T is a semiprimary ring. Put

$$e = \begin{pmatrix} 1 & 0 & 0 \\ 0 & 0 & 0 \\ 0 & 0 & 0 \end{pmatrix} + S, \qquad f = \begin{pmatrix} 0 & 0 & 0 \\ 0 & 1 & 0 \\ 0 & 0 & 0 \end{pmatrix} + S.$$

Then Rf_{fRf} and $_{eRe}eR$ are artinian and $(eR; Rf)$ is an i-pair. Therefore both eR_R and $_RRf$ are injective. If F is infinite dimensional over D, then R is neither left nor right artinian.

2.2 *M*-Simple-Injective and Quasi-Simple-Injective Modules

In this section and Sections 2.3-2.5, we further develop Fuller's Theorem in a more general setting.

An R-module M is called *local* (resp. *colocal*) if the Jacobson radical $J(M)$ of M is small and the top $T(M)$ of M is simple (resp. the socle $S(M)$ of M is simple and essential in M). A bimodule $_RM_S$ is called colocal if both $_RM$ and M_S are colocal.

The following proposition gives a particular instance of when M-simple-injectivity implies M-injectivity.

Proposition 2.2.1. *Let M and N be right R-modules with $S(N_R) \cong T(fR_R)$ for some $f \in Pi(R)$. If N is M-simple-injective and either $L(Nf_{fRf}) < \infty$ or $L(Mf_{fRf}) < \infty$ holds, then N is M-injective.*

Proof. Suppose that $L(Nf_{fRf}) < \infty$ (resp. $L(Mf_{fRf}) < \infty$). Let M' be a submodule of M and $\varphi \in \text{Hom}_R(M', N)$. Put $N_i = \varphi^{-1}(S_i(Nf_{fRf})R)$ (resp. $M_i = S_i(M'f_{fRf})R)$ for any $i \in \mathbb{N}$.

By assumption, we have $\tilde{\varphi}_1 \in \text{Hom}_R(M, N)$ such that $\tilde{\varphi}_1|_{N_1} = \varphi|_{N_1}$ (resp. $\tilde{\varphi}_1|_{M_1} = \varphi|_{M_1}$). Next set $\varphi_2 = \varphi|_{N_2} - \tilde{\varphi}_1|_{N_2} \in \text{Hom}_R(N_2, N)$ (resp. $\varphi_2 = \varphi|_{M_2} - \tilde{\varphi}_1|_{M_2} \in \text{Hom}_R(M_2, N)$). If $\varphi_2 \neq 0$, there exists $\tilde{\varphi}_2 \in \text{Hom}_R(M, N)$ such that $\tilde{\varphi}_2|_{N_2} = \varphi_2$ (resp. $\tilde{\varphi}_2|_{M_2} = \varphi_2$). If $\varphi_2 = 0$, set $\tilde{\varphi}_2 = 0$. Furthermore, let $\varphi_3 = \varphi|_{N_3} - (\tilde{\varphi}_1 + \tilde{\varphi}_2)|_{N_3}$ (resp. $\varphi_3 = \varphi|_{M_3} - (\tilde{\varphi}_1 + \tilde{\varphi}_2)|_{M_3}$) and repeat this argument.

As a consequence, we get $\tilde{\varphi} = \tilde{\varphi}_1 + \cdots + \tilde{\varphi}_n \in \text{Hom}_R(M, N)$ with $\tilde{\varphi}|_{M'} = \varphi$.   □

Next we give a lemma which plays an important role in showing M-simple-injectivity in §2.2.

Lemma 2.2.2. *Let M and N be right R-modules with $S(N)$ essential in N and f an idempotent of R with $S(N)fR = S(N)$. If $\varphi, \psi \in \text{Hom}_R(M, N)$ such that $\varphi|_{Mf} = \psi|_{Mf}$ and $\text{Im}\,\psi$ is semisimple, then $\varphi = \psi$.*

Proof. Let $f = f_1 + \cdots + f_n$ be a decomposition into orthogonal primitive idempotents of R and let $\{e_i\}_{i=1}^m$ be a complete set of orthogonal primitive idempotents of R. We may assume that there exists $m' \in \{1, \ldots, m\}$

with $\{e_i\}_{i=1}^{m'} = \{e_i \mid {}_RRe_i \cong {}_RRf_j$ for some $j \in \{1, ..., n\}\}$. Put $e = \sum_{i=1}^{m'} e_i$. Then $\varphi|_{Me} = \psi|_{Me}$ since $Me \subseteq MfR$. We now claim that $\varphi(x)(1 - e) = 0$ for any $x \in M$. Assume on the contrary that $\varphi(x)(1-e) \neq 0$. Then there is an $r \in R$ with $0 \neq \varphi(x)(1-e)r \in S(N)$ since $S(N) \subseteq_e N$. Further $S(N)e = S(N)$ because $S(N)fR = S(N)$. Therefore $0 \neq \varphi(x)(1 - e)re = \varphi(x(1 - e)re) = \psi(x(1 - e)re) = \psi(x)(1 - e)re \in S(N)(1-e)re = S(N)e(1-e)re = 0$, a contradiction. Hence, for any $x \in M$, $\varphi(x) = \varphi(x)e + \varphi(x)(1 - e) = \varphi(x)e = \varphi(xe) = \psi(xe) = \psi(x)e = \psi(x)$ since $\mathrm{Im}\,\psi \subseteq S(N) = S(N)e$. $\square$

Theorem 1.1.23 provided characterizations of quasi-injective modules and quasi-projective modules. Now we note two further results on quasi-projective modules. For their details, see, for instance, [64, 5.4.10 Proposition, 5.4.11 Theorem and 5.4.12 Theorem].

(1) Let R be a ring, g an idempotent of R and X a left gRg-right R-subbimodule of gR. Then gR/X is a quasi-projective right R-module.

(2) Conversely, let M be a quasi-projective right R-module with a projective cover $\varphi : gR \to M$, where g is an idempotent of R. Then $\mathrm{Ker}\,\varphi$ is a left gRg-right R-subbimodule of gR and, if M is an indecomposable module, $g \in Pi(R)$.

An R-module M is called *quasi-simple-injective* if M is M-simple-injective.

Now we give a characterization of indecomposable quasi-projective quasi-simple-injective modules.

Lemma 2.2.3. *Let M be an indecomposable quasi-projective quasi-simple-injective right R-module with a projective cover $\varphi : eR \to M$, where e is an idempotent of R. If $S(M_R) \neq 0$, then $S(M_R)$ is simple.*

Proof. Put $K = \mathrm{Ker}\,\varphi$. Then K is a left eRe-right R-subbimodule of

eR, $\text{End}(M_R) \cong eRe/Ke$ and $\text{End}(M_R)$ is a local ring. Let S be a simple submodule of eR/K_R ($\cong M_R$). Then we may show that $S = S(eR/K_R)$. Suppose that $S \subsetneq S(eR/K_R)$. Let $\pi : S(eR/K_R) \to S$ be a projection. Since $eR/K_R \cong M_R$ is quasi-simple-injective, there exists $x \in eRe$ with $(x)_L|_{S(eR/K_R)} = \pi$, where we consider $(x)_L \in \text{End}(eR/K_R)$. Then $(x)_L$ is not isomorphic because π is not monic. Hence $x \in eJe$. Therefore $(e - x) + Ke$ is a unit element in eRe/Ke. But we consider $(e - x)_L \in \text{End}(eR/K_R)$, and $(e - x)_L(S) = 0 + K$, a contradiction. $\qquad\square$

Finally in this section, we establish a basic property of the quasi-projective right R-module $eR/l_{eR}(Rf)$, where $e, f \in Pi(R)$. In the next section, we study the injectivity of this module.

Lemma 2.2.4. *Let $e, f \in Pi(R)$ with eRf a colocal right fRf-module. Then $eR/l_{eR}(Rf)_R$ is a local right R-module with $S(eR/l_{eR}(Rf)_R) \cong T(fR_R)$.*

Proof. Take $0 \neq s \in S(eRf_{fRf})$. We have a right R-epimorphism $\psi : sR \to T(fR_R)$. Let $\tilde{\psi} : eR \to E(T(fR_R))$ be an extension homomorphism of ψ. Then we claim that $\text{Ker}\,\tilde{\psi} = l_{eR}(Rf)$. Assume that there exists $x \in \text{Ker}\,\tilde{\psi} - l_{eR}(Rf)$. Then $0 \neq xRf \subseteq eRf$ and hence $s \in S(eRf_{fRf}) \subseteq xRf \subseteq \text{Ker}\,\tilde{\psi}$ since $S(eRf_{fRf})$ is simple and essential in eRf. But $\psi(s) \neq 0$, i.e., $s \notin \text{Ker}\,\tilde{\psi}$, a contradiction. Conversely assume that there exists $x \in l_{eR}(Rf) - \text{Ker}\,\tilde{\psi}$. We have $r \in R$ with $0 \neq \tilde{\psi}(x)r \in T(fR_R)$. Then we may assume that $xr \in eRf$, so xr does not annihilate f. But $xr \in l_{eR}(Rf)$, a contradiction. Hence $S(eR/l_{eR}(Rf)_R)$ is simple and essential in $eR/l_{eR}(Rf)$ with $S(eR/l_{eR}(Rf)_R) \cong T(fR_R)$. $\qquad\square$

2.3 Simple-Injectivity and the Condition $\alpha_r[e, g, f]$

In this section, we characterize M-simple-injective modules and quasi-simple-injective modules.

For any e, $f \in Pi(R)$ and any idempotent g of R, we say that R satisfies $\alpha_r[e, g, f]$ (or $\alpha_l[e, g, f]$) if the following condition $\alpha_r[e, g, f]$ (or $\alpha_l[e, g, f]$) holds:

$\alpha_r[e, g, f]$: $r_{gRf} l_{eRg}(X) = X$ for any right fRf-module X with $r_{gRf}(eRg) \subseteq X \subseteq gRf$.

$\alpha_l[e, g, f]$: $l_{eRg} r_{gRf}(X) = X$ for any left eRe-module X with $l_{eRg}(gRf) \subseteq X \subseteq eRg$.

We may easily obtain the following characterization of $\alpha_r[e, g, f]$ (resp. $\alpha_l[e, g, f]$).

Lemma 2.3.1. *Let e, $f \in Pi(R)$ and g an idempotent of R. Then the following are equivalent:*

(1) *R satisfies $\alpha_r[e, g, f]$ (resp. $\alpha_l[e, g, f]$).*
(2) *There exists $a \in eRg$ such that $a\overline{X} = 0$ but $a\overline{Y} \neq 0$ for any right fRf-modules $\overline{X}$ and $\overline{Y}$ with $\overline{X} \subsetneqq \overline{Y} \subseteq gRf/r_{gRf}(eRg)$ (resp. there exists $a \in gRf$ such that $\overline{X}a = 0$ but $\overline{Y}a \neq 0$ for any left eRe-modules $\overline{X}$ and $\overline{Y}$ with $\overline{X} \subsetneqq \overline{Y} \subseteq eRg/l_{eRg}(gRf)$).*

Let e, $f \in Pi(R)$. We say that $(eR; Rf)$ is a *colocal pair* (abbreviated *c-pair*) if $_{eRe}eRf_{fRf}$ is a colocal bimodule.

In the following proposition, we further characterize $\alpha_r[e, g, f]$ using simple-injectivity.

Proposition 2.3.2. *Let $(eR; Rf)$ be a c-pair and g an idempotent of R.*

(1) *Consider the following two conditions:*

   (a) *R satisfies $\alpha_r[e, g, f]$.*
   (b) *The quasi-projective module $eR/l_{eR}(Rf)_R$ is $gR/r_{gR}(eRg)$-simple-injective.*

   Then $(a) \Rightarrow (b)$ holds and if the ring fRf is right or left perfect, the converse also holds.

(2) *The following are equivalent:*

(a)   The quasi-projective module $eR/l_{eR}(Rf)_R$ is $gR/l_{gR}(Rf)$-simple-injective.

(b)   Condition (1) (b) holds and $r_{gRf}(eRg) = 0$.

Proof. (1). $(a) \Rightarrow (b)$. Let $\bar{I}$ be a right R-submodule of $gR/r_{gR}(eRg)$ and let φ be a homomorphism $\bar{I}_R \to eR/l_{eR}(Rf)_R$ with $\operatorname{Im} \varphi$ simple. Consider the restriction map $\varphi|_{\bar{I}f} : \bar{I}f \to S(eR/l_{eR}(Rf)_R)f = S(eRf_{fRf})$. Since R satisfies $\alpha_r[e, g, f]$, we have $y \in eRg$ such that $y \operatorname{Ker}(\varphi|_{\bar{I}f}) = 0$ but $y\bar{I} \neq 0$. Then, since the left eRe-module $S(eRf_{fRf})$ $(= S(_{eRe}eRf))$ is simple and essential in $_{eRe}eRf$ by Lemma 2.1.1 (3) and its proof, there is $y' \in eRe$ such that $\varphi|_{\bar{I}f} = (y'y)_L$ as right fRf-homomorphisms $\bar{I} \to S(eRf_{fRf})$. Consider $(y'y)_L \in \operatorname{Hom}_R(gR/r_{gR}(eRg), eR/l_{eR}(Rf))$. Then $\varphi = (y'y)_L|_{\bar{I}}$ by Lemmas 2.2.2 and 2.2.4. Therefore $eR/l_{eR}(Rf)$ is $gR/r_{gR}(eRg)$-simple-injective.

$(b) \Rightarrow (a)$. Let X and Y be right fRf-modules with $r_{gRf}(eRg) \subseteq X \subsetneq Y \subseteq gRf$. We must show that there exists $r \in eRg$ such that $rX = 0$ but $rY \neq 0$ by Lemma 2.3.1. We may assume that Y/X is a simple right fRf-module since the ring fRf is right or left perfect (see, for instance, [5, 28.4.Theorem]). Then we have a right fRf-epimorphism $\varphi : Y \to S(eR/l_{eR}(Rf)_R)f$ with $\operatorname{Ker} \varphi = X$ since $S(eR/l_{eR}(Rf)_R)f = S(eRf_{fRf})$ is a simple right fRf-module. We claim that we can define a right R-epimorphism $\tilde{\varphi} : YR/r_{gRf}(eRg)R \to S(eR/l_{eR}(Rf)_R)$ by $\sum_{i=1}^{n} a_ir_i + r_{gRf}(eRg)R \mapsto \sum_{i=1}^{n} \varphi(a_i)r_i$, where $a_i \in Y$ and $r_i \in fR$. Assume that $\sum_{i=1}^{n} \varphi(a_i)r_i \neq \bar{0}$. There exists $s \in Rf$ with $\bar{0} \neq (\sum_{i=1}^{n} \varphi(a_i)r_i)s \in S(eR/l_{eR}(Rf)_R)f$ by Lemma 2.2.4. Then $(0 \neq)$ $(\sum_{i=1}^{n} \varphi(a_i)r_i)s = \sum_{i=1}^{n} \varphi(a_i)r_is = \varphi(\sum_{i=1}^{n} a_ir_is) = \varphi((\sum_{i=1}^{n} a_ir_i)s)$. Hence $\sum_{i=1}^{n} a_ir_i \notin r_{gRf}(eRg)R$ because $\operatorname{Ker} \varphi = X \supseteq r_{gRf}(eRg)$. Further we have a right R-isomorphism $\eta : (YR + r_{gR}(eRg))/r_{gR}(eRg) \to YR/r_{gRf}(eRg)R$ since $(YR + r_{gR}(eRg))/r_{gR}(eRg) \cong YR/(YR \cap r_{gR}(eRg))$ and $YR \cap r_{gR}(eRg) = r_{gRf}(eRg)R$. Therefore there exists $r \in eRg$ with $(r)_L = \tilde{\varphi}\eta$ because $eR/l_{eR}(Rf)_R$ is $gR/r_{gR}(eRg)$-simple-injective. Then $rX = 0$ but $rY \neq 0$.

(2). $(a) \Rightarrow (b)$. Let I be a right R-submodule of gR with $I \supseteq r_{gR}(eRg)$ and let $\varphi \in \operatorname{Hom}_R(I/r_{gR}(eRg), S(eR/l_{eR}(Rf)_R))$. By Lemma 2.2.4 we may define a right R-homomorphism $\psi : (I + l_{gR}(Rf))/l_{gR}(Rf) \rightarrow S(eR/l_{eR}(Rf)_R)$ by $x + l_{gR}(Rf) \mapsto \varphi(x + r_{gR}(eRg))$ for $x \in I$. Then, because $eR/l_{eR}(Rf)_R$ is $gR/l_{gR}(Rf)$-simple-injective, there exists $a \in eRg$ with $(a)_L |_{(I+l_{gR}(Rf))/l_{gR}(Rf)} = \psi$, where we consider $(a)_L : gR/l_{gR}(Rf) \rightarrow eR/l_{eR}(Rf)$. Now define a right R-homomorphism $\tilde{\varphi} : gR/r_{gR}(eRg) \rightarrow eR/l_{eR}(Rf)$ by $g + r_{gR}(eRg) \mapsto a + l_{eR}(Rf)$. Then $\tilde{\varphi}(x + r_{gR}(eRg)) = ax + l_{eR}(Rf) = (a)_L(x + l_{gR}(Rf)) = \psi(x + l_{gR}(Rf)) = \varphi(x + r_{gR}(eRg))$ for any $x \in I$. Therefore $eR/l_{eR}(Rf)_R$ is $gR/r_{gR}(eRg)$-simple-injective.

Suppose that there exists a non-zero element $x \in r_{gRf}(eRg)$. Then we have a right R-epimorphism $\xi : (xR + l_{gR}(Rf))/l_{gR}(Rf) \rightarrow S(eR/l_{eR}(Rf)_R)$ since $T(xR_R) \cong T(fR_R)$. Therefore, because $eR/l_{eR}(Rf)_R$ is $gR/l_{gR}(Rf)$-simple-injective, there exists $a \in eRg$ with $(a)_L = \xi$, where we consider $(a)_L : (xR + l_{gR}(Rf))/l_{gR}(Rf) \rightarrow S(eR/l_{eR}(Rf)_R)$. Then $ax \neq 0$. This contradicts the fact that $x \in r_{gRf}(eRg)$.

$(b) \Rightarrow (a)$. Let I be a right R-submodule of gR with $I \supseteq l_{gR}(Rf)$ and let $\psi \in \operatorname{Hom}_R(I/l_{gR}(Rf)_R, S(eR/l_{eR}(Rf)_R))$. Then we may define a right R-homomorphism $\varphi : (I + r_{gR}(eRg))/r_{gR}(eRg) \rightarrow S(eR/l_{eR}(Rf)_R)$ by $x + r_{gR}(eRg) \mapsto \psi(x + l_{gR}(Rf))$ for $x \in I$ because the assumptions that $r_{gRf}(eRg) = 0$ and $S(eR/l_{eR}(Rf)_R) \cong T(fR_R)$ induce $\psi(y + l_{gR}(Rf)) = 0$ for any $y \in I \cap r_{gR}(eRg)$. Since $eR/l_{eR}(Rf)_R$ is $gR/r_{gR}(eRg)$-simple-injective, there exists $a \in eRg$ with $(a)_L |_{(I+r_{gR}(eRg))/r_{gR}(eRg)} = \varphi$, where we consider $(a)_L : gR/r_{gR}(eRg) \rightarrow eR/l_{eR}(Rf)$. Next define the right R-homomorphism $\tilde{\psi} : gR/l_{gR}(Rf) \rightarrow eR/l_{eR}(Rf)$ by $g + l_{gR}(Rf) \mapsto a + l_{eR}(Rf)$. For any $x \in I$, $\tilde{\psi}(x + l_{gR}(Rf)) = ax + l_{eR}(Rf) = (a)_L(x + r_{gR}(eRg)) = \varphi(x + r_{gR}(eRg)) = \psi(x + l_{gR}(Rf))$. Therefore $eR/l_{eR}(Rf)_R$ is $gR/l_{gR}(Rf)$-simple-injective. $\square$

The following lemma is used to simplify later proofs.

Lemma 2.3.3. *Let* $h \in Pi(R)$, *g an idempotent of R and H a right R-submodule of gR. If I is a gR/H-simple-injective right R-module with $S(I_R) \cong T(hR_R)$, then, for each non-zero element $t \in gRh - H$ and each non-zero element $s \in S(I_R)h$, there is an $x \in I$ such that $xt = s$.*

    Proof. $T(hR_R) \cong tR/tJ$ since $t \in Rh$. Thus we have a right R-epimorphism $\varphi : (gR + H)/H \to S(I_R)$. Define an automorphism $\eta : S(I_R) \to S(I_R)$ by $\varphi(t + H) \mapsto s$. Then we have an extension right R-homomorphism $\tilde{\varphi} : gR/H \to I$ of $\eta\varphi$. Put $x = \tilde{\varphi}(g + H)$. Then $xt = \tilde{\varphi}(g + H)t = \tilde{\varphi}(t + H) = \eta\varphi(t + H) = s$.   □

    Now we have a characterization of indecomposable quasi-projective quasi-simple-injective modules. Here $\alpha_r[e, e, f]$ and $\alpha_l[e, f, f]$ play an important role. By their definition and Lemma 2.3.1, it follows that R satisfies $\alpha_r[e, e, f]$ (resp. $\alpha_l[e, f, f]$) if and only if $r_{eRf}l_{eRe}(X) = X$ for any right fRf-submodule X of eRf (resp. $l_{eRf}r_{fRf}(Y) = Y$ for any left eRe-submodule Y of eRf), or equivalently, there exists $a \in eRe$ such that $aX = 0$ and $aY \neq 0$ for any right fRf-modules X and Y with $X \subsetneq Y \subseteq eRf$ (resp. there exists $a \in fRf$ such that $Xa = 0$ and $Ya \neq 0$ for any left eRe-submodules X and Y with $X \subsetneq Y \subseteq eRf$).

    Now for the promised characterization. It will used later to establish more important results.

Proposition 2.3.4. *Let R be a left perfect ring and $e, f \in Pi(R)$ with $eRf \neq 0$. The following are equivalent:*

    (1)   *The quasi-projective module $eR/l_{eR}(Rf)_R$ is quasi-simple-injective.*

    (2)   (a) *$(eR; Rf)$ is a c-pair, and*

           (b) *R satisfies $\alpha_r[e, e, f]$.*

    Proof. (1) $\Rightarrow$ (2). (a). $S(eR/l_{eR}(Rf)_R) \cong T(fR_R)$ by Lemma 2.2.3 since $eRf \neq 0$. Take $s \in eRf$ with $l_{eR}(Rf) \neq s + l_{eR}(Rf) \in$

$S(eR/l_{eR}(Rf)_R)$. Then $sfRf$ is a simple right fRf-submodule of eRf. Moreover $sfRf$ is an essential right fRf-submodule of eRf since $(s + l_{eR}(Rf))R = S(eR/l_{eR}(Rf)_R)$ is an essential simple right R-submodule of $eR/l_{eR}(Rf)$. Therefore $sfRf = S(eRf_{fRf})$ (and it is a simple right fRf-module). Further $S(eRf_{fRf})$ is also a left eRe-submodule of eRf. Also, for any non-zero $t \in eRf$, applying Lemma 2.3.3 (to $I = eR/l_{eR}(Rf)$, $H = l_{eR}(Rf)$, $h = f$ and $g = e$) we have $x \in eRe$ such that $xt = s$, i.e., $S(eRf_{fRf}) = sfRf$ is an essential simple left eRe-submodule of eRf. Therefore $S(_{eRe}eRf) = S(eRf_{fRf})$ (and it is a simple left eRe-module).

(b). This follows from Proposition 2.3.2 and (a) above.

(2) $\Rightarrow$ (1). This follows from Proposition 2.3.2. $\square$

Next we characterize indecomposable projective quasi-simple-injective modules and indecomposable quasi-projective R-simple-injective modules.

Theorem 2.3.5.

(1)   *The following are equivalent for a right perfect ring R and $f \in Pi(R)$:*

   (a) $_R Rf$ *is quasi-simple-injective.*

   (b)   *There exists $e \in Pi(R)$ such that*

      (i) $S(_R Rf)$ *is simple and essential in $_R Rf$ with $S(_R Rf) \cong T(_R Re)$,*

      (ii) $S(eRf_{fRf})$ *is simple and essential in eRf_{fRf}, and*

      (iii) R *satisfies $\alpha_l[e, f, f]$.*

(2)   *The following are equivalent for a left perfect ring R and $e, f \in Pi(R)$:*

   (a)   *The quasi-projective module $eR/l_{eR}(Rf)_R$ is R-simple-injective.*

   (b)   (i) $S(_R Rf)$ *is simple and essential in $_R Rf$ with $S(_R Rf) \cong T(_R Re)$,*

      (ii) $S(eRf_{fRf})$ *is simple and essential in eRf_{fRf}, and*

(iii) R *satisfies* $\alpha_r[e, e, f]$.

Proof. (1). This follows from Proposition 2.3.4 and Lemma 2.2.3.

(2). (a) $\Rightarrow$ (b). $eR/l_{eR}(Rf)_R$ is quasi-simple-injective since it is R-simple-injective. Thus (ii) and (iii) hold and $S(_{eRe}eRf)$ is also simple and essential in eRf by Proposition 2.3.4. Therefore $S(_{eRe}eRf) = S(eRf_{fRf})$ by Lemma 2.1.1 (3). Further $S(eR/l_{eR}(Rf)_R)f = S(eRf_{fRf})$ by Lemma 2.1.1 (1) because $S(eR/l_{eR}(Rf)_R) \cong T(fR_R)$ by Lemma 2.2.4. Let s be a non-zero element of $S(_{eRe}eRf)$. Then, for any $t \in Rf$, applying Lemma 2.3.3 (to $I = eR/l_{eR}(Rf), H = 0, h = f$ and $g = 1$), we have a non-zero $x \in S(eRf_{fRf})$ such that $xt = s$ since $s \in S(eR/l_{eR}(Rf)_R)f$. Therefore $R\,S(_{eRe}eRf)$ is an essential simple left R-submodule of Rf, establishing (i).

(b) $\Rightarrow$ (a).   Let I be a right ideal of R and let $\varphi : I \to eR/l_{eR}(Rf)$ be a right R-homomorphism with Im φ simple. Consider the right fRf-epimorphism $\varphi|_{If} : If \to S(eR/l_{eR}(Rf)_R)f = S(eRf_{fRf})$. Now $eIf \neq 0$ since $S(_RRf) \cong T(_RRe)$. Therefore we have $y \in eRe$ such that $y \cdot \mathrm{Ker}(\varphi|_{If}) = 0$ and $yIf \neq 0$ by Lemma 2.3.1. Then there exists $y' \in eRe$ such that $(y'y)_L = \varphi|_{If}$ because $S(eRf_{fRf}) = S(_{eRe}eRf)$ is a simple left eRe-module. Consider $(y'y)_L \in \mathrm{Hom}_R(R_R, eR_R)$ and put $\tilde\varphi = \pi(y'y)_L \in \mathrm{Hom}_R(R_R, eR/l_{eR}(Rf)_R)$, where $\pi : eR \to eR/l_{eR}(Rf)$ is the natural epimorphism. Then $\varphi|_{If} = \tilde\varphi|_{If}$ since $\varphi|_{If} = (y'y)_L$. Therefore $\varphi = \tilde\varphi|_I$ by Lemma 2.2.2. Hence $eR/l_{eR}(Rf)_R$ is R-simple-injective.$\quad\square$

The following important characterization of indecomposable projective R-simple-injective modules now follows from Theorem 2.3.5 (2) and Lemma 2.2.3.

Corollary 2.3.6.   *Let R be a left perfect ring and $e \in Pi(R)$. Then the following are equivalent:*

(1) eR_R *is R-simple-injective.*

(2)   *There exists $f \in Pi(R)$ such that*

(a) $(eR; Rf)$ *is an i-pair, and*

(b) *R satisfies* $\alpha_r[e, 1, f]$.

Using Corollary 2.3.6 and Proposition 2.2.1, we have the following corollary, which establishes Theorem 2.B (a).

Corollary 2.3.7. *Let R be a semiprimary ring and let $e \in Pi(R)$. Then the following are equivalent:*

(1) eR_R *is injective.*

(2) *There exists $f \in Pi(R)$ such that*

    (a) $(eR; Rf)$ *is an i-pair, and*

    (b) *R satisfies* $\alpha_r[e, 1, f]$.

Remark 2.3.8. If R is a semiprimary ring then, using Proposition 2.2.1, we may replace the terms "quasi-simple-injective", "M-simple-injective" and "R-simple-injective" in Propositions 2.3.2 and 2.3.4 and Theorem 2.3.5 by "quasi-injective", "M-injective" and "injective", respectively.

We now present a proposition which is frequently used in later chapters.

Proposition 2.3.9. (cf. Lemma 2.1.1 (4)) *Let R be a left perfect ring and $e, f \in Pi(R)$ such that eR_R is colocal and fR_R is a projective cover of $S(eR_R)$. If eR_R is R-simple-injective, then the following hold:*

(1) $(eR; Rf)$ *is an i-pair,*

(2) $S(eR_R)f = S(eRf_{fRf}) = S(_{eRe}eRf) = eS(_RRf)$.

Proof. (1). This follows from Corollary 2.3.6.

(2). From (a), we see that $(eR; Rf)$ is a c-pair. Therefore $S(eRf_{fRf}) = S(_{eRe}eRf)$ by Lemma 2.1.1 (3). In consequence, we have $S(eR_R)f = S(eRf_{fRf}) = S(_{eRe}eRf) = eS(_RRf)$ from Lemma 2.1.1 (2). □

2.4 ACC on Right Annihilator Ideals and the Condition $\alpha_r[e, g, f]$

In Proposition 2.3.4, Theorem 2.3.5 and Corollary 2.3.6, we studied the simple-injectivity of modules using the condition $\alpha_r[e, g, f]$ (or $\alpha_l[e, g, f]$). We say that a ring R satisfies ACC on right annihilator ideals if ACC holds on $\{\, r_R(X) \mid X$ is a subset of $R\,\}$. In this section we show that, if R is a semiprimary ring which satisfies ACC on right annihilator ideals, then we can remove the conditions $\alpha_r[e, g, f]$ and $\alpha_l[e, g, f]$ from Theorem 2.3.5 and its Corollary. This will be established in Corollaries 2.4.4, 2.4.5 and 2.4.6.

First we note the following lemma.

Lemma 2.4.1. *If R is a right perfect ring and satisfies ACC on right annihilator ideals, then R is a semiprimary ring.*

Proof.   We must show that there exists $n \in \mathbb{N}$ with $J^n = 0$. Consider the ascending chain $r_R(J) \subseteq r_R(J^2) \subseteq r_R(J^3) \subseteq \cdots$. By assumption, there exists $n \in \mathbb{N}$ with $r_R(J^n) = r_R(J^{n+1})$. Suppose that $r_R(J^n) \neq R$. Then $S(_R R/r_R(J^n)) \neq 0$ since R is a right perfect ring. Thus we have a left ideal $I \ (\supsetneq r_R(J^n))$ with $I/r_R(J^n) = S(_R R/r_R(J^n))$. Then $I \subseteq r_R(J^{n+1})$ $(= r_R(J^n))$, a contradiction. Therefore $r_R(J^n) = R$, i.e., $J^n = 0$.   □

For $e,\ f \in Pi(R)$ and an idempotent g of R, we put

$$A_r[e, g, f] = \{\, X_{fRf}\ (\subseteq gRf) \mid r_{gRf}l_{eRg}(X) = X \,\},$$

and

$$A_l[e, g, f] = \{\, _{eRe}X\ (\subseteq eRg) \mid l_{eRg}r_{gRf}(X) = X \,\}.$$

We note that, by the definitions of $\alpha_r[e, g, f]$ and $\alpha_l[e, g, f]$, the following hold:

(1) R satisfies $\alpha_r[e, g, f]$ if and only if $A_r[e, g, f] = \{\, X_{fRf} \mid r_{gRf}(eRg) \subseteq X \subseteq gRf \,\}$,

(2) R satisfies $\alpha_l[e, g, f]$ if and only if $A_l[e, g, f] = \{\, _{eRe}X \mid l_{eRg}(gRf) \subseteq X \subseteq eRg \,\}$.

Lemma 2.4.2. *Let $(eR; Rf)$ be a c-pair, g an idempotent of R and X, Y right fRf-modules with $r_{gRf}(eRg) \subseteq X \subsetneq Y \subseteq gRf$. If the right fRf-module Y/X is simple and $X \in A_r[e, g, f]$, then the following hold:*

   (1) $_{eRe}\, l_{eRg}(X)/l_{eRg}(Y)$ *is simple.*

   (2) $Y \in A_r[e, g, f]$.

Proof.   (1).   Take $y \in Y - X$. Then $Y = yfRf + X$. We claim that $l_{eRg}(X)yfRf = S(eRf_{fRf})$. Since $Y \supsetneq X = r_{gRf}l_{eRg}(X)$, $l_{eRg}(Y) \subsetneq l_{eRg}(X)$, and hence $0 \neq l_{eRg}(X)Y = l_{eRg}(X)yfRf$. Further $YfJf \subseteq X$ since Y/X_{fRf} is simple, and hence $l_{eRg}(X)yfJf = 0$. Therefore $l_{eRg}(X)yfRf = S(eRf_{fRf})$. Hence $l_{eRg}(X)y = S(_{eRe}eRf)$ by Lemma 2.1.1 (3). Hence $l_{eRg}(X)/l_{eRg}(Y)$ is simple as a left eRe-module since $(y)_R : l_{eRg}(X)/l_{eRg}(Y) \rightarrow l_{eRg}(X)y$ is an isomorphism.

   (2). $l_{eRg}(Y) \subsetneq l_{eRg}(X)$ with $_{eRe}(l_{eRg}(X)/l_{eRg}(Y))$ simple by (1). Then, since $l_{eRg}(Y) \in A_l[e, g, f]$, $(r_{gRf}l_{eRg}(Y)/r_{gRf}l_{eRg}(X))_{fRf}$ is simple by (1). Hence we see that $Y = r_{gRf}l_{eRg}(Y)$ since $r_{gRf}l_{eRg}(X) = X \subsetneq Y \subseteq r_{gRf}l_{eRg}(Y)$ and Y/X_{fRf} is simple.   □

By Lemma 2.4.2 we see that, if $(eR; Rf)$ is a c-pair, every right fRf-submodule X $(\supseteq r_{gRf}(eRg))$ of gRf with $|X/r_{gRf}(eRg)_{fRf}| < \infty$ is an element of $A_r[e, g, f]$. If we further assume that ACC holds on $\{ r_{gRf}(I) \mid I$ is a left eRe-submodule of $eRg \}$, the following proposition shows that every right fRf-submodule X $(\supseteq r_{gRf}(eRg))$ of gRf is an element of $A_r[e, g, f]$, i.e., R satisfies $\alpha_r[e, g, f]$.

Proposition 2.4.3. *Let $(eR; Rf)$ be a c-pair and g an idempotent of R. If ACC holds on $\{ r_{gRf}(I) \mid I$ is a left eRe-submodule of $eRg \}$ (equivalently, DCC holds on $\{ l_{eRg}(I') \mid I'$ is a right fRf-submodule of $gRf \}$) and fRf is a left perfect ring, then the following hold:*

   (1) $|(gRf/r_{gRf}(eRg))_{fRf}| < \infty$.

   (2) R *satisfies* $\alpha_r[e, g, f]$.

Moreover, if eRe is a right perfect ring, then R also satisfies $\alpha_l[e, g, f]$.

Proof. Put $B = \{ X_{fRf} \mid r_{gRf}(eRg) \subseteq X \subseteq gRf$ and $|(X/r_{gRf}(eRg))_{fRf}| < \infty \}$. Then $B \subseteq A_r[e, g, f]$ by Lemma 2.4.2. Therefore we have a maximal element M in B because ACC holds on $\{ r_{gRf}(I) \mid I$ is a left eRe-submodule of $eRg \}$. We claim that $M = gRf$. If not, we have a submodule Y of gRf_{fRf} such that $M \subsetneq Y$ and Y/M_{fRf} is simple since fRf is a left perfect ring. Then $Y \in B$. This contradicts the maximality of M, and hence (1) holds. Moreover $A_r[e, g, f] = \{ X_{fRf} \mid r_{gRf}(eRg) \subseteq X \subseteq gRf \}$. Hence (2) also holds.

We further assume that eRe is a right perfect ring. By (1) DCC holds on $\{ r_{gRf}(I) \mid I$ is a left eRe-submodule of $eRg \}$. Then $\{ l_{eRg}(I') \mid I'$ is a right fRf-submodule of $gRf \}$ satisfies ACC. Hence, by applying the above argument with left and right interchanged, we see that R satisfies $\alpha_l[e, g, f]$. $\square$

We now give some useful corollaries. Corollaries 2.4.5 and 2.4.6, in particular, play important roles in §3.3.

Corollary 2.4.4. *Let R be a semiprimary ring which satisfies ACC on right annihilator ideals and $e, f \in Pi(R)$ with $eRf \neq 0$. Then the following are equivalent:*

(1) $_R Rf/r_{Rf}(eR)$ *is quasi-injective.*
(2) $eR/l_{eR}(Rf)_R$ *is quasi-injective.*
(3) $(eR; Rf)$ *is a c-pair.*

Proof. (1) or (2) $\Rightarrow$ (3). These follow from Proposition 2.3.4.

(3) $\Rightarrow$ (1) and (2). Since ACC holds on right annihilator ideals, R satisfies both $\alpha_l[e, f, f]$ and $\alpha_r[e, e, f]$ by Proposition 2.4.3. Hence (1) and (2) hold by Proposition 2.3.4 and Remark 2.3.8. $\square$

Corollary 2.4.5. *Let R be a semiprimary ring which satisfies ACC on* *right annihilator ideals and e, f $\in$ Pi(R). Then the following are equivalent:*

(1) $_RRf$ *is quasi-injective with* $S(_RRf) \cong T(_RRe)$.

(2) $eR/l_{eR}(Rf)_R$ *is injective, i.e.,* $eR/l_{eR}(Rf)_R \cong E(T(fR_R))$.

(3) *(i)* $S(_RRf) \cong T(_RRe)$, *and*

  (ii) $S(eRf_{fRf})$ *is simple.*

Proof. This follows from Proposition 2.4.3, Theorem 2.3.5 and Remark 2.3.8. □

Corollary 2.4.6. *Let R be a semiprimary ring which satisfies ACC on right annihilator ideals and e, f $\in$ Pi(R). Then the following are equivalent:*

(1) $_RRf$ *is injective with* $S(_RRf) \cong T(_RRe)$.

(2) eR_R *is injective with* $S(eR_R) \cong T(fR_R)$.

(3) $(eR; Rf)$ *is an i-pair.*

Proof. This follows from Proposition 2.4.3, Corollary 2.3.6 and Remark 2.3.8. □

2.5 Injectivity and Composition Length

In the previous sections, we observed the relationships between M-(quasi-)injectivity and the notions of a c-pair, an i-pair and an intermediate condition. In this section, we assume that $(eR; Rf)$ is a c-pair or an i-pair and consider the relationships between the (M-)injectivity of both $eR/l_{eR}(Rf)$ and $Rf/r_{Rf}(eR)$ and the composition length of these.

First we give two lemmas.

Lemma 2.5.1. *Let $(eR; Rf)$ be a c-pair and g an idempotent of R. Then,* *for each $n \in \mathbb{N}$, $r_{gRf}(eJ^n g)/r_{gRf}(eJ^{n-1}g)$ is either 0 or the essential socle of the right fRf-module $gRf/r_{gRf}(eJ^{n-1}g)$.*

Proof. Suppose that $r_{gRf}(eJ^n g) \neq r_{gRf}(eJ^{n-1}g)$ and choose x in the former. Then $0 \neq eJ^{n-1}gx \subseteq S(_{eRe}eRf)$ $(= r_{eRf}(eJe))$, so $eJ^{n-1}gx \subseteq S(eRf_{fRf})$ by Lemma 2.1.1 (3). Therefore $eJ^{n-1}gxfJf = 0$, i.e., $xfJf \subseteq r_{gRf}(eJ^{n-1}g)$, i.e., $r_{gRf}(eJ^n g)/r_{gRf}(eJ^{n-1}g)_{fRf}$ is semisimple. Further, for any $x \in gRf - r_{gRf}(eJ^{n-1}g)$, there exists $r \in fRf$ with $0 \neq eJ^{n-1}gxr \in S(eRf_{fRf})$ $(= S(_{eRe}eRf))$. Therefore $eJeJ^{n-1}gxr = 0$, i.e., $xr \in r_{gRf}(eJ^n g) - r_{gRf}(eJ^{n-1}g)$. Hence $r_{gRf}(eJ^n g)/r_{gRf}(eJ^{n-1}g)$ is the essential socle of $gRf/\, r_{gRf}(eJ^{n-1}g)_{fRf}$.   □

Lemma 2.5.2. *Let $(eR; Rf)$ be a c-pair, g an idempotent of R and X_{fRf} and Y_{fRf} submodules of gRf such that $r_{gRf}(eRg) \subseteq X \subsetneq Y$ and Y/X is the essential socle of gRf/X_{fRf}. Suppose that $eR/l_{eR}(Rf)_R$ is $gR/r_{gR}(eRg)$-simple-injective and $_RRf/r_{Rf}(eR)$ is a $Rg/l_{Rg}(gRf)$-simple-injective. Then $|Y/X_{fRf}| < \infty$.*

Proof. Suppose that $|Y/X_{fRf}| = \infty$. Then there is an infinite subset $\{y_\lambda\}_{\lambda \in \Lambda}$ of $Y - X$ such that $\oplus_{\lambda \in \Lambda}(y_\lambda + X)fRf = Y/X$. For each $\lambda \in \Lambda$, put $M_\lambda = y_\lambda J + \sum_{\lambda' \in \Lambda - \{\lambda\}} y_{\lambda'} R + XR$. Each M_λ is a maximal right R-submodule of YR such that $YR/M_\lambda \cong T(fR_R)$ $(\cong S(eR/l_{eR}(Rf)_R))$. Therefore there exists $z_\lambda \in eRg$ with $z_\lambda y_\lambda \neq 0$ but $z_\lambda M_\lambda = 0$ for each λ since $eR/l_{eR}(Rf)_R$ is $gR/r_{gR}(eRg)$-simple-injective. Then $z_\lambda \in l_{eRg}(X) - l_{eRg}(Y)$. Moreover we claim that $\{Rz_\lambda\}_{\lambda \in \Lambda}$ is a set of independent left ideals modulo $l_{Rg}(Y)$. To this end, suppose that $\sum_{i=1}^n r_i z_{l_i} \in l_{Rg}(Y)$, where $r_i \in R$ and $l_i \in \Lambda$. For each j, $r_j z_{l_j} y_{l_j} = (\sum_{i=1}^n r_i z_{l_i})y_{l_j} \in l_{Rg}(Y)Y = 0$. Hence $r_j z_{l_j} \in l_{Rg}(Y)$ since $z_{l_j} M_{l_j} = 0$, justifying the claim.

Now take $l \in \Lambda$ and set $T = \sum_{\lambda \in \Lambda} Rz_\lambda$ and $W = Jz_l + \sum_{\lambda' \in \Lambda - \{l\}} R(z_{\lambda'} - z_l)$. Then $_R(T + r_{Rg}(Y))/(W + r_{Rg}(Y)) \cong {}_RT/W \cong T(_RRe) \cong S(_RRf/r_{Rf}(eR))$ since $\{Rz_\lambda\}_{\lambda \in \Lambda}$ is a set of independent left ideals modulo $l_{Rg}(Y)$. Thus there is an $a \in gRf$ with $Ta \neq 0$ but $Wa = 0$ because $_RRf/r_{Rf}(eR)$ is $Rg/l_{Rg}(gRf)$-simple-injective. We claim that $a \in Y$. We show this claim. If not, then $afJf \not\subseteq X$ since $Y/X = S(gRf/X_{fRf})$. Then there is an $r \in fJf$ with $ar \notin X$. We may assume that $ar = y_{l'}$ for some $l' \in \Lambda$ because Y/X is the essential socle of gRf/X_{fRf}. Then $z_{l'}ar \neq 0$. On the other hand, $z_la r = 0$ since $z_la + r_{Rf}(eR) \in S(_RRf/r_{Rf}(eR))$ induces $z_la \in eS(_RRf/r_{Rf}(eR)) = S(_{eRe}eRf) = S(eRf_{fRf})$ and $r \in fJf$. Therefore $z_\lambda ar = 0$ for any $\lambda \in \Lambda$ because $Wa = 0$. This is a contradiction. Hence we can represent $a = \sum_{i=1}^{m} y_{l_i'}r_i + x$, where $l_i' \in \Lambda$, $r_i \in R$ and $x \in X$. Then $z_la = z_{\lambda'}a$ for any $\lambda' \in \Lambda - \{l\}$ since $Wa = 0$ and we can take $l'' \in \Lambda - \{l_i'\}_{i=1}^{m}$ because Λ is an infinite set. Therefore $0 \neq z_la = z_{l''}a = z_{l''}(\sum_{i=1}^{m} y_{l_i'}r_i + x) = 0$, a contradiction. Hence $|Y/X_{fRf}| < \infty$. $\square$

Using Lemmas 2.5.1 and 2.5.2 we easily have the following proposition.

Proposition 2.5.3. *Let $(eR; Rf)$ be a c-pair and g an idempotent of R. If fRf is a left perfect ring, $eR/l_{eR}(Rf)_R$ is $gR/r_{gR}(eRg)$-simple-injective and $_RRf/r_{Rf}(eR)$ is $Rg/l_{Rg}(gRf)$-simple-injective, then $|gRf/r_{gRf}(eRg)_{fRf}| < \infty$ and $|_{eRe}eRg/l_{eRg}(gRf)| < \infty$.*

Proof. Since fRf is left perfect, $gRf/r_{gRf}(eRg)_{fRf}$ is artinian by Lemma 2.5.2 or, for instance, [5, 10.10.Proposition]. Therefore there exists $n \in \mathbb{N}$ with $gJ^nf \subseteq r_{gRf}(eRg)$. On the other hand, $|_{eRe}l_{eRg}(gJ^if)/l_{eRg}(gJ^{i-1}f)| < \infty$ for any $i = 1, \ldots, n$ by Lemmas 2.5.1 and 2.5.2. Therefore $|_{eRe}eRg/l_{eRg}(gRf)| < \infty$. Hence $|gRf/r_{gRf}(eRg)_{fRf}| < \infty$ by Lemma 2.5.1. $\square$

We can now give the following theorem.

Theorem 2.5.4. *Let $(eR; Rf)$ be a c-pair and g an idempotent of R. Suppose that fRf is a left perfect ring. Then the following are equivalent:*

(1) (a) $eR/l_{eR}(Rf)_R$ is $gR/r_{gR}(eRg)$-injective, and

   (b) $_RRf/r_{Rf}(eR)$ is $Rg/l_{Rg}(gRf)$-injective.

(2) (a) $eR/l_{eR}(Rf)_R$ *is* $gR/r_{gR}(eRg)$*-simple-injective, and*

(b) $_RRf/r_{Rf}(eR)$ *is* $Rg/l_{Rg}(gRf)$*-simple-injective.*

(3) $|(gRf/r_{gRf}(eRg))_{fRf}| < \infty$.

(4) $|_{eRe}(eRg/l_{eRg}(gRf))| < \infty$.

(5) *ACC holds on* $\{r_{gRf}(I) \mid {}_{eRe}I \subseteq eRg\}$.

(6) *DCC holds on* $\{l_{eRg}(I') \mid I'_{fRf} \subseteq gRf\}$.

Proof. (1) $\Rightarrow$ (2). Obvious.

(2) $\Rightarrow$ (3), (4). This follows from Proposition 2.5.3.

(2) $\Rightarrow$ (1). Since we have already shown that (2) $\Rightarrow$ (3), (4), this follows from Proposition 2.2.1.

(3) $\Leftrightarrow$ (4). This follows from Lemma 2.4.2.

(3) $\Rightarrow$ (2). R satisfies $\alpha_r[e, g, f]$ by Lemma 2.4.2. Similarly R also satisfies $\alpha_l[e, g, f]$ since we've already shown (3) $\Leftrightarrow$ (4). Therefore (2) holds by Proposition 2.3.2 (1).

(3) $\Rightarrow$ (5). Obvious.

(5) $\Rightarrow$ (3). This follows from Proposition 2.4.3.

(5) $\Leftrightarrow$ (6). Obvious. $\square$

The following corollaries are easily deduced from Theorem 2.5.4.

Corollary 2.5.5. *Let* $(eR; Rf)$ *be a c-pair. If* fRf *is a left perfect ring, then the following are equivalent:*

(1) $eR/l_{eR}(Rf)_R$ *and* $_RRf/r_{Rf}(eR)$ *are injective.*

(2) $eR/l_{eR}(Rf)_R$ *and* $_RRf/r_{Rf}(eR)$ *are R-simple-injective.*

(3) $|Rf/r_{Rf}(eR)_{fRf}| < \infty$.

(4) $|_{eRe}eR/l_{eR}(Rf)| < \infty$.

(5) *ACC holds on* $\{r_{Rf}(I) \mid {}_{eRe}I \subseteq eR\}$.

Proof. Obviously (3), (4), (5) and the following (1′) and (2′) are equivalent by Theorem 2.5.4 and Proposition 2.3.2 (2).

(1′) $eR/l_{eR}(Rf)_R$ is $R/l_R(Rf)$-injective and $_RRf/r_{Rf}(eR)$ is

$R/r_R(eR)$-injective.

(2') $eR/l_{eR}(Rf)_R$ is $R/l_R(Rf)$-simple-injective and $_RRf/r_{Rf}(eR)$ is $R/r_R(eR)$-simple-injective.

Clearly (1') (resp. (2')) is equivalent to (1) (resp. (2)). □

Corollary 2.5.6. *Let $(eR; Rf)$ be an i-pair and g an idempotent of R. If fRf is a left perfect ring, then the following are equivalent:*

(1) eR_R and $_RRf$ are injective.

(2) eR_R and $_RRf$ are R-simple-injective.

(3) $|Rf_{fRf}| < \infty$.

(4) $|_{eRe}eR| < \infty$.

(5) ACC holds on $\{r_{Rf}(I) \mid _{eRe}I \subseteq eR\}$.

Finally in this section, we give another proposition on the conditions $\alpha_r[e, 1, f]$ and $\alpha_l[e, 1, f]$. This gives a proof for $(3) \Rightarrow (2)$ in Theorem 2.A.

Proposition 2.5.7. *Let R be a basic semiprimary ring and e, $f \in Pi(R)$. If R satisfies both $\alpha_r[e, 1, f]$ and $\alpha_l[e, 1, f]$, then $(eR; Rf)$ is an i-pair.*

Proof. Since R satisfies $\alpha_r[e, 1, f]$, $0 = r_{Rf}l_{eR}(0) = r_{Rf}(eR)$. Thus $S(_RRf) \cong T(_RRe)^n$ for some $n \in \mathbb{N}_0$. We let $S(_RRf) = \oplus_{i=1}^n S_i$, where $S_i \cong T(_RRe)$. Then $l_{eR}(S_1) = l_{eR}(\oplus_{i=1}^n S_i)$ because R is a basic semiprimary ring. It follows that $S_1 = r_{Rf}l_{eR}(S_1) = r_{Rf}l_{eR}(\oplus_{i=1}^n S_i) = \oplus_{i=1}^n S_i$. Hence $n = 1$, i.e., $S(_RRf) \cong T(_RRe)$. By a similar argument, we also have $S(eR_R) \cong T(fR_R)$. □

COMMENTS

The criterion for an indecomposable projective module to be injective given in Theorem 2.A is due to Fuller [54]. Baba and Oshiro improved Theorem 2.A to Theorem 2.B in [20]. From this point of view, in [11], Baba further generalized Theorem 2.B and gave criterions for an indecomposable projective to be quasi-injective and an indecomposable injective module to be quasi-projective. In [118] Morimoto and Sumioka generalized Theorem 2.B to module theory. Meanwhile Theorem 2.B was generalized to perfect rings in Hoshino-Sumioka [79] and Xue [186]. More investigation about the results in [11] was undertaken in Baba [15], from where most of material in Sections 2.2-2.5 is taken.

Chapter 3

Harada Rings

In this chapter, we introduce two classes of artinian rings, namely, left Harada rings and left co-Harada rings. However, although these rings are defined using mutually dual notions, these classes coincide. Indeed, it is shown that a ring is a "left" Harada ring if and only if it is a "right" co-Harada ring. By this fact we see that left Harada rings have rich structures for their left ideals and also for their right ideals. The class of these rings contains QF-rings and Nakayama rings. Moreover, as we will show in later chapters, the three classes of left Harada rings, QF-rings and Nakayama rings are very closely interrelated.

3.1 Definition of Harada Rings

A right R-module M is called a *small* module if $M \ll E(M)$ and is called *non-small* if M is not small.

First we give a useful lemma.

Lemma 3.1.1. *Let M be a right R-module and N a submodule of M.*

(1) *If N is non-small, then so is M.*

(2) *If M/N is non-small, then so is M.*

Proof. (1). This is straightforward to show.

(2). We put $\overline{M} = M/N$ and let $\pi : M \to \overline{M}$ be the natural epimorphism and $\tilde{\pi} : E(M) \to E(\overline{M})$ an extension map of π. Since $\overline{M}$ is non-small and $\overline{M} \subseteq \operatorname{Im}\tilde{\pi} \subseteq E(\overline{M})$, we see that $\overline{M} \not\ll \operatorname{Im}\tilde{\pi}$. Hence we have a proper submodule $\overline{K}$ of $\operatorname{Im}\tilde{\pi}$ with $\overline{M} + \overline{K} = \operatorname{Im}\tilde{\pi}$. Then $\tilde{\pi}^{-1}(\overline{K})$ is a proper submodule of $E(M)$ with $M + \tilde{\pi}^{-1}(\overline{K}) = E(M)$. $\qquad\square$

A faithful right R-module M is called *minimal faithful* if, for any faithful right R-module N, there exists a direct summand M' of N such that $M' \cong M$. A ring R is called a *right (left) QF-3* ring if R has a minimal faithful right (left) R-module and is called a *QF-3 ring* if it is both left and right *QF-3*.

Let R be a semiperfect ring. We say that a set $\{f_i\}_{i=1}^m$ of primitive idempotents of R is *pairwise non-isomorphic* if $f_i R \not\cong f_j R$ for any distinct $i, j \in \{1, \ldots, m\}$. If $\{Rf_i\}_{i=1}^m$ is an irredundant set of representatives of indecomposable projective left R-modules, we say that $\{f_i\}_{i=1}^m$ is *basic*. Any basic set of primitive idempotents is pairwise non-isomorphic.

Recall that a right R-module M is called *uniform* if every non-zero submodule of M is essential. We note that, if R is left perfect, M_R is uniform if and only if M_R is colocal.

The *uniform dimension* of a module M is the infimum of those cardinal numbers c such that $\#I \leq c$ for every independent set $\{N_i\}_{i \in I}$ of non-zero submodules of M. We denote the uniform dimension of M by unif.dim$M$.

Proposition 3.1.2. *Let R be a ring. We consider the following three conditions.*

    (a) R is right QF-3.

    (b) R contains a faithful injective right ideal.

    (c) $E(R_R)$ is projective.

Then the following hold.

    (1) $(a) \Rightarrow (b)$ holds. Further, if R is a left perfect ring, then $(b) \Rightarrow (a)$ also holds.

(2) *If DCC holds on right annihilator ideals, then $(b) \Rightarrow (c)$ does. And if either ACC or DCC holds on right annihilator ideals, then $(c) \Rightarrow (b)$ does.*

Proof. (1). (a) $\Rightarrow$ (b). Let M be a minimal faithful right R-module. Then there exists a right ideal I of R with $I \cong M_R$ since R_R is faithful. And M is injective because $E(R_R)$ is faithful. Hence R has a faithful injective right ideal I.

(b) $\Rightarrow$ (a). We assume that there exists a faithful injective right ideal I of R. Then $I = \oplus_{i=1}^{n} e_i R$ for some primitive idempotent $e_1, \ldots, e_n$ of R since R is left perfect. We may assume that $\{e_i\}_{i=1}^{k}$ is a basic set of $\{e_i\}_{i=1}^{n}$. Then we note that $\{S(e_i R_R)\}_{i=1}^{k}$ is a set of pairwise non-isomorphic essential simple socles since R is left perfect. Let N be a faithful right R-module. For each $i = 1, \ldots, k$, we have a monomorphism $\varphi_i : e_i R \to N$ since $S(e_i R_R)$ is an essential simple socles. Then $\sum_{i=1}^{k} \operatorname{Im} \varphi_i = \oplus_{i=1}^{k} \operatorname{Im} \varphi_i$ since $S(e_1 R_R), \ldots, S(e_k R_R)$ are pairwise non-isomorphic. Therefore we have a split monomorphism : $\oplus_{i=1}^{k} e_i R \to N$. Hence $\oplus_{i=1}^{k} e_i R$ is a minimal faithful right R-module.

(2). First we note that the following Claim 1.

Claim 1. *Let I be a right ideal of R. If there exists a monomorphism $\varphi : R/I \to R^S$, where S is a set, then $I = r_R(X)$ for some $X \subseteq R$.*

Proof of Claim 1. Let $\eta : R \to R/I$ be the natural epimorphism. And, for each $s \in S$, let $\pi_s : R^S \to R$ be the projection and put $x_s = \pi_s \varphi \eta(1)$. We further put $X = \{x_s\}_{s \in S}$. Then $r_R(X) = \cap_{s \in S} r_R(x_s) = \cap_{s \in S} \operatorname{Ker} \pi_s \varphi \eta = \operatorname{Ker} \eta = I$.

(b) $\Rightarrow$ (c). Let eR be a faithful injective right ideal of R, where e is an idempotent of R. Since eR_R is faithful, there exists a monomorphism $f : R_R \to (eR)^S$, where S is a set. From Claim 1, we have a subset X of R such that $r_R(\{x_s\}_{s \in S}) = 0$. Then we have a finite subset F of S with $r_R(\{x_s\}_{s \in F}) = 0$ because DCC holds on right annihilator ideals. Now we may assume that $F = \{1, 2, \ldots, n\}$. Then a monomorphism $\varphi' :$

$R_R \to (eR)^n$ is defined by $\varphi'(r) = (x_1 r, x_2 r, \ldots, x_n r)$ for any $r \in R$. Since $(eR)^n$ is injective, φ' is extended to a monomorphism $\tilde{\varphi}' : E(R_R) \to (eR)^n$. Therefore $E(R_R)_R$ is isomorphic to a direct summand of $(eR)^n_R$. Hence $E(R_R)_R$ is projective.

(c) $\Rightarrow$ (b). First we show the following Cliam 2.

Claim 2.

(1) *For any right ideal A of R,* $\mathrm{Hom}_R(r_R l_R(A)/A, E(R_R)) = 0$.

(2) *For any non-zero right ideals A, B of R with $A + B = A \oplus B$,*
$r_R l_R(A) < r_R l_R(A \oplus B)$.

(3) *unif.dim $R_R < \infty$.*

Proof of Claim 2. (1). Suppose that $\mathrm{Hom}_R(r_R l_R(A)/A, E(R_R)) \neq 0$ for some right ideal A of R. Let $0 \neq \varphi \in \mathrm{Hom}_R(r_R l_R(A)/A, E(R_R))$. φ is extended to $\tilde{\varphi} \in \mathrm{Hom}_R(R/A, E(R_R))$. And, since $E(R_R)$ is projective, there exists $\psi \in \mathrm{Hom}_R(E(R_R), R)$ with $\psi\varphi \neq 0$. We put $a = \psi\tilde{\varphi}(\bar{1})$. Then $aA = 0$. Hence $a \cdot r_R l_R(A) = 0$ since $a \in l_R(A) = l_R r_R l_R(A)$. Hence $\psi\varphi = (a)_L|_{r_R l_R(A)/A} = 0$, a contradiction.

(2). Suppose that $r_R l_R(A) = r_R l_R(A \oplus B)$. Then $A \oplus B \leq r_R l_R(A)$. On the other hand, $\mathrm{Hom}_R((A \oplus B)/A, E(R_R)) \neq 0$ since $(A \oplus B)/A \cong B$ is isomorphic to a submodule of $E(R_R)$. Hence there exists an epimorphism: $\mathrm{Hom}_R(r_R l_R(A)/A, E(R_R)) \to \mathrm{Hom}_R((A \oplus B)/A, E(R_R))$ because $E(R_R)$ is injective. This contradicts (1).

(3). This follows from (2).

Claim is shown.

From Claim 2 (3), we see that $E(R_R)$ has a finite direct sum decomposition $E(R_R) = \oplus_{i=1}^n E_i$ of indecomposable submodules. Then, for each $i = 1, \ldots, n$, E_i has the exchange property because $\mathrm{End}(E_i)$ is a local ring. On the other hand, since $E(R_R)$ is projective, we have a (split) monomorphism $\iota : E(R_R) \to R^{(S)}$ for some set S. So $E_i \cong e_i R$ for some primitive idempotent e_i of R. Now we may consider that $\{E_1, E_2, \ldots, E_m\}$ is an irredundant subset of $\{E_i\}_{i=1}^n$. Then $\sum_{i=1}^m e_i R = \oplus_{i=1}^m e_i R$ is a faithful injective right ideal of R. □

The following fundamental result is due to Faith.

Theorem 3.1.3. ([47, 20. 3 A and 20. 6 A] or [5, 25.1. Theorem]) *If a right R-module M is injective, then the following are equivalent:*

(1) M is $\sum$-*injective.*

(2) *ACC holds on* $\{ r_R(X) \mid X$ *is a subset of* $M \}$.

(3) $M^{(\mathbb{N})}$ *is injective.*

Next we quote a characterization of *QF-3* rings due to Colby and Rutter ([32, 1.1 Proposition and 1.2 and 1.3 Theorems]).

Theorem 3.1.4. *For a ring R, the following are equivalent:*

(1) *R is right perfect and contains a faithful $\sum$-injective right ideal.*

(2) *R is right perfect and injective hulls of every projective right R-module are projective.*

(3) *R is right perfect and projective covers of every injective right R-module are injective.*

(4) *R is perfect and contains minimal faithful right and left ideals.*

If the equivalent conditions hold, then R is a semiprimary QF-3 ring and satisfies ACC on right, and also left, annihilator ideals.

Lemma 3.1.5. *Let R be a left perfect ring. If ACC holds on right annihilator ideals, then ACC holds on* $\{ r_R(X') \mid X'$ *is a subset of* $eR/S_k(eR_R) \}$ *for any idempotent e of R and $k \in \mathbb{N}_0$.*

Proof. Assume that ACC holds on $\{ r_R(X') \mid X'$ is a subset of $eR/S_k(eR_R) \}$ but ACC does not hold on $\{ r_R(X'') \mid X''$ is a subset of $eR/S_{k+1}(eR_R) \}$. Then we have a strictly ascending chain $r_R(X_1'') \subsetneqq r_R(X_2'') \subsetneqq \cdots$ in $\{ r_R(X'') \mid X''$ is a subset of $eR/S_{k+1}(eR_R) \}$. For any $i \in \mathbb{N}$ take $r_i \in r_R(X_{i+1}'') - r_R(X_i'')$. We may assume that each X_i'' is contained in the set $(S_{k+1}(eR_R) \subseteq) X_i (\subseteq eR)$. Then $r_R(X_i'') = \{ r \in R \mid X_i r \subseteq S_{k+1}(eR_R) \} = \{ r \in R \mid X_i r J^{k+1} = 0 \}$ since R is left perfect. Therefore $X_i r_i J^{k+1} \neq 0$ but $X_{i+1} r_i J^{k+1} = 0$. Take $t_i \in J$ with $X_i r_i t_i J^k \neq 0$ and let X_i' be the subset of $eR/S_k(eR_R)$ induced from X_i. Then we have an ascending chain $r_R(X_1') \subseteq r_R(X_2') \subseteq \cdots$ in $\{ r_R(X') \mid X' \subseteq eR/S_k(eR_R) \}$. Moreover $r_i t_i \in r_R(X_{i+1}') - r_R(X_i')$ since $r_R(X_i') = \{ r \in R \mid X_i r l_{eR}(J^k) = 0 \}$. This shows that $r_R(X_1') \subsetneq r_R(X_2') \subsetneq \cdots$, i.e., ACC does not hold on $\{ r_R(X') \mid X' \subseteq eR/S_k(eR_R) \}$, a contradiction. $\qquad \square$

Let $S \subseteq T$ be sets of R-modules. We say that S is *a set of representatives* of T if each $M \in T$ is isomorphic to some element in S. Furthermore we say that a set S of representatives of T is *irredundant* if no two elements in S are isomorphic. Note that, for a right perfect ring, the irredundant set of representatives of simple right R-modules is finite, and hence so is that of indecomposable injective modules.

Proposition 3.1.6. *Let R be a right perfect ring. Suppose that the family of all injective right R-modules is closed under taking small covers, i.e., given any epimorphism $\varphi : M \to E$, where E is an injective right R-module and $\mathrm{Ker}\, \varphi$ is small in M, then M is also injective. Then the following hold:*

(1) *R is semiprimary QF-3.*

(2) *Any indecomposable injective right R-module is isomorphic to $eR/S_k(eR_R)$ for some $e \in Pi(R)$ and some $k \in \mathbb{N}_0$.*

(3) *Every local non-small right R-module is injective.*

Proof. (1). This follows from Theorem 3.1.4.

(2). Let E be an indecomposable injective right R-module and let $\varphi : P \to E$ be a projective cover. We may express P as $P = \oplus_{i \in I} e_i R$, where $e_i \in Pi(R)$. Then P is injective by assumption. Suppose that E is not local. Then we claim that $\mathrm{Ker}\, \varphi \supseteq S(P)$. If not, then $\mathrm{Ker}\, \varphi \not\supseteq S(e_j R_R)$ for some $j \in I$. Then, since $0 \neq \varphi(S(e_j R_R)) \subseteq E$ and E is indecomposable injective, $E \cong E(S(e_j R_R)) = e_j R$. Therefore E is local, a contradiction.

Thus φ induces an epimorphism $\varphi_1 : P/S_1(P) = \oplus_{i \in I} e_i R/S_1(e_i R_R) \to E$. Then $\text{Ker}\,\varphi_1 \ll P/S_1(P)$ since $S_1(P) \subseteq \text{Ker}\,\varphi \ll P$. Therefore $P/S_1(P)$ is injective by assumption. Hence $e_i R/S_1(e_i R_R)$ is injective for any $i \in I$. We now claim that $\text{Ker}\,\varphi_1 \supseteq S_2(P)/S_1(P)$. If not, then $\text{Ker}\,\varphi_1 \not\supseteq (S_2(e_j R_R) + S_1(P))/S_1(P)$ for some $j \in I$. Then $E \cong E(e_j R/S_1(e_j R_R)) = e_j R/S_1(e_j R_R)$. Therefore E is local, a contradiction. Hence $\text{Ker}\,\varphi_1 \supseteq S_2(P)/S_1(P)$.

Hence $\text{Ker}\,\varphi \supseteq S_2(P)$ and φ induces an epimorphism $\varphi_2 : P/S_2(P) = \oplus_{i \in I} e_i R/S_2(e_i R_R) \to E$. Inductively we see that $\text{Ker}\,\varphi \supseteq S_i(P)$ for any $i \in \mathbb{N}$. Hence $\text{Ker}\,\varphi = P$ since R is a semiprimary ring by (1). This is a contradiction. Therefore E is local.

Thus $P = eR$ for some $e \in Pi(R)$. Let $k \in \mathbb{N}_0$ such that $\text{Ker}\,\varphi \supseteq S_k(eR_R)$ but $\text{Ker}\,\varphi \not\supseteq S_{k+1}(eR_R)$. Then φ induces an epimorphism $\varphi' : eR/S_k(eR_R) \to E$. Since eR is local, we have $\text{Ker}\,\varphi' \ll eR/S_k(eR_R)$. Then $eR/S_k(eR_R)$ is injective by assumption. Hence $S(eR/S_k(eR_R))$ is simple. Therefore φ' is an isomorphism since $\text{Ker}\,\varphi \not\supseteq S_{k+1}(eR_R)$.

(3). Let M be a local non-small right R-module. Note that M is a cyclic module. Put $E = E(M)$ and let us express $S(E) = \oplus_{i \in I} S_i$, where each S_i is a simple module. For each $i \in I$, put $E_i = E(S_i)$. Then $E_i \cong e_i R/S_{k(i)}(e_i R_R)$ for some $e_i \in Pi(R)$ and some $k(i) \in \mathbb{N}_0$ by (2). Now ACC holds on right annihilator ideals by Theorem 1.1.13 and hence ACC holds on $\{\, r_R(X') \mid X' \text{ is a subset of } e_i R/S_{k(i)}(e_i R_R) \,\}$ by Lemma 3.1.5. Therefore each E_i is $\sum$-injective by Theorem 1.1.13. Hence we see that $E = \oplus_{i \in I} E_i$ since any irredundant set of representatives of indecomposable injective right R-modules is finite. Then, since M is cyclic, I is a finite set, say $I = \{1, \ldots, m\}$. Let $\pi_i : E = \oplus_{j=1}^{m} E_j \to E_i$ be the projection. Then there exists $i \in \{1, \ldots, m\}$ with $\pi_i(M) = E_i$ since M is non-small and R is right perfect. Hence M is injective by assumption. $\qquad\square$

We say that a ring R is a *right Harada ring* (abbreviated *right H-ring*) if it is a perfect ring satisfying the following condition:

(∗) Every non-small right R-module contains a non-zero injective submodule.

We have the following structure theorem for right H-rings.

Theorem 3.1.7. *For a perfect ring R, the following are equivalent:*

(1) *R is a right H-ring.*

(2) *For any $e \in Pi(R)$ with eR_R non-small, there exists $n(e) \in \mathbb{N}_0$ such that*

    (a) *$eR/S_{i-1}(eR_R)$ is injective for each $i = 1, \ldots, n(e)$, and*

    (b) *$eR/S_{n(e)}(eR_R)$ is small.*

Moreover, if R is a right H-ring, every indecomposable injective right R-module is of the form $eR/S_{k-1}(eR_R)$ for some $e \in Pi(R)$ with eR_R injective and $1 \le k \le n(e)$.

Proof. (1) $\Rightarrow$ (2). Let $e \in Pi(R)$ with eR_R injective. If $eR/S(eR_R)$ is non-small, it is also injective by (1). Repeating this argument, we obtain injective modules eR, $eR/S_1(eR_R)$, $\ldots$, $eR/S_t(eR_R)$ if they are non-small. However $eR/S_i(eR_R)$ is small for some i because any irredundant set of representatives of simple right R-modules is finite.

(2) $\Rightarrow$ (1) and the last comment. Let M be a non-small right R-module. Then $M \nsubseteq E(M)J$ since R is right perfect. Take $x \in M - E(M)J$. Then xR is a non-small right R-module. Let $\{e_1, \ldots, e_m, g_1, \ldots, g_n\}$ be a complete set of orthogonal primitive idempotents of R such that $e_1R, \ldots, e_mR$ are injective and $g_1R, \ldots, g_nR$ are not injective. We claim that xe_sR is non-small for some $s \in \{1, \ldots, m\}$. Note that xg_jR is small for any $j = 1, \ldots, n$ by Lemma 3.1.1 (2) since the left multiplication by x gives an epimorphism: $g_jR \to xg_jR$. Suppose that xe_iR is small for any $i = 1, \ldots, m$. Then $xR = \sum_{i=1}^m xe_iR + \sum_{j=1}^n xg_jR \ll E(xR)$, a contradiction. Put $e = e_s$ and consider an epimorphism $(x)_L : eR \to xeR$. Then either $\mathrm{Ker}(x)_L = S_{k-1}(eR_R)$ for some $k \le n(e)$ or we have $\mathrm{Ker}(x)_L \supseteq S_{n(e)}(eR_R)$ since $S_{n(e)}(eR_R)$ is uniserial by (2). Suppose that $\mathrm{Ker}(x)_L \supseteq S_{n(e)}(eR_R)$

holds. Then we have an epimorphism: $eR/S_{n(e)}(eR_R) \to eR/\operatorname{Ker}(x)_L$. Hence $eR/\operatorname{Ker}(x)_L$ ($\cong xeR$) is small by Lemma 3.1.1 (2), a contradiction. Therefore $\operatorname{Ker}(x)_L = S_{k-1}(eR_R)$ for some $k \leq n(e)$. Thus xeR ($\cong eR/S_{k-1}(eR_R)$) is injective by (2), i.e., M has a non-zero injective submodule.

Hence, if M is an indecomposable injective right R-module, $M = xeR$, i.e., $M \cong eR/S_{k-1}(eR_R)$, verifying the last comment. $\square$

Later in Corollary 4.3.6, we show that right H-rings are two-sided artinian. As a first step, we show the following

Proposition 3.1.8. *Right H-rings are right artinian.*

Proof. Let E be an indecomposable injective right R-module. First we claim that E is $\sum$-injective. We show this claim. Since R is left perfect, we may consider $E = E(S)$ for some simple right R-module S. Put $M = \oplus_{i \in I} E_i$, where $E_i \cong E$ for any i. Then $(M + E(M)J)/E(M)J$ is a direct summand of $E(M)/E(M)J$. We show that $(M + E(M)J)/E(M)J = E(M)/E(M)J$. If not, there is an $x \in E(M) - (M + E(M)J)$. Then xR is non-small and we may assume that xR is local, since xR/xJ is semisimple. Then xR is injective by condition (∗). Hence $S(xR_R)$ is simple and $xR \cong E$.

We therefore have a finite subset F of I such that $S(xR_R) \subseteq \oplus_{i \in F} E_i$. Put $E_F = \oplus_{i \in F} E_i$ and $E(M) = E_F \oplus E'$ for some E'. Let $\pi : E(M) = E_F \oplus E' \to E'$ be the projection. We claim that $\pi(x) \notin E'J$. If not, then $\pi(x) \in E'J \subseteq E(M)J$, so $x \in E_F + E(M)J \subseteq M + E(M)J$. But $x \notin M + E(M)J$, a contradiction. Thus $\pi(x)R \cong E$ in the same way as xR is since E' is injective. On the other hand, $\pi(S(xR_R)) = 0$ since $S(xR_R) \subseteq E_f$. Hence $\pi|_{xR}$ induces an epimorphism: $E \to E$ which is not injective. Because $E \cong eR/S_{k-1}(eR_R)$ for some $e \in Pi(R)$ and $1 \leq k \leq n(e)$ by Theorem 3.1.7, we have an epimorphism $\varphi : eR/S_{k-1}(eR_R) \to eR/S_{k-1}(eR_R)$ which is not injective. Let N be a submodule of eR with $N/S_{k-1}(eR_R) = \operatorname{Ker}\varphi$. Then by the last comment of Theorem 3.1.7 we see that $N = S_{l-1}(eR_R)$ for some $k < l \leq n(e)$ and therefore we have an isomorphism $\tilde{\varphi} : eR/S_{l-1}(eR_R) \to eR/S_{k-1}(eR_R)$. There then exists a unit u in a ring eRe with $(u)_L = \tilde{\varphi}$. Then $uS_{l-1}(eR_R) = S_{k-1}(eR_R)$ and $uS_{l-1}(eR_R) = S_{l-1}(eR_R)$ since u is a unit and $S_{n(e)-1}(eR_R)$ is uniserial by Theorem 3.1.7 (2)(a). But $S_{l-1}(eR_R) \supsetneqq S_{k-1}(eR_R)$ because $k < l$. This is a contradiction. Therefore $M = E(M)$, i.e., E is $\sum$-injective. In consequence, every indecomposable injective right R-module is $\sum$-injective.

Let E' be an injective right R-module. We express $S(E') = \oplus_{i \in L} S_i$, where S_i is simple for any i. Now, since R is left perfect, the irredundant set of representatives of indecomposable injective right R-modules is finite. Since every indecomposable injective right R-module is $\sum$-injective, this implies that $\oplus_{i \in L} E(S_i)$ is injective, so $E' = \oplus_{i \in L} E(S_i)$. Hence every injective right R-module can be expressed as a direct sum of cyclic indecomposable modules. Therefore we see that R is right artinian by Faith-Walker's theorem in [45] (see Faith [47, 20.17]).   □

In Theorem 3.1.7 we showed that, if R is a right H-ring, every indecomposable injective right R-module is of the form $eR/S_{k-1}(eR_R)$ for some $e \in Pi(R)$ with eR_R injective and $1 \leq k \leq n(e)$. In the following theorem we show that, over an artinian ring R, this description of indecomposable injectives characterizes right H-rings.

Theorem 3.1.9. *For a ring R, the following are equivalent:*

(1) *R is a right H-ring.*

(2) *R is a right artinian ring and every indecomposable injective right R-module is of the form $eR/S_k(eR_R)$ for some $e \in Pi(R)$ and some $k \in \mathbb{N}_0$.*

Proof. (1) $\Rightarrow$ (2). This follows from Theorem 3.1.7 and Proposition 3.1.8.

(2) $\Rightarrow$ (1). We show that R satisfies condition (2) of Theorem 3.1.7. Let $e \in Pi(R)$ with eR_R injective. First we show the following claim.

Claim. *If $eR/S_k(eR_R)$ is non-small, then it is injective.*

Proof of Claim. Suppose that $eR/S_k(eR_R)$ is non-small. We express $S(eR/S_k(eR_R)) = \oplus_{i=1}^n S_i$, where each S_i is simple. Then $E(S_i) \cong e_i R/S_{k_i}(e_i R_R)$ for some $e_i \in Pi(R)$ and $k_i \in \mathbb{N}_0$ by assumption. Therefore we have a homomorphism $\zeta_i :$ $eR/S_k(eR_R) \to e_i R/S_{k_i}(e_i R_R)$ with $\zeta_i(S_i) = S(e_i R/S_{k_i}(e_i R_R))$. Put $\zeta = \oplus_{i=1}^n \zeta_i eR/S_k(eR_R) \to \oplus_{i=1}^n e_i R/S_{k_i}(e_i R_R)$. Then ζ is a monomorphism. Therefore $E(eR/S_k(eR_R)) \cong \oplus_{i=1}^n e_i R/S_{k_i}(e_i R_R)$. Then ζ_j is surjective for some j since $eR/S_k(eR_R)$ is non-small. We may assume $e_j = e$ and show that $k_j = k$. Let N be a submodule of eR with $N/S_k(eR_R) = \mathrm{Ker}\,\zeta_j$. Then $N = S_l(eR_R)$ for some l by assumption. Since $\zeta_j(S_j) \neq 0$ and $S_k(eR_R) \subseteq N$, $l = k$, and hence $eR/S_k(eR_R) \cong eR/S_{k_j}(eR_R)$. This isomorphism is of the form $(u)_L$, where u is a unit element of eRe. Therefore $k_j = k$, and hence $eR/S_k(eR_R)$ is injective. This has established the claim.

If $eR/S_1(eR_R)$ is non-small, it is injective by the claim. Further, if $eR/S_2(eR_R)$ is non-small, it is also injective. Repeating this argument, we see that Theorem 3.1.7 (2) holds. $\square$

Let M be a module and N a submodule of M with $N = \oplus_{i \in I} N_i$. Then we say that N is a *locally direct summand* of M (with respect to the decomposition $N = \oplus_{i \in I} N_i$) if $\oplus_{i \in F} N_i$ is a direct summand of M for any finite subset F of I.

The following lemma due to Harada [60] is a crucial result.

Lemma 3.1.10. *If R satisfies $(*)$, then every injective right R-module contains a cyclic injective submodule and R contains a non-zero injective right ideal.*

Proof. Let E_R be injective. Consider the homomorphism $\varphi :$ $\oplus_{x \in E} E_x \to E$, where $E_x = E$ and $\varphi|_{E_x} = 1_E$ for each $x \in E$. Then $\varphi(\oplus_{x \in E} xR) = E$ and hence $\oplus_{x \in E} xR$ is non-small by Lemma 3.1.1. Hence $\oplus_{x \in E} xR$ contains an injective submodule F. Since F satisfies the exchange property, some xR contains an injective submodule isomorphic to a direct summand of F, as required.

Replacing $\oplus_{x \in E} E_x$ in the above argument by a free R-module, we obtain the last part of the lemma. □

Proposition 3.1.11. *If every injective right module over a ring R is a lifting module, then R is a right artinian ring.*

Proof. By Lemma 3.1.10, every indecomposable injective right R-module is cyclic. Then, by [45, Theorem 3.1] or [47, 20,17], it suffices to show that any injective right R-module E can be expressed as a direct sum of indecomposable submodules. First we show the following claim.

Claim 1. *For any independent family $\{A_i\}_{i \in I}$ of submodules of E, if $\oplus_{i \in I} A_i$ is a locally direct summand of E, then $\oplus_{i \in I} A_i$ is a direct summand of E.*

Proof of Claim 1. Put $A = \oplus_{i \in I} A_i$. Since E is lifting, there exists a direct decomposition $E = E_1 \oplus E_2$ such that $E_1 \subseteq A$ and $A \cap E_2 \ll E_2$. Then it suffices to show that $A \cap E_2 = 0$. Suppose not and take $0 \neq x \in A \cap E_2$. Then $x \in \oplus_{i \in F} A_i$ for some finite subset F of I and there exists a direct summand X of $\oplus_{i \in F} A_i$ with $xR \subseteq_e X$ since $\oplus_{i \in F} A_i$ is an extending module.

Then $X \cap E_1 = 0$. Thus if $\pi : A = E_1 \oplus (A \cap E_2) \to A \cap E_2$ is the projection map then $\pi |_X$ is a monomorphism. Let $\iota : A \cap E_2 \to E$ be the inclusion map. Then, since E is injective, we have a homomorphism $\varphi : E \to E$ with $\varphi \iota \pi |_X = 1_X$. Now $E = X \oplus X'$ for some X' since X is a direct summand of $\oplus_{i \in F} A_i$ and $\oplus_{i \in F} A_i$ is a direct summand of E. Let $\pi' : E = X \oplus X' \to X$ be the projection. Then $(\pi' \varphi)(\iota \pi |_X) = 1_X$. Therefore $E = \pi(X) \oplus \operatorname{Ker} \pi' \varphi$, i.e., $A \cap E_2$ contains a non-zero direct summand $\pi(X)$ of E. This contradicts $A \cap E_2 \ll E_2$. Hence $A = E_1$. Claim 1 is shown.

Next we show another claim.

Claim 2. *Every non-zero direct summand of E contains a non-zero indecomposable direct summand of E.*

Proof of Claim 2. Assume that there exists a non-zero direct summand E' of E which does not contain a non-zero indecomposable direct summand of E. Take $0 \neq x \in E'$. We have a maximal independent family $\{B_i\}_{i \in L}$ of submodules of E' such that $x \notin B = \oplus_{i \in L} B_i$ and B is a locally direct summand of E'. Then $E' = B \oplus B'$ for some B' by Claim 1. Now $B' \neq 0$ since $x \notin B$, and there exists an infinite set $\{B'_i\}_{i \in L'}$ of non-zero submodules of B' with $B' = \oplus_{i \in L'} B'_i$ since B' does not contain any non-zero indecomposable direct summand of E. Then we have a finite subset F' of L' with $x \in \oplus_{i \in F'} B'_i$. Therefore $x \notin B \oplus (\oplus_{i \in L' - F'} B'_i)$. This contradicts the maximality of $\{B_i\}_{i \in L}$. Claim 2 is shown.

Now let $\{E_i\}_{i \in K}$ be a maximal independent family of non-zero indecomposable direct summands of E such that $\oplus_{i \in K} E_i$ is a locally direct summand of E. Then $K \neq \varphi$ by Claim 2 and $\oplus_{i \in K} E_i$ is a direct summand of E by Claim 1. Hence $E = \oplus_{i \in K} E_i$ by Claim 2.   □

In the following theorem we give several characterizations of a right H-ring.

Theorem 3.1.12. *For a ring R the following are equivalent:*

(1) *R is a right H-ring.*

(2) *Every injective right R-module is a lifting module.*

(3) (a) *R is right perfect, and*

  (b) *the family of all injective right R-modules is closed under taking small covers.*

(4) *Every right R-module can be expressed as a direct sum of an injective module and a small module.*

If a ring R satisfies the above conditions, it is a right artinian QF-3 ring by Propositions 3.1.6 and 3.1.11.

Proof. $(1) \Rightarrow (2)$. Let E be an injective right R-module and let N be a non-small submodule of E. Take a maximal independent family $\{N_i\}_{i \in I}$ of non-zero injective submodules of N. Put $E_1 = \oplus_{i \in I} N_i$. Since R is right noetherian by Proposition 3.1.8, E_1 is also injective. Thus $E = E_1 \oplus E_2$ for some submodule E_2 of E. Then $N = E_1 \oplus (N \cap E_2)$. Suppose that $N \cap E_2 \not\ll E_2$. Then $N \cap E_2$ is non-small and hence, by (1), it has a non-zero injective submodule. This contradicts the maximality of $\{N_i\}_{i \in I}$. Hence $N \cap E_2 \ll E_2$. Thus E is lifting.

$(2) \Rightarrow (1)$. R is right artinian by Proposition 3.1.8. Moreover it is obvious that R satisfies $(*)$. Hence R is a right H-ring.

$(2) \Rightarrow (3)$.   Let E be an injective right R-module and let $\varphi : M \to E$ be a small cover. We have an extension epimorphism $\tilde{\varphi} : E(M) \to E$. Since $E(M)$ is lifting, there exists a direct decomposition $E(M) = E_1 \oplus E_2$ with $E_1 \subseteq M$ and $M \cap E_2 \ll E_2$. Then $M = E_1 \oplus (M \cap E_2)$ and $M \cap E_2 \ll E(M)$. Thus $\varphi(M \cap E_2) = \tilde{\varphi}(M \cap E_2) \ll E$. Therefore $E = \varphi(E_1)$ since $E = \varphi(M) = \varphi(E_1) + \varphi(M \cap E_2)$. Hence $M = E_1 + \mathrm{Ker}\,\varphi$ and so $M = E_1$ since $\mathrm{Ker}\,\varphi \ll M$. Thus M is injective.

$(2) \Leftrightarrow (4)$ is obvious.

$(3) \Rightarrow (1)$. R is semiprimary by Proposition 3.1.6 (1).

Let $e \in Pi(R)$ with eR_R injective. If $eR/S_1(eR_R)_R$ is non-small, it is injective by Proposition 3.1.6 (3). Further, if $eR/S_2(eR_R)$ is non-small, it is also injective by Proposition 3.1.6 (3). Repeating this argument, we eventually satisfy Theorem 3.1.7 (2).   □

By the proof of Theorem 3.1.12 we have the following remark.

Remark 3.1.13. For a right artinian ring R whose indecomposable injective right R-modules are finitely generated, the following are equivalent:

(1) R is a right H-ring.

(2)   Every finitely generated injective right R-module is a lifting module.

(3)   The family of all finitely generated injective right R-modules is

closed under taking small covers.

(4)   Every finitely generated right R-module can be expressed as a direct sum of an injective module and a small module.

3.2   A Dual Property of Harada Rings

In this section we consider a dual property of right H-rings. First we define a dual notion of non-small modules.

A right R-module M is called *cosmall* if there exist a projective module P and an epimorphism $\varphi : P \to M$ with $\operatorname{Ker} \varphi \subseteq_e P$ or, equivalently, $\operatorname{Ker} \psi \subseteq_e L$ holds for any epimorphism $\psi : L \to M$.

If M is not a cosmall module, i.e., for any projective module P and an epimorphism $\varphi : P \to M$, $\operatorname{Ker} \varphi \not\subseteq_e P$, then M is called a *non-cosmall* module.

The following two lemmas are easily obtained from the definition of non-cosmall modules.

Lemma 3.2.1. *Let M be a right R-module and N a submodule of M.*

(1)   *If N is non-cosmall, then so is M.*

(2)   *If M/N is non-cosmall, then so is M.*

Lemma 3.2.2. *Let M be a right R-module. Then M is non-cosmall if and only if $M \neq Z(M)$.*

Proof.   This follows from the well-known fact that a module M is singular if and only if there exists an epimorphism $\varphi : L \to M$ with $\operatorname{Ker} \varphi \subseteq_e L$ (see, for instance, Goodearl [55, Proposition 1.20 (b)]).   □

Now we consider a dual of the condition $(*)$:

$(*)^*$ Every non-cosmall module contains a non-zero projective direct
summand.

In order to characterize condition $(*)^*$, we give two preparatory lemmas.

Lemma 3.2.3. *If R satisfies $(*)^*$, then every indecomposable projective
right R-module is uniform.*

Proof.   Let P be an indecomposable projective right R-module and
suppose that there exists a submodule N of P with $N \not\subseteq_e P$. Then P/N
is non-cosmall and hence we have a direct decomposition $P/N = X \oplus Y$
with X non-zero projective by $(*)^*$. Let $\varphi : P \to P/N$ be the natural
epimorphism and let $\psi : P/N = X \oplus Y \to X$ be the projection. Then
$\psi\varphi$ splits and we have a direct decomposition $P = \mathrm{Ker}(\psi\varphi) \oplus P'$, where
$P' \neq 0$. Therefore $N \subseteq \mathrm{Ker}(\psi\varphi) = 0$ since P is indecomposable.   □

Lemma 3.2.4. *Suppose that R satisfies $(*)^*$. Let P be a uniform projective
right R-module and N a submodule of P. Then N is projective if and only
if $N \not\subseteq Z(P)$.*

Proof. $(\Rightarrow)$ is obvious.

$(\Leftarrow)$.   Suppose that $N \not\subseteq Z(P)$. Then N is non-cosmall by Lemma
3.2.2. Moreover N is indecomposable since it is a submodule of the uniform
module P. Thus N is projective since R satisfies $(*)^*$.   □

A module M is called *completely indecomposable* if $\mathrm{End}(M)$ is a local
ring. If M is completely indecomposable projective, it is a local module
(see, for instance, [5, 17.19. Proposition]).

Now we characterize condition $(*)^*$.

Theorem 3.2.5. *Let R be a ring with a complete set of orthogonal primitive idempotents. Then the following are equivalent:*

(1) R satisfies $(*)^*$.

(2) For any $f \in Pi(R)$, there exist $e \in Pi(R)$ with eR_R injective and $j \in \mathbb{N}_0$ with $fR \cong eJ^j$.

Moreover if these equivalent conditions hold, R is a semiperfect ring and, for any $e \in Pi(R)$ with eR_R injective, we have $n(e) \in \mathbb{N}_0$ such that

(a) eJ^{i-1} is completely indecomposable projective for any $1 \leq i \leq n(e)$, and

(b) $eJ^{n(e)}$ is singular.

Furthermore the integer j in (2) satisfies $j \leq n(e)$.

Proof. (1) $\Rightarrow$ (2) and the last comment. Let $e \in Pi(R)$ with eR_R injective. Then eR_R is completely indecomposable by Harada [64, 5.4.9 Proposition] or [5, 25.4. Lemma]. If eJ is not singular, then it is projective by Lemma 3.2.4. Further eJ is completely indecomposable again by [64, 5.4.9 Proposition] or the proof of [5, 25.4. Lemma] since it is quasi-injective. Therefore eJ^2 is a unique maximal submodule of eJ. If eJ^2 is not singular, then it is also completely indecomposable projective. Repeating this argument, we claim that there exists $n(e) \in \mathbb{N}$ such that eJ^{i-1} is completely indecomposable projective for any $1 \leq i \leq n(e)$ and $eJ^{n(e)}$ is singular. Assume that eJ^s is completely indecomposable projective for any $s \in \mathbb{N}_0$. Then there exists an isomorphism $\eta : eJ^s \to eJ^t$ for some distinct $s, t \in \mathbb{N}_0$ since there exists a complete set of orthogonal primitive idempotents of R. Moreover, η or η^{-1} is extendable to a monomorphism $eR \to eR$. This is also an isomorphism since eR is injective. Therefore $s = t$, a contradiction. Hence the second part of the last comment holds.

Let $f \in Pi(R)$ with fR_R not injective. Then fR is uniform by Lemma 3.2.3 and hence $E(fR_R)$ is indecomposable. Furthermore $Z(E(fR_R)) \neq E(fR_R)$ since fR is not singular. Therefore $E(fR_R)$ is indecomposable projective by Lemma 3.2.2 because R satisfies $(*)^*$. Say $E(fR_R) \cong eR_R$, where $e \in Pi(R)$. Moreover there exists $j \in \{1, \ldots, n(e)\}$ with $fR \cong eJ^{j-1}$ since fR is not singular and we've already established the second part of the last comment. Thus fR is also completely indecomposable. Hence R is a semiperfect ring (see, for instance, [5, 27.6. Theorem]).

(2) $\Rightarrow$ (1). Every indecomposable projective module is quasi-injective by (2). Thus R is semiperfect by [64, 5.4.9 Proposition] or the proof of [5, 25.4. Lemma]. Next we show the following

Claim. *If eJ^s is projective for some $e \in Pi(R)$ with eR_R injective and $s \in \mathbb{N}$, then eJ^i is also projective for any $0 \leq i \leq s$. Consequently we have the unique chain $eR \supsetneq eJ \supsetneq \cdots \supsetneq eJ^s$ of submodules of eR.*

Proof of Claim. Let eJ^s be as described. Take $x \in eJ^{s-1} - eJ^s$ with xR local and let $\pi : gR \to xR$ be a projective cover, where $g \in Pi(R)$. Then eJ^s is a local right R-module since it is indecomposable projective and R is semiperfect. Thus $\pi(gJ) = eJ^s$ and since gR is colocal by (2), π is an isomorphism. Hence xR is projective and there exists an isomorphism $\eta : xR \to eJ^j$ for some $j \in \mathbb{N}_0$ by (2). Then η or η^{-1} is extendable to an endomorphism of eR and this is also an isomorphism. Therefore $xR = eJ^{s-1}$ holds since $x \in eJ^{s-1} - eJ^s$. Hence eJ^{s-1} is projective. The claim now follows by induction.

Now let M be a non-cosmall right R-module. Then there exist $m \in M$ and $g \in Pi(R)$ such that mgR_R is non-cosmall by Lemma 3.2.2. Consider an epimorphism $(m)_L : gR \to mgR$. Then $\text{Ker}(m)_L \not\subseteq_e gR$. Since gR is uniform by (2), it follows that $(m)_L$ is an isomorphism, i.e., $mgR \cong gR$. Furthermore, there exists an isomorphism $\varphi : gR \to eJ^t$ for some $e \in Pi(R)$ with eR_R injective and $t \in \mathbb{N}_0$ by (2). Hence we have $\psi : M \to eR$ such that $\psi|_{mgR} = \varphi(m)_L^{-1}$. Then $\text{Im}\,\psi = eJ^{t'}$ for some $t' \leq t$ since $\text{Im}\,\psi \supseteq eJ^t$ and $eR \supsetneq eJ \supsetneq \cdots \supsetneq eJ^t$ by the above claim. Therefore $\text{Im}\,\psi$ is projective, again by the claim, and hence ψ splits. Hence M contains a projective direct summand. $\qquad\square$

We say that a ring R is a *right co-Harada ring* (abbreviated *right co-H-ring*) if it satisfies $(*)^*$ and ACC on right annihilator ideals.

Before giving an important characterization of right co-H-rings, we present a well-known fact and three lemmas.

A quasi-injective module M is called $\sum$-*quasi-injective* if the direct sum of any number of copies of M is quasi-injective.

First we recall the following theorem from [47, 20.2 A, 20.6A].

Theorem 3.2.6. *For a quasi-injective right R-module M, the following are equivalent.*

(1) M *is* $\sum$-*quasi-injective.*

(2) $M^{(\mathbb{N})}$ *is quasi-injective.*

(3) $E(M)$ *is* $\sum$-*injective.*

(4) $E(M)^{(\mathbb{N})}$ *is injective.*

(5) *ACC holds on* $\{\, r_R(X) \mid X \subseteq E(M)\,\}$.

Further, if these equivalent conditions hold, M can be expressed as a direct sum of completely indecomposable modules.

Lemma 3.2.7. *Let R be a semiperfect ring and $\{g_i\}_{i=1}^n$ a complete set of orthogonal primitive idempotents of R. If $g_i R$ is $\sum$-quasi-injective for any $i = 1, \ldots, n$, then R is right perfect.*

Proof. By Theorem 1.2.17, it suffices to show that every projective right R-module has the exchange property. Let P be a projective right R-module. We may assume that $P = \oplus_{i=1}^n (g_i R)^{(I_i)}$. For each $i = 1, \ldots, n$, $(g_i R)^{(I_i)}$ is quasi-injective by assumption and hence it has the exchange property. We claim that $P = \oplus_{i=1}^n (g_i R)^{(I_i)}$ also has the exchange property.

Let L be a right R-module containing P and let $L = (\oplus_{i=1}^n (g_i R)^{(I_i)}) \oplus N = \oplus_{i \in I} L_i$ be direct decompositions. Since $(g_1 R)^{(I_1)}$ has the exchange property, there exists a submodule L_i' of L_i for any $i \in I$ satisfying $L = (g_1 R)^{(I_1)} \oplus (\oplus_{i \in I} L_i')$. Further, since $(g_2 R)^{(I_2)}$ has the exchange property, for any $i \in I$ there exist submodules M_1 and L_i'' of $(g_1 R)^{(I_1)}$ and L_i', respectively, satisfying $L = (g_2 R)^{(I_2)} \oplus M_1 \oplus (\oplus_{i \in I} L_i'')$. Then M_1 and L_i'' are direct summands of $(g_1 R)^{(I_1)}$ and L_i', respectively. Let

$(g_1 R)^{(I_1)} = M_1 \oplus \overline{M_1}$ and $L'_i = L''_i \oplus \overline{L_i}''$. Then $\overline{M_1} \oplus (\oplus_{i \in I} \overline{L_i}'') \cong (g_2 R)^{(I_2)}$ because $L = (g_1 R)^{(I_1)} \oplus (\oplus_{i \in I} L'_i) = (M_1 \oplus \overline{M_1}) \oplus (\oplus_{i \in I}(L''_i \oplus \overline{L_i}''))$ and $L = (g_2 R)^{(I_2)} \oplus M_1 \oplus (\oplus_{i \in I} L''_i)$. Therefore $\overline{M_1} = 0$ by the Krull-Remak-Schmidt-Azumaya Theorem (see, for instance, [64, 7.1.2 Theorem]) since $(g_1 R)^{(I_1)}$ and $(g_2 R)^{(I_2)}$ have no non-zero direct summands which are isomorphic to each other. Hence $L = (g_1 R)^{(I_1)} \oplus (g_2 R)^{(I_2)} \oplus (\oplus_{i \in I} L''_i)$. Proceeding inductively, we see that $L = (\oplus_{i=1}^{n}(g_i R)^{(I_i)}) \oplus (\oplus_{i \in I} L'''_i)$ for some submodule L'''_i of L_i. $\square$

Lemma 3.2.8. *Let M be a right R-module with $M = \oplus_{i=1}^{n} M_i$, where each M_i is uniform, let U be a uniform submodule of M, and set $I = \{1, \ldots, n\}$. Suppose that, for any $\varphi \in \mathrm{Hom}_R(A_i, M_j)$, where $i, j \in I$ and A_i is a submodule of M_i, either of the following (1) or (2) holds:*

(1) *φ can be extended to an element of $\mathrm{Hom}_R(M_i, M_j)$.*

(2) *$\mathrm{Ker}\,\varphi = 0$ and an isomorphism $\varphi^{-1} : \mathrm{Im}\,\varphi \to A_i$ can be extended to an element of $\mathrm{Hom}_R(M_j, M_i)$.*

Then there exists a direct summand M' of M with $U \subseteq_e M'$.

Proof. Let $\pi_j : M = \oplus_{i \in I} M_i \to M_j$ be the projection for any $j \in I$. Put $K = \{\, k \in I \mid \mathrm{Ker}(\pi_k \mid_U) = 0 \,\}$. Then $K \neq \emptyset$ since U is uniform. We claim that there exists $k \in K$ such that $\pi_i(\pi_k \mid_U)^{-1} : \pi_k(U) \to \pi_i(U)$ can be extended to $\psi_i \in \mathrm{Hom}_R(M_k, M_i)$ for any $i \in I$. For any $i, j \in K$, either the following (1) or (2) holds by assumption:

(1) *$\pi_j(\pi_i \mid_U)^{-1} : \pi_i(U) \to \pi_j(U)$ can be extended to an element of $\mathrm{Hom}_R(M_i, M_j)$.*

(2) *$\pi_i(\pi_j \mid_U)^{-1} : \pi_j(U) \to \pi_i(U)$ can be extended to an element of $\mathrm{Hom}_R(M_j, M_i)$.*

Assume that there exist $i, j, l \in K$ such that $\pi_j(\pi_i \mid_U)^{-1}$ and $\pi_l(\pi_j \mid_U)^{-1}$ are extended to $\psi_{j,i}$ and $\psi_{l,j}$, respectively. Then $\psi_{l,j}\psi_{j,i}$ is an extended homomorphism of $\pi_l(\pi_i \mid_U)^{-1}$. Therefore there exists $k \in K$ such that $\pi_i(\pi_k|_U)^{-1}$ is extended to $\psi_i \in \mathrm{Hom}_R(M_k, M_i)$ for any $i \in K$. Then $\pi_i(\pi_k|_U)^{-1}$ is also extended to $\psi_i \in \mathrm{Hom}_R(M_k, M_i)$ for any $i \in I - K$ by assumption. Put $M' = \{\, x + \sum_{i \in I - \{k\}} \psi_i(x) \mid x \in M_k \,\}$. Then $M = M' \oplus (\oplus_{i \in I - \{k\}} M_i)$ and $U \subseteq_e M'$ because $U = \{\, y + \sum_{i \in I - \{k\}} \pi_i(\pi_k|_U)^{-1}(y) \mid y \in \pi_k(U) \,\}$. $\qquad \square$

Lemma 3.2.9. *Let R be a right perfect ring, M a projective right R-module and $N = \oplus_{i \in I} N_i$ a locally direct summand of M. Then N is a direct summand of M.*

Proof. Since M_R is lifting by Theorem 1.2.17, we have a direct decomposition $M = M_1 \oplus M_2$ such that $M_1 \subseteq N$ and $N \cap M_2 \ll M_2$. We show that $N \cap M_2 = 0$, i.e., $M = N \oplus M_2$. Given $x \in N \cap M_2$, there exists a finite subset F of I with $x \in \oplus_{i \in F} N_i$. Put $N' = \oplus_{i \in F} N_i$. Then there exists a direct decomposition $M = N' \oplus N''$ since $N = \oplus_{i \in I} N_i$ is a locally direct summand of M. Now let $\pi_i : M = M_1 \oplus M_2 \to M_i$ be the projection for $i = 1, 2$. Because M_1 is projective, we have a direct decomposition $M_1 = M_1' \oplus M_1''$ such that $M_1' \subseteq \pi_1(N')$ and $\pi_1(N') \cap M_1'' \ll M_1''$. Put $X = N' \cap (M_1'' \oplus M_2)$. Then, since $M = N' + (M_1'' \oplus M_2) = N' \oplus N'' = M_1' \oplus M_1'' \oplus M_2$, we see that $M/X = N'/X \oplus (M_1'' \oplus M_2)/X = N'/X \oplus (N'' \oplus X)/X = (M_1' \oplus X)/X \oplus (M_1'' \oplus M_2)/X$. Thus $N'/X \cong (M_1 \oplus X)/X \cong M_1 \oplus M$ and $(M_1'' \oplus M_2)/X \cong (N'' \oplus X)/X \cong N'' \oplus M$. Therefore N'/X and $(M_1'' \oplus M_2)/X$ are projective modules. Hence M/X is projective and X is a direct summand of M. On the other hand, since $N = N \cap M = N \cap (M_1 \oplus M_2) = M_1 \oplus (N \cap M_2)$, we have $\pi_2(N) = N \cap M_2 \ll M_2$. Thus $\pi_2(N) \ll M$. Let $\pi'' : M_1 = M_1' \oplus M_1'' \to M_1''$ be the projection. Then, because $\pi_1(N') = \pi_1(N') \cap M_1 = \pi_1(N') \cap (M_1' \oplus M_1'') = M_1' \oplus (\pi_1(N) \cap M_1'')$, we further see that $\pi'' \pi_1(N') = \pi_1(N') \cap M_1'' \ll M_1''$. Hence $\pi'' \pi_1(N') \ll M$. Consequently $X \subseteq \pi'' \pi_1(N') + \pi_2(N') \ll M$. Therefore $X = 0$ because X is a direct summand of M. Hence $x = 0$ since $x \in N' \cap M_2 \subseteq X$. $\qquad \square$

Now we characterize right *co-H*-rings.

Theorem 3.2.10. *For a ring R, the following are equivalent:*

(1) *R is a right co-H-ring.*

(2) *Every projective right R-module is an extending module.*

(3) *The family of all projective right R-modules is closed under taking essential extensions.*

(4) *Every right R-module can be expressed as a direct sum of a projective module and a singular module.*

Moreover, if these equivalent conditions hold, R is a semiprimary QF-3 ring which satisfies ACC on right annihilator ideals.

Proof. (1) $\Rightarrow$ (2). We show this by the following Claims 1–7.

Claim 1. *R has a complete set of orthogonal primitive idempotents.*

Proof of Claim 1. Assume that, for any $k \in \mathbb{N}$, we have a decomposition $1 = \sum_{i=1}^{k} f_i + f_k'$ of orthogonal idempotents of R such that $f_k' = f_{k+1} + f_{k+1}'$. Then $Rf_1' \supsetneq Rf_2' \supsetneq Rf_3' \supsetneq \cdots$ and hence $r_R(Rf_1') \subseteq r_R(Rf_2') \subseteq r_R(Rf_3') \subseteq \cdots$. Therefore there exists $n \in \mathbb{N}$ such that $r_R(Rf_n') = r_R(Rf_{n+1}')$ since ACC holds on right annihilator ideals. But $r_R(Rf_n') = \sum_{i=1}^{n} f_i R \neq \sum_{i=1}^{n+1} f_i R = r_R(Rf_{n+1}')$, a contradiction.

Claim 2.

(a) *R is a semiprimary ring.*

(b) *For any projective right R-module P, $Z(P)$ is $\sum$-quasi-injective.*

Proof of Claim 2. (a). For any $f \in Pi(R)$, there exists $e \in Pi(R)$ such that eR_R is injective and $fR \cong eJ^j$ for some $j \in \mathbb{N}$ by Theorem 3.2.5. Therefore fR is $\sum$-quasi-injective by Theorem 3.2.6 since ACC holds on right annihilator ideals. On the other hand, R is semiperfect by Theorem 3.2.5. Hence R is right perfect by Lemma 3.2.7. Thus the Jacobson radical of R is nilpotent by [5, 29.1. Proposition]. Hence R is semiprimary.

(b). Let $\{f_i\}_{i=1}^{n}$ be a complete set of orthogonal primitive idempotents of R. By Theorem 3.2.5, we have $Z(E(f_j R_R)) \cong Z(f_j R)$ for each

$j = 1, \ldots, n$. Then $Z(f_j R)$ is $\oplus_{i=1}^{n} Z(f_i R)$-injective since $\oplus_{i=1}^{n} Z(f_i R) = Z(\oplus_{i=1}^{n} f_i R)$. Therefore $\oplus_{i=1}^{n} Z(f_i R)$ is quasi-injective (see, for instance, [5, 16.10. Proposition]). Hence it is $\sum$-quasi-injective by Theorem 3.2.6. Now we may assume that P is expressed as $P = \oplus_{i=1}^{n} (f_i R)^{(I_i)}$ for some I_i. Then, since $Z(P) = \oplus_{i=1}^{n} Z(f_i R)^{(I_i)}$, $Z(P)$ is a direct summand of $(\oplus_{i=1}^{n} Z(f_i R))^{(I)}$ for some I. Therefore $Z(P)$ is $\sum$-quasi-injective because $\oplus_{i=1}^{n} Z(f_i R)$ is $\sum$-quasi-injective.

Claim 3. *Let P be a projective right R-module and $P = \oplus_{i \in I} P_i$, where each P_i is indecomposable. Then, for each uniform submodule U of P, there exists a finite subset F of I with $U \subseteq \oplus_{i \in F} P_i$.*

Proof of Claim 3. For each $i \in I$, $E(P_i)$ is local $\sum$-injective by Theorems 3.2.5 and 3.2.6. Therefore $E(P) = \oplus_{i \in I} E(P_i)$ since any irredundant set of representatives of indecomposable injective right R-modules is finite by Claim 2 (a). Moreover there exists $j \in I$ with $E(P) = E(U) \oplus (\oplus_{i \in I - \{j\}} E(P_i))$ since $E(U)$ has the exchange property. Then $U \oplus (\oplus_{i \in I - \{j\}} P_i) \subseteq_e P$. Put $P' = \oplus_{i \in I - \{j\}} P_i$ and let $\pi_j : P = P_j \oplus P' \to P_j$ and $\pi' : P = P_j \oplus P' \to P'$ be the projections. Then, since $U \cap P' = 0$, we may define a right R-homomorphism $\varphi : \pi_j(U) \to \pi'(U)$ by $\pi_j(x) \mapsto \pi'(x)$ for any $x \in U$. Let $\tilde{\varphi} : E(P_j) \to E(P')$ be an extension map of φ. Then there exists a finite subset F of $I - \{j\}$ with $\tilde{\varphi}(E(P_j)) \subseteq \oplus_{i \in F} E(P_i)$ because $E(P_j)$ is local. Then $\varphi(\pi_j(U)) \subseteq \oplus_{i \in F} P_i$. Now $U = \{ y + \varphi(y) \mid y \in \pi_j(U) \}$ by the definition of φ. Hence $U \subseteq P_j \oplus (\oplus_{i \in F} P_i)$.

Claim 4. *Let P be a projective right R-module and U a uniform submodule of P. Then there exists a direct summand P' of P with $U \subseteq_e P'$.*

Proof of Claim 4. By Claim 2 (a), we have a direct decomposition $P = \oplus_{i \in I} P_i$, where each P_i is indecomposable. We may assume that I is finite by Claim 3. Therefore, by Lemma 3.2.8, it suffices to show that, for any $\varphi \in \mathrm{Hom}_R(A_i, P_j)$, where $i, j \in I$ and A_i is a submodule of P_i, either the following (1) or (2) holds:

(1) φ can be extended to an element of $\mathrm{Hom}_R(P_i, P_j)$.

(2) $\mathrm{Ker}\,\varphi = 0$ and an isomorphism $\varphi^{-1} : \mathrm{Im}\,\varphi \to A_i$ can be extended to an element of $\mathrm{Hom}_R(P_j, P_i)$.

Let $\psi : E(P_i) \to E(P_j)$ be an extension map of φ. Suppose that φ is a monomorphism. Then ψ is an isomorphism. Moreover we have $e \in Pi(R)$ with $eR \cong E(P_i) \cong E(P_j)$ and $k_i, k_j \in \mathbb{N}_0$ with $P_i \cong eJ^{k_i}$ and $P_j \cong eJ^{k_j}$ by Theorem 3.2.5 and Claim 1. If $k_i \geq k_j$, then $\psi(P_i) = \psi(E(P_i)J^{k_i}) \subseteq E(P_j)J^{k_j} = P_j$. Therefore $\psi|_{P_i} \in \mathrm{Hom}_R(P_i, P_j)$. If $k_i < k_j$, then $\psi^{-1}|_{P_j} \in \mathrm{Hom}_R(P_j, P_i)$ by the same argument.

Now suppose that φ is not a monomorphism. Then we claim that $\mathrm{Im}\,\psi \subseteq Z(E(P_j))$. If not, then $\mathrm{Im}\,\psi$ is projective by Lemma 3.2.4 and Theorem 3.2.5. Then the epimorphism $\psi : E(P_i) \to \mathrm{Im}\,\psi$ splits and it is a monomorphism since $E(P_i)$ is uniform. Hence φ is a monomorphism, a contradiction. Thus $\mathrm{Im}\,\psi \subseteq Z(E(P_j)) = Z(P_j) \subseteq P_j$ by Theorem 3.2.5. Hence $\psi|_{P_i} \in \mathrm{Hom}_R(P_i, P_j)$.

Claim 5. Let P be a projective right R-module and A a submodule of P. Then there exist direct decompositions $P = \tilde{P} \oplus Q$ and $A = \tilde{A} \oplus Z$ such that $\tilde{A}$ is projective with $\tilde{A} \subseteq_e \tilde{P}$ and Z is a singular module with $Z \subseteq Q$.

Proof of Claim 5. Let $\{f_i\}_{i=1}^n$ be a complete set of orthogonal primitive idempotents of R and let $\{e_i\}_{i=1}^m$ be a subset of $\{f_i\}_{i=1}^n$ such that $\{e_iR\}_{i=1}^m$ is an irredundant set of representatives of indecomposable projective injective right R-modules. By Theorem 3.2.5, for each $i = 1, \ldots, m$, we have $n(i) \in \mathbb{N}_0$ such that e_iJ^{j-1} is projective for each $j = 1, \ldots, n(i)$ and $e_iJ^{n(i)} = Z(e_iR)$. Next note that $\{ e_iJ^{j-1} \mid 1 \leq i \leq m,\ 1 \leq j \leq n(i) \}$ is an irredundant set of representatives of indecomposable projective right R-modules.

If $A = Z(A)$, then $P = 0 \oplus P$ and $A = 0 \oplus A$ are the desired direct decompositions. Suppose that $A \neq Z(A)$. Then A is non-cosmall by Lemma 3.2.2, and hence A has a non-zero projective direct summand by $(*)^*$. Now we may assume that there exist $k \in \{1, \ldots, m\}$ and $l \in \{1, \ldots, n(k)\}$ such that

(†) A contains a direct summand which is isomorphic to $e_k J^{l-1}$ and

(††) if A contains a direct summand which is isomorphic to $e_i J^{j-1}$,
      then

    (a) $i = k$ and $l \leq j \leq n(k)$ or

    (b) $k + 1 \leq i \leq m$.

By (†), there exists a direct decomposition $A = A' \oplus A''$ with $A' \cong e_k J^{l-1}$. Then A' is uniform. Hence we have a direct decomposition $P = P' \oplus P''$ with $A' \subseteq_e P'$ by Claim 4. Then P' is uniform, and so is $P' \cap A$. Therefore $P' \cap A = P' \cap (A' \oplus A'') = A' \oplus (P' \cap A'') = A'$. In consequence, P' is an indecomposable direct summand of P such that $P' \cap A \, (= A')$ is a projective direct summand of A satisfying $P' \cap A \cong e_k J^{l-1}$ and $P' \cap A \subseteq_e P'$. Hence, using Zorn's Lemma, we obtain a maximal family $\{P_i\}_{i \in I}$ of independent indecomposable direct summands of P such that

(† † †) $\oplus_{i \in I} P_i$ is a locally direct summand of P and

(† † ††) $A_i = P_i \cap A$ is a projective direct summand of A with $A_i \cong e_k J^{l-1}$
          and $A_i \subseteq_e P_i$ for any $i \in I$.

By Lemma 3.2.9 and († † †), $\oplus_{i \in I} P_i$ is a direct summand of P. Put $P = (\oplus_{i \in I} P_i) \oplus Q^{(1)}$ and let $\pi_i : P = (\oplus_{j \in I} P_j) \oplus Q^{(1)} \to P_i$ be the projection for any $i \in I$. We show that $\pi_i(A) = A_i$ for any $i \in I$. The inclusion $\pi_i(A) \supseteq A_i$ is obvious. For the converse, suppose that $\pi_i(A) \supsetneq A_i$. Then $\pi_i(A)$ is projective with $\pi_i(A) \cong e_k J^{j-1}$ for some $j \leq l - 1$ by Theorem 3.2.5 (a) since A_i is isomorphic to the projective module $e_k J^{l-1}$ by († † ††) and $\pi_i(A)$ is a submodule of an indecomposable module P_i. Thus $\pi_i |_A : A \to \pi_i(A)$ splits, and hence A contains a direct summand of A which is isomorphic to $e_k J^{j-1}$ because $\pi_i(A) \cong e_k J^{j-1}$. This contradicts (††) since $j \leq l - 1$. Hence $\pi_i(A) = A_i$. Therefore $A = A \cap P = A \cap ((\oplus_{i \in I} P_i) \oplus Q^{(1)}) = (\oplus_{i \in I} A_i) \oplus (A \cap Q^{(1)})$ since $\pi_i(A) = A_i \subseteq A$.

Put $Z^{(1)} = A \cap Q^{(1)}$. We show that $Z^{(1)}$ does not contain a direct summand which is isomorphic to $e_k J^{l-1}$. Assume, on the contrary, that there does exist a direct summand Z' of $Z^{(1)}$ with $Z' \cong e_k J^{l-1}$. Then Q' is a direct summand of $Q^{(1)}$ with $Z' \subseteq_e Q'$ by Claim 4, and we see that $Z' = A \cap Q'$. Thus $\{P_i\}_{i \in I} \cup \{Q'\}$ satisfies ($\dagger\dagger\dagger$) and ($\dagger\dagger\dagger\dagger$), contradicting the maximality of $\{P_i\}_{i \in I}$. Put $P^{(1)} = \oplus_{i \in I} P_i$ and $A^{(1)} = \oplus_{i \in I} A_i$. In consequence, we obtain direct decompositions $P = P^{(1)} \oplus Q^{(1)}$ and $A = A^{(1)} \oplus Z^{(1)}$ such that $A^{(1)}$ is projective with $A^{(1)} \subseteq_e P^{(1)}$, $Z^{(1)}$ does not contain any projective direct summand which is isomorphic to $e_k J^{l-1}$, and $Z^{(1)} \subseteq Q^{(1)}$.

If $Z^{(1)}$ is singular, $P = P^{(1)} \oplus Q^{(1)}$ and $A = A^{(1)} \oplus Z^{(1)}$ are the desired direct decompositions. If $Z^{(1)}$ is not singular, the same argument works on $Z^{(1)} \subseteq Q^{(1)}$ instead of $A \subseteq P$. Repeating this argument, we obtain direct decompositions $P = P^{(1)} \oplus P^{(2)} \oplus \cdots \oplus P^{(k)} \oplus Q^{(k)}$ and $A = A^{(1)} \oplus A^{(2)} \oplus \cdots \oplus A^{(k)} \oplus Z^{(k)}$ such that each $A^{(i)}$ is projective with $A^{(i)} \subseteq_e P^{(i)}$ and $Z^{(k)}$ is singular with $Z^{(k)} \subseteq Q^{(k)}$.

Claim 6. Let P be a projective right R-module, A a direct summand of $Z(P)$ and $\oplus_{i \in I} A_i$ a locally direct summand of A. Then $\oplus_{i \in I} A_i$ is a direct summand of A.

Proof of Claim 6. Since A is quasi-injective by Claim 2 (b), it is extending by, for instance, [64, 9.9.1 Proposition]. Thus there exists a direct summand A' of A with $\oplus_{i \in I} A_i \subseteq_e A'$. Now each A_i is $\sum$-quasi-injective by Claim 2 (b) since A_i is a summand of A, and hence $E(A') = E(\oplus_{i \in I} A_i) = \oplus_{i \in I} E(A_i)$ by Theorem 3.2.6 and Claim 2 (a). Hence $Z(E(A')) = \oplus_{i \in I} Z(E(A_i))$. Furthermore, if we let $\{e_i\}_{i=1}^m$ and $\{n(i)\}_{i=1}^m$ be as in the proof of Claim 5, then $Z(P) \cong \oplus_{i=1}^m (e_i J^{n(i)})^{(K_i)}$ for some $K_1, \ldots, K_m$ by Theorem 3.2.5. Thus, since the quasi-injective module A' has the exchange property, $A' \cong \oplus_{i=1}^m (e_i J^{n(i)})^{(K_i')}$ for some $K_1', \ldots, K_m'$. Therefore $A' = Z(E(A'))$ because $E(A') \cong \oplus_{i=1}^m (e_i R)^{(K_i')}$ by Claim 2 (b) and Theorem 3.2.6. In consequence, $A' = Z(E(A')) = \oplus_{i \in I} Z(E(A_i))$. Then, for any $i \in I$, $A' = A_i \oplus (\oplus_{j \in I - \{i\}} Z(E(A_j)))$ because the quasi-injective module A_i has the exchange property. Thus $A_i = Z(E(A_i))$ since $A_i \subseteq Z(E(A_i))$. Hence $A' = \oplus_{i \in I} A_i$.

Claim 7. Every projective right R-module P is extending. This establishes (2).

Proof of Claim 7. Let A be a submodule of P. By Claim 5, we may assume that A is singular, i.e., $A \subseteq Z(P)$. Since $Z(P)$ is quasi-injective by Claim 2 (b), it is extending. Hence we may further assume that A is a direct summand of $Z(P)$. Now $Z(P)$ may be expressed as a direct sum of completely indecomposable uniform submodules by Theorems 3.2.5, 3.2.6 and Claim 2 (a), (b). Therefore there exists a non-zero direct summand A' of A which contains a completely indecomposable uniform direct summand A'' by, for instance, [5, 12.6 Theorem]. Then there exists a direct summand P'' of P with $A'' \subseteq_e P''$ by Claim 4. Hence, using Zorn's Lemma, we obtain maximal independent families $\{P_i\}_{i \in I}$ of indecomposable direct summands of P and $\{A_i\}_{i \in I}$ of indecomposable direct summands of A such that $\oplus_{i \in I} A_i \subseteq_e \oplus_{i \in I} P_i$ and $\oplus_{i \in I} P_i$ and $\oplus_{i \in I} A_i$ are locally direct summands of P and A, respectively. Then, by Lemma 3.2.9 and Claim 6, $\oplus_{i \in I} P_i$ and $\oplus_{i \in I} A_i$ are direct summands of P and A, respectively, say $P = (\oplus_{i \in I} P_i) \oplus Q$ and $A = (\oplus_{i \in I} A_i) \oplus B$.

We complete the proof by showing that $B = 0$. Suppose that $B \neq 0$. Then there exists a uniform direct summand B' of B since $Z(P)$ is expressed as a direct sum of completely indecomposable uniform submodules and A (and so B) is a direct summand of $Z(P)$. Let $\pi : P = (\oplus_{i \in I} P_i) \oplus Q \to Q$ be the projection. Then $\pi|_{B'}$ is a monomorphism since $\mathrm{Ker}(\pi|_{B'}) = B' \cap (\oplus_{i \in I} P_i)$, $B' \cap (\oplus_{i \in I} A_i) = 0$ and $\oplus_{i \in I} A_i \subseteq_e \oplus_{i \in I} P_i$. Hence $\pi(B')$ is uniform. Moreover we have a uniform direct summand Q' of Q with $\pi(B') \subseteq_e Q'$ by Claim 4. Then $(\oplus_{i \in I} A_i) \oplus B' \subseteq_e (\oplus_{i \in I} P_i) \oplus B' = (\oplus_{i \in I} P_i) \oplus \pi(B') \subseteq_e (\oplus_{i \in I} P_i) \oplus Q'$. Therefore the existence of the families $\{P_i\}_{i \in I} \cup \{Q'\}$ and $\{A_i\}_{i \in I} \cup \{B'\}$ contradicts the maximality of $\{P_i\}_{i \in I}$ and $\{A_i\}_{i \in I}$. Hence $B = 0$, and hence (2) is attained.

$(2) \Rightarrow (4)$. Let M be a right R-module and let $\varphi : F \to M$ be an epimorphism, where F is a free right R-module. There exists a direct decomposition $F = P_1 \oplus P_2$ with $\mathrm{Ker}\,\varphi \subseteq_e P_1$ since F is extending by

(2). Then $M = \varphi(P_1) \oplus \varphi(P_2)$, where $\varphi(P_1)$ ($\cong P_1/\operatorname{Ker}\varphi$) is singular and $\varphi(P_2)$ ($\cong P_2$) is projective.

(4) $\Rightarrow$ (3). Let P be a non-zero projective right R-module and let M be a right R-module with $P \subseteq_e M$. By (4), we may write $M = P' \oplus Z'$, where P' is projective and Z' is singular. Let $\pi : M = P' \oplus Z' \to P'$ be the projection. By (4), we may write $\pi(P) = P'' \oplus Z''$, where P'' is projective and Z'' is singular. Let $\pi' : \pi(P) = P'' \oplus Z'' \to P''$ be the projection. Then $\pi'\pi|_P : P \to P''$ splits, and hence $\operatorname{Ker}(\pi'\pi|_P)$ is projective. So we see that $\operatorname{Ker}(\pi'\pi|_P) = 0$ since $\operatorname{Ker}(\pi'\pi|_P) \subseteq \operatorname{Ker}(\pi'\pi) = \pi^{-1}(Z'') = Z' \oplus Z''$ and $Z' \oplus Z''$ is singular. Therefore $Z' \cap P = \operatorname{Ker}\pi \cap P = 0$. Hence, because $P \subseteq_e M$, we have $Z' = 0$, i.e., $M = P'$ is projective.

(3) $\Rightarrow$ (1) and the last comment. We show this by Claims 1–4 as follows.

Claim 1.

(a)   *Every projective injective right R-module contains a completely indecomposable $\sum$-injective submodule.*

(b)   *Let $\{P_l\}_{l\in L}$ be a family of indecomposable projective injective right R-modules. Then $\oplus_{l\in L}P_l$ is $\sum$-injective.*

Proof of Claim 1.   (a).   Let P be a projective injective right R-module. Choose a cardinal number τ such that $\tau > \max\{\aleph_0, \#R\}$ and put $E = E(P^{(\tau)})$. Then E is projective by (3), and hence we may consider E as a direct summand of $R^{(I)}$ for some set I. Then, since the injective module E has the exchange property, $R^{(I)} = E \oplus (\oplus_{i\in I}e_iR)$ for some $e_i \in Pi(R)$. Hence $E \cong \oplus_{i\in I}(1 - e_i)R$, and we can write $E = \oplus_{i\in I}x_iR$. Hence $\tau \le$ unif.dim$P^{(\tau)} = $ unif.dim$E = \sum_{i\in I}$ unif.dim$(x_iR) \le \#I \times \#R$. Then $\tau \le \#I$ since $\tau > \#R$. Hence there exists $J \subseteq I$ such that $\#J \ge \aleph_0$ and $x_iR \cong x_jR$ for any $i, j \in J$ because, if γ denotes the cardinality of an irredundant set of representatives of $\{xR \mid x \in R\}$, then $\gamma \le \#R < \tau \le \#I$. Then x_jR is $\sum$-injective for any $j \in J$ since $(x_jR)^{(\aleph_0)}$ may be considered as a direct summand of $\oplus_{i\in I}x_iR = E$ which is injective. Therefore x_jR can be expressed as a direct sum of completely indecomposable modules by

Theorem 3.2.6. Choose a completely indecomposable direct summand E' of $x_j R$. Then E' is uniform, since it is completely indecomposable $\sum$-injective, and hence it may be considered as a submodule of $P^{(\tau)}$ because P is injective. Since E' has the exchange property, we have $P^{(\tau)} = E' \oplus (\oplus_{i \in K} P'_i)$, where K is a set with $\#K = \tau$ and $P = P'_i \oplus P''_i$ for any $i \in K$. Thus $E' \cong \oplus_{i \in K} P''_i$, and hence, since E' is indecomposable, $E' \cong P''_k$ for some $k \in K$. Therefore P contains a completely indecomposable $\sum$-injective submodule P''_k, showing (a).

(b). Each P_l is $\sum$-injective by (a). Hence we may assume that $P_l \not\cong P_m$ for any distinct l, m. Put $P = \oplus_{l \in L} P_l$ and $E = E(P^{(\tau)})$, where we let τ be as in the proof of (a). Then, as in the proof of (a), we can write $E = \oplus_{i \in I} x_i R$, where each $x_i R \cong f_i R$ for some idempotent f_i of R. For any $i \in I$, there exists a non-zero element $y_i \in x_i R \cap P^{(\tau)}$ since $P^{(\tau)} \subseteq_e E = \oplus_{i \in I} x_i R$. Then $E(y_i R) \subseteq x_i R$ since $x_i R$ is injective. Further $E(y_i R) \cong \oplus_{l \in L_i} P_l^{k_{i,l}}$ for some finite subset L_i of L and $k_{i,l} \in \mathbb{N}$ because $E(y_i R) \subseteq P^{(\tau)} = (\oplus_{l \in L} P_l)^{(\tau)}$, each P_l is indecomposable and $E(y_i R)$ has the exchange property. Thus we may assume that $E(y_i R) \cong P_{l_i}$ for some $l_i \in L$. Therefore each $x_i R$ contains a direct summand $E(y_i R)$ which is isomorphic to some P_{l_i}.

Now we consider the partitions $I = \cup_{\gamma \in \Gamma} I_\gamma$ and $\Gamma = \Gamma_1 \cup \Gamma_2$, where, for any $i \in I_\gamma$ and $j \in I_{\gamma'}$,

$$\begin{cases} x_i R \cong x_j R \text{ if } \gamma = \gamma', \\ x_i R \not\cong x_j R \text{ if } \gamma \neq \gamma', \end{cases}$$

and

$$\begin{cases} \#I_\gamma \geq \chi_0 \text{ if } \gamma \in \Gamma_1, \\ \#I_\gamma < \infty \text{ if } \gamma \in \Gamma_2. \end{cases}$$

Put $I_n = \cup_{\gamma \in \Gamma_n} I_\gamma$ for $n = 1, 2$. For any $i \in I_1$, $(x_i R)^{(\aleph_0)}$ is isomorphic to a direct summand of $E = \oplus_{j \in I} x_j R$ since $\#I_\gamma \geq \aleph_0$ for any $\gamma \in \Gamma_1$. Then $(x_i R)^{(\chi_0)}$ is injective. Now $x_i R$ can be expressed as a direct sum of indecomposable modules by Theorem 3.2.6. Hence we may assume that $x_i R$ is indecomposable and is isomorphic to P_{l_i}. Put $E_n = \oplus_{i \in I_n} x_i R$ for $n = 1, 2$.

On the other hand, for any $l \in L$ we now show that there exists $i_l \in I_1$ with $P_l \cong x_{i_l} R$. Assume, on the contrary, that there exists $l \in L$ such that $P_l \not\cong x_i R$ for any $i \in I_1$. Now $P_l^{(\tau)} \subseteq E = E_1 \oplus E_2$. If there exists $0 \neq z \in P_l^{(\tau)} \cap E_1$, then $E(zR) \subseteq P_l^{(\tau)} \cap E_1$ since $P_l^{(\tau)}$ and E_1 are direct sums of injective modules. Therefore $E(zR)$ contains a direct summand which is isomorphic to both P_l and some direct summand of E_1, i.e., $P_l \cong x_i R$ for some $i \in I_1$, a contradiction. Hence $P_l^{(\tau)} \cap E_1 = 0$. Thus, if $\pi : E = E_1 \oplus E_2 \to E_2$ is the projection, then $\pi|_{P_l^{(\tau)}}$ is a monomorphism. Therefore it follows that $\#E_2 \geq \tau$. Hence $\#\Gamma_2 \geq \tau$ since $\tau > \#R$ and $\#I_\gamma < \infty$ for any $\gamma \in \Gamma_2$. But $\#\Gamma_2 \leq \#\Gamma \leq \#R$ because each $x_i R$ is isomorphic to some $f_i R$ for some idempotent f_i of R. This contradicts the definition of τ. Hence, for any $l \in L$, there exists $i_l \in I_1$ with $P_l \cong x_{i_l} R$. As a consequence, E_1 contains a direct summand which is isomorphic to P since $P_l \not\cong P_m$ for any distinct l, m.

Moreover $E_1^{(\aleph_0)} = \oplus_{\gamma \in \Gamma_1}(\oplus_{i \in I_\gamma} x_i R)^{(\aleph_0)} \cong \oplus_{\gamma \in \Gamma_1}(\oplus_{i \in I_\gamma} x_i R) = E_1$ by the definition of Γ_1. Therefore E_1 is $\sum$-injective. Hence P is also $\sum$-injective.

Claim 2. *Let P be a projective right R-module. Then $E(P)$ is projective $\sum$-injective.*

Proof of Claim 2. By (3), $E(P)$ is projective. Thus we have a non-empty maximal independent set $\{P_i\}_{i \in I}$ of indecomposable $\sum$-injective submodules of $E(P)$ by Claim 1 (a). Then $\oplus_{i \in I} P_i$ is $\sum$-injective by Claim 1 (b). We have a direct decomposition $E(P) = (\oplus_{i \in I} P_i) \oplus E'$. Then $E' = 0$ by Claim 1 (a) and the maximality of $\{P_i\}_{i \in I}$.

Claim 3. *R is a semiperfect ring which satisfies $(*)^*$.*

Proof of Claim 3. By Claim 2 and Theorem 3.2.6, $E(R_R)$ can be expressed as a (finite) direct sum of indecomposable right R-modules and hence unif.dim$(R_R) < \infty$. Therefore there exists a complete set $\{f_j\}_{j=1}^n$ of orthogonal primitive idempotents of R.

Let $f \in \{f_j\}_{j=1}^n$. We may express $E(fR_R)$ as a direct sum $E(fR_R) =$

$\oplus_{i=1}^{m} y_i R$ of completely indecomposable projective modules by Claim 2 and Theorem 3.2.6. We claim that $m = 1$. Suppose that $m \geq 2$. Let $\pi_k : \oplus_{i=1}^{m} y_i R \to y_k R$ be the projection for each $k = 1, \ldots, m$. Then $fR \subseteq_e \oplus_{i=1}^{m} \pi_i(fR)$ since $fR \subseteq_e E(fR_R) = \oplus_{i=1}^{m} \pi_i(E(fR_R))$. Therefore $\pi_i(fR)$ is projective for each $i = 1, \ldots, m$ by (3). Then $\pi_1 \mid_{fR}$ is monic because $\pi_1 \mid_{fR} : fR \to \pi_1(fR)$ is a split epimorphism and fR is indecomposable. Therefore $fR \cap (\oplus_{i=2}^{m} y_i R) = \mathrm{Ker}(\pi_1 \mid_{fR}) = 0$, i.e., $\oplus_{i=2}^{m} y_i R = 0$, a contradiction. Hence $m = 1$.

Consequently $E(fR_R)$ is an indecomposable projective injective module, and hence we have an isomorphism $\varphi : E(fR_R) \to eR$ for some $e \in \{f_j\}_{i=1}^{m}$ since $E(fR_R)$ has the exchange property. Then eJ^i is indecomposable quasi-injective for any $i \in \mathbb{N}_0$. Since $E(eJ^i) = eR$ and it is uniform, eJ^i is quasi-injective and uniform. Then $eJ^i \oplus eJ^i$ is quasi-injective and satisfies the exchange property, and hence eJ^i is completely indecomposable by Theorem 1.1.14. Now there exists $s \in \mathbb{N}_0$ such that $\varphi(fR) \subseteq eJ^s$ but $\varphi(fR) \not\subseteq eJ^{s+1}$ because $eJ^i \not\cong eJ^j$ for any distinct $i, j \in \mathbb{N}$. Then eJ^s is completely indecomposable and projective by (3), and hence it is local. Therefore, since $\varphi(fR) \not\subseteq eJ^{s+1}$, we have $\varphi(fR) = eJ^s$. Hence R is a semiperfect ring which satisfies $(*)^*$ by Theorem 3.2.5.

Claim 4. *R is a semiprimary QF-3 ring which satisfies ACC on right annihilator ideals.*

Proof of Claim 4. By (3) and Theorem 3.1.4, it suffices to show that R is a right perfect ring.

Let $f \in Pi(R)$. Then fR_R is quasi-injective by Theorem 3.2.5 (2) and hence $\sum$-quasi-injective by Claim 2 and Theorem 3.2.6. Therefore R is a right perfect ring by Lemma 3.2.7 and Claim 3. □

In view of the arguments in the proof of Theorem 3.2.10, we have the following remark.

Remark 3.2.11. For a right noetherian ring R whose indecomposable injective right R-modules are finitely generated, the following are equivalent:

(1) R is a right *co-H-ring*.

(2) Every finitely generated projective right R-module is an extending module.

(3) The family of all finitely generated projective right R-modules is closed under taking essential extensions.

(4) Every finitely generated right R-module can be expressed as a direct sum of a projective module and a singular module.

3.3   The Relationships between Harada Rings and Co-Harada Rings

In this section we present the important fact that a ring is a left H-ring if and only if it is a right *co-H-ring*.

We begin with the following lemma.

Lemma 3.3.1. *Let R be a semiprimary ring which satisfies ACC on right annihilator ideals and $e_1, \ldots, e_n, f \in Pi(R)$. Then the following are equivalent:*

(1) $Rf/S_{j-1}(_RRf) \cong E(T(_RRe_j))$ *holds for each $j = 1, \ldots, n$.*

(2) (a) e_1R_R *is injective with $S(e_1R_R) \cong T(fR_R)$, and*

     (b) $e_jR \cong e_1J^{j-1}$ *holds for each $j = 1, \ldots, n$.*

Proof. (1) $\Rightarrow$ (2). $_RRf$ is injective with $S(_RRf) \cong T(_RRe_1)$ by (1). Therefore (a) holds by Corollary 2.4.6.

Next we show (b). First note that $\{e_j\}_{j=1}^n$ is pairwise non-isomorphic by (1). We now use induction on j. Let $k \in \{1, \ldots, n\}$ be such that $e_jR \cong e_1J^{j-1}$ holds for any $j \leq k - 1$.

Claim. *Then e_1J^{k-1}/e_1J^k is isomorphic to a direct sum of copies of $T(e_kR_R)$.*

Proof of Claim. Choose $x \in e_1 J^{k-1} - J^k$ with $xg = x$ for some $g \in Pi(R)$. We show that $Rg \cong Re_k$. Since $e_{k-1}R \cong e_1 J^{k-2}$ by our inductive assumption, $e_1 J^{k-2} = yR$ for some $y \in e_1 J^{k-2} e_{k-1}$. Thus there exists $r \in e_{k-1} Jg - J^2$ with $yr = x$. Then we have an epimorphism $\varphi : Rr \to S(_R Rf / S_{k-2}(_R Rf))$ because $S(_R Rf / S_{k-2}(_R Rf)) \cong T(_R Re_{k-1})$ by (1). Moreover there exists an extension map $\tilde{\varphi} : Rg \to Rf / S_{k-2}(_R Rf)$ of φ by (1). Then $0 \neq (g)\tilde{\varphi} \in S_k(_R Rf) / S_{k-2}(_R Rf) - S_{k-1}(_R Rf) / S_{k-2}(_R Rf)$ since $r \in Jg - J^2$. Hence $g(S_k(_R Rf) / S_{k-1}(_R Rf)) \neq 0$. Therefore $Rg \cong Re_k$ because $S_k(_R Rf) / S_{k-1}(_R Rf) \cong T(_R Re_k)$ by (1). This shows the claim.

Now $_R Rf / S_{k-1}(_R Rf)$ is quasi-projective injective with $S(_R Rf / S_{k-1}(_R Rf)) \cong T(_R Re_k)$ by (1). Further $S_{k-1}(_R Rf) = r_{Rf}(e_k R)$ by (1) since $\{e_j\}_{j=1}^n$ is pairwise non-isomorphic. Hence $e_k R$ is a quasi-injective right R-module with $S(e_k R_R) \cong T(f R_R)$ by Corollary 2.4.5. On the other hand, $e_1 Re_k \subseteq e_1 J^{k-1}$ by our inductive assumption since $\{e_j\}_{j=1}^n$ is pairwise non-isomorphic. Hence $e_1 J^{k-1} \cong e_k R$ by the claim above because $e_1 R \cong E(T(f R_R))$ and $e_k R_R$ is quasi-injective with $S(e_k R_R) \cong T(f R_R)$. Thus (2) holds.

(2) $\Rightarrow$ (1). We use induction on j. Firstly $Rf \cong E(T(_R Re_1))$ follows from (2)(i) and Corollary 2.4.6. Let $k \in \{1, \ldots, n\}$ be such that $Rf / S_{j-1}(_R Rf) \cong E(T(_R Re_j))$ for any $1 \leq j \leq k - 1$.

Claim. $S(Rf / S_{k-1}(_R Rf))$ *is isomorphic to a direct sum of copies of* $T(_R Re_k)$.

Proof of Claim. Take $x \in S_k(_R Rf) - S_{k-1}(_R Rf)$ with $gx = x$ for some $g \in Pi(R)$. We show that $gR \cong e_k R$. Since $S(_R Rf / S_{k-2}(_R Rf)) \cong T(_R Re_{k-1})$ by our inductive assumption, we have $S_{k-1}(_R Rf) = Ry$ for some $y \in e_{k-1} S_{k-1}(_R Rf)$. Therefore there exists $r \in e_{k-1} Jg - J^2$ with $rx = y$. Hence $gR \cong e_k R$ because $e_{k-1} J \cong e_1 J^{k-1} \cong e_k R$ by (2)(ii). This shows the claim.

Now $e_k R$ is a quasi-injective right R-module with $S(e_k R_R) \cong T(f R_R)$

by (2)(ii). Thus $Rf/r_{Rf}(e_k R)$ is an injective right R-module with $S({}_R Rf/r_{Rf}(e_k R)) \cong T({}_R Re_k)$ by Corollary 2.4.5. Further $r_{Rf}(e_k R) = S_{k-1}({}_R Rf)$ by our inductive assumption and the claim above since $\{e_j\}_{j=1}^n$ is pairwise non-isomorphic by (2)(ii). Hence $Rf/S_{k-1}({}_R Rf) \cong E(T({}_R Re_k))$, i.e., (1) holds.   □

Using Lemma 3.3.1, we can easily show the following proposition.

Proposition 3.3.2. *Let R be a semiprimary ring which satisfies ACC on right annihilator ideals and let $\{e_{ij}\}_{i=1,j=1}^{m\ \ n(i)}$ and $\{f_i\}_{i=1}^m$ be two sets of primitive idempotents of R. Then the following are equivalent:*

(1) *$\{f_i\}_{i=1}^m$ satisfies the following conditions:*

    (a) *$\{Rf_i\}_{i=1}^m$ is an irredundant set of representatives of indecomposable projective injective left R-modules,*

    (b) *$Rf_i/S_{j-1}({}_R Rf_i) \cong E(T({}_R Re_{ij}))$,*

    (c) *${}_R Rf_i/S_{n(i)}({}_R Rf_i)$ is small*

    for any $i = 1, \ldots, m$, $j = 1, \ldots, n(i)$.

(2) *$\{e_{ij}\}_{i=1,j=1}^{m\ \ n(i)}$ is a basic set of primitive idempotents of R which satisfies the following conditions:*

    (d) *$\{e_{i1}R\}_{i=1}^m$ is an irredundant set of representatives of indecomposable projective injective right R-modules with $S(e_{i1}R_R) \cong T(f_i R_R)$,*

    (e) *$e_{ij}R \cong e_{i1}J^{j-1}$,*

    (f) *$e_{i1}J_R^{n(i)}$ is singular*

    for any $i = 1, \ldots, m$, $j = 1, \ldots, n(i)$.

Proof. (1) $\Rightarrow$ (2). $S({}_R Rf_i) \cong T({}_R Re_{i1})$ holds for each $i = 1, \ldots, m$ by (b). Therefore (d) follows from (a) and Corollary 2.4.6. Moreover $\{e_{ij}\}_{i=1,j=1}^{m\ \ n(i)}$ is a basic set of primitive idempotents of R by (a), (b), (c) and Theorem 3.1.7. Further (e) follows from Lemma 3.3.1. Thus (f) also follows from Theorem 3.2.5 since $\{e_{ij}\}_{i=1,j=1}^{m\ \ n(i)}$ is a basic set of primitive idempotents of R.

(2) $\Rightarrow$ (1). We can show this using the dual of the argument for (1) $\Rightarrow$ (2). However we give a proof for the benefit of the reader.

From (d) and Corollary 2.4.6, (a) holds and $S(_RRf_i) \cong T(_RRe_{i1})$ for each $i = 1, \ldots, m$. Hence, in particular, we have $E(T(_RRe_{i1})) \cong Rf_i$ for each $i = 1, \ldots, m$. Further (b) holds by Lemma 3.3.1. Thus (c) also holds since $\{e_{ij}\}_{i=1,j=1}^{m\ \ n(i)}$ is a basic set of primitive idempotents of R. $\square$

The following theorem is the main theorem of this section.

Theorem 3.3.3. *A ring is a left H-ring if and only if it is a right co-H-ring.*

Proof.   Assume that R is a left H-ring. Then R is right artinian by Proposition 3.1.8. Further condition (1) of Proposition 3.3.2 holds by Theorem 3.1.7. Thus R satisfies $(*)^*$ by Proposition 3.3.2 (d), (e) and Theorem 3.2.5, and hence R is a right co-H-ring.

Conversely assume that R is a right co-H-ring. Then R is a semiprimary ring by Theorem 3.2.10. Moreover condition (2) of Proposition 3.3.2 holds by Theorem 3.2.5. Thus R is a left H-ring by Proposition 3.3.2 (a), (b), (c) and Theorem 3.1.7. $\square$

Let R be a left H-ring with $Pi(R)$. Then, by Theorems 3.2.5 and 3.3.3, $Pi(R)$ can be arranged as $Pi(R) = \{e_{ij}\}_{i=1,j=1}^{m\ \ n(i)}$, where

(1) $e_{i1}R$ is an injective right R-module for each $i = 1, \ldots, m$, and

(2) $e_{ij}R \cong e_{i,j-1}R$ or $e_{ij}R \cong e_{i,j-1}J$ for any $i = 1, \ldots, m$, $j = 2, \ldots, n(i)$.

In case R is a basic left H-ring, the condition (2) can be replaced by

(2′) $e_{ij}R \cong e_{i1}J^{j-1}$ for any $i = 1, \ldots, m$, $j = 2, \ldots, n(i)$.

We say that such an arranged $Pi(R) = \{e_{ij}\}_{i=1,j=1}^{m\ \ n(i)}$ is a *left well-indexed set*. Symmetrically, we can also define a *right well-indexed set* for a right H-ring.

For later use, we say that the left H-ring R has *m-row blocks* if R has a left well-indexed set $Pi(R) = \{e_{ij}\}_{i=1,j=1}^{m\ \ n(i)}$.

By Proposition 3.3.2, we have the following theorem.

Theorem 3.3.4. *Let R be a basic left H-ring with a left well-indexed set $Pi(R) = \{e_{ij}\}_{i=1,j=1}^{m\ \ n(i)}$. Then the following hold:*

(1) $Re_{\sigma(i)\rho(i)}/S_{j-1}(_RRe_{\sigma(i)\rho(i)}) \cong E(T(_RRe_{ij}))$ *for any* $i = 1,\ldots,m,\ k = 1,\ldots,n(i)$.

(2) $\oplus_{i=1}^m \oplus_{j=1}^{n(i)} Re_{\sigma(i)\rho(i)}/S_{j-1}(_RRe_{\sigma(i)\rho(i)}) \cong E(R/J)$.

(3) $\oplus_{i=1}^m \oplus_{j=1}^{n(i)} Re_{\sigma(i)\rho(i)}/S_{j-1}(_RRe_{\sigma(i)\rho(i)})$ *is a finitely generated injective cogenerator in R-FMod.*

We close this section with the following example which shows that in general left H-rings need not be right H-rings.

Example 3.3.5. Consider the local QF-ring $Q = K[x,y]/(x^2, y^2)$, where K is a commutative field. Then $J(Q) = (x,y)/(x^2, y^2)$. Put

$$R = \begin{pmatrix} Q & \overline{Q} \\ J(Q) & \overline{Q} \end{pmatrix} = \begin{pmatrix} Q & Q \\ J(Q) & Q \end{pmatrix} / \begin{pmatrix} 0 & S(Q) \\ 0 & S(Q) \end{pmatrix}.$$

Then R is a left H-ring. But, since the second column of R does not have simple socle, R is not left QF-2. Therefore R is not right Harada.

COMMENTS

The conditions $(*)$ and $(*)^*$ were introduced in Harada [60]. The nomenclature "Harada ring" and "co-Harada ring" are due to Oshiro [147] and almost all the results of this chapter are taken from this paper and Oshiro [150], [151] and [152]. It should be noted that the study of lifting modules and extending modules originated in the works Harada [63] and Harada-Oshiro [73] in the early 1980s. In these papers, the lifting property was first studied for simple factor rings while the extending property was considered for simple submodules. However it soon became apparent that these notions were important more generally in module theory. Indeed many ring theorists have studied continuous modules and discrete modules using the extending and lifting properties without their specific mention.

In [18], Baba and Iwase called a right artinian ring a right quasi-Harada ring when every indecomposable injective right module is quasi-projective. In this paper, basic results of quasi-Harada rings are given and it is shown that quasi-Harada rings are constructed by QF-rings. Further in Baba [12] and [13], quasi-Harada rings are well characterized using the structure theorem of Tachikawa-Ringel on QF-3 rings. Furthermore, in Koike [100], the self-duality of quasi-Harada rings is investigated.

Chapter 4

The Structure Theory of Left Harada Rings

This chapter is concerned with the matrix representation of left H-rings. We show that, for a given basic left H-ring R, there exists a QF-subring $F(R)$ of R and R is represented as an upper staircase factor ring of a block extension of $F(R)$. The subring $F(R)$ is uniquely determined for a given $Pi(R)$ and is called the frame QF-subring of R. As we see in Chapter 7, Nakayama rings are H-rings and, therefore, this reprentation theory holds for any basic Nakayama rings, namely, any basic Nakayama ring R has a frame Nakayama QF-subring $F(R)$ and R is represented as an upper staircase factor ring of its suitable block extension. The terminologies of the frame subrings, block extensions and upper staircase factor rings are new and are introduced in this chapter.

4.1 Left Harada Rings of Types ($\sharp$) and ($*$)

Before we develop the structure of left H-rings, we introduce left H-rings of type ($\sharp$) and of type ($*$). Let R be a left H-ring with a left well-index set $Pi(R) = \{e_{ij}\}_{i=1,j=1}^{m,\ n(i)} Pi(R)$. We recall that

(1) $e_{i1}R$ is an injective right R-module for each $i = 1, \ldots, m$, and

(2) $e_{ij}R \cong e_{i,j-1}R$ or $e_{ij}R \cong e_{i,j-1}J$ for any $i = 1, \ldots, m$, $j = 2, \ldots, n(i)$.

In case R is a basic left H-ring, the condition (2) can be replaced by

(2′) $e_{ij}R \cong e_{i1}J^{j-1}$ for any $i = 1, \ldots, m$, $j = 2, \ldots, n(i)$.

For each $i = 1, \ldots, m$, since $e_{i1}R$ is injective, there exist $s \in \{1, \ldots, m\}$ and $t \in \{1, \ldots, n(i)\}$ such that $(e_{i1}R; Re_{st})$ is an i-pair. This s is uniquely determined, and if R is a basic ring, t is also uniquely determined. If R is not basic, we let this t as the largest one. Then we may define two maps $\sigma, \rho : \{1, \ldots, m\} \to \mathbb{N}$ by the rule: $\sigma(i) = s$ and $\rho(i) = t$, so that $(e_{i1}R; Re_{\sigma(i)\rho(i)})$ is an i-pair and $1 \le \rho(i) \le n(\sigma(i))$.

When σ is a permutation of $\{1, \ldots, m\}$, we say that R is a *left H-ring of type* ($\sharp$), and when σ is a permutation of $\{1, \ldots, m\}$ and $\rho(i) = n(\sigma(i))$, i.e., $(e_{i1}R; Re_{\sigma(i)n(\sigma(i))})$ is an i-pair for each $i = 1, \ldots, m$, we say that R is a *left H-ring of type* ($*$).

4.2   A Construction of Left Harada Rings as Upper Staircase Factor Rings of Block Extensions of QF-Rings

Let F be a basic semiperfect ring with $Pi(F) = \{e_1, \ldots, e_y\}$. We put $A_{ij} = e_i F e_j$ for any i, j and, in particular, put $Q_i = A_{ii}$ for any i. Then we may represent F as

$$F = \begin{pmatrix} A_{11} & A_{12} & \cdots & A_{1y} \\ A_{21} & A_{22} & \cdots & A_{2y} \\ \cdots & \cdots & \cdots & \cdots \\ A_{y1} & A_{y2} & \cdots & A_{yy} \end{pmatrix} = \begin{pmatrix} Q_1 & A_{12} & \cdots & A_{1y} \\ A_{21} & Q_2 & \cdots & A_{2y} \\ \cdots & \cdots & \cdots & \cdots \\ A_{y1} & \cdots & A_{y,y-1} & Q_y \end{pmatrix}.$$

For $k(1), \ldots, k(y) \in \mathbb{N}$, we define the *block extension* $F(k(1), \ldots, k(y))$ of F as follows: For any $i, s \in \{1, \ldots, y\}$, $j \in \{1, \ldots, k(i)\}$, $t \in \{1, \ldots, k(s)\}$, let

$$P_{ij,st} = \begin{cases} Q_i & \text{if } i = s, \ j \le t, \\ J(Q_i) & \text{if } i = s, \ j > t, \\ A_{is} & \text{if } i \ne s \end{cases}$$

and put

$$P(i,s) = \begin{pmatrix} P_{i1,s1} & P_{i1,s2} & \cdots & P_{i1,sk(s)} \\ P_{i2,s1} & P_{i2,s2} & \cdots & P_{i2,sk(s)} \\ \cdots & \cdots & \cdots & \cdots \\ P_{ik(i),s1} & P_{ik(i),s2} & \cdots & P_{ik(i),sk(s)} \end{pmatrix}.$$

Consequently, when $i = s$, we have a $(k(i), k(i))$-matrix

$$P(i,i) = \begin{pmatrix} Q_i & \cdots & & Q_i \\ J(Q_i) & \ddots & & \vdots \\ \vdots & \ddots & \ddots & \vdots \\ J(Q_i) & \cdots & J(Q_i) & Q_i \end{pmatrix}$$

which we denote by $Q(i)$ and, when $i \neq s$, we have a $(k(i), k(s))$-matrix

$$P(i,s) = \begin{pmatrix} A_{is} & \cdots & A_{is} \\ & \cdots & \\ A_{is} & \cdots & A_{is} \end{pmatrix}.$$

Furthermore we put

$$P = F(k(1), \ldots, k(y)) = \begin{pmatrix} P(1,1) & P(1,2) & \cdots & P(1,y) \\ P(2,1) & P(2,2) & \cdots & P(2,y) \\ \cdots & \cdots & \cdots & \cdots \\ P(y,1) & P(y,2) & \cdots & P(y,y) \end{pmatrix}$$

$$= \begin{pmatrix} Q(1) & P(1,2) & \cdots & P(1,y) \\ P(2,1) & Q(2) & \cdots & P(2,y) \\ \cdots & \cdots & \cdots & \cdots \\ P(y,1) & P(y,2) & \cdots & Q(y) \end{pmatrix}.$$

Since F is a basic semiperfect ring, P is also a basic semiperfect ring with the matrix size $k(1)+\cdots+k(y)$. We say that $F(k(1), \ldots, k(y))$ is a *block extension* of F for $\{k(1), \ldots, k(y)\}$. In more detail, this matrix representation is given as

$$P = F(k(1), \ldots, k(y)) = \begin{pmatrix} P_{11,11} & \cdots & P_{11,1k(1)} & \cdots & P_{11,y1} & \cdots & P_{11,yk(y)} \\ \vdots & & \vdots & & \vdots & & \vdots \\ P_{1k(1),11} & \cdots & P_{1k(1),1k(1)} & \cdots & P_{1k(1),y1} & \cdots & P_{1k(1),yk(y)} \\ \vdots & & \vdots & & \vdots & & \vdots \\ P_{y1,11} & \cdots & P_{y1,1k(1)} & \cdots & P_{y1,y1} & \cdots & P_{y1,yk(y)} \\ \vdots & & \vdots & & \vdots & & \vdots \\ P_{yk(y),11} & \cdots & P_{yk(y),1k(1)} & \cdots & P_{yk(y),y1} & \cdots & P_{yk(y),yk(y)} \end{pmatrix}$$

Put

$$p_{ij} = \langle 1 \rangle_{ij,ij}$$

for each $i = 1, \ldots, y$, $j = 1, \ldots, k(i)$, then $\{p_{ij}\}_{i=1,j=1}^{y \quad k(i)}$ is a complete set of orthogonal primitive idempotents of $P = F(k(1), \ldots, k(y))$. To distinguish it from any other $Pi(F(k(1), \ldots, k(y)))$, we denote it by $Pi^*(F(k(1), \ldots, k(y)))$ or $Pi^*(P)$.

Remark 4.2.1. For $Pi^*(P)$, we note that

$$p_{ij}P_P \cong p_{i1}J(P)_P^{j-1}$$

for any $i = 1, \ldots, y$ and $j = 1, \ldots, k(i)$.

Theorem 4.2.2. *Given the situation above, the following are equivalent:*

(1) *F is QF with the Nakayama permutation:*

$$\begin{pmatrix} e_1 & \cdots & e_y \\ e_{\sigma(1)} & \cdots & e_{\sigma(y)} \end{pmatrix}.$$

(2) *$P = F(k(1), \ldots, k(y))$ is a basic left H-ring of type $(*)$ with a left well-indexed set $Pi^*(P) = \{p_{ij}\}_{i=1,j=1}^{y \quad k(i)}$.*

Proof. (1) $\Rightarrow$ (2). Since $(e_iF; Fe_{\sigma(i)})$ is an i-pair, we see that $(p_{i1}P; Pp_{\sigma(i)n(\sigma(i))})$ is an i-pair for each $i = 1, \ldots, y$. Therefore $p_{i1}P_P$ is injective. As we noted above, $p_{ij}P_P \cong p_{i1}J(P)_P^{j-1}$ for any $i = 1, \ldots, y$, $j = 1, \ldots, k(i)$. Hence P is a basic left H-ring of type $(*)$.

(2) $\Rightarrow$ (1).   Since P is a left H-ring of type $(*)$, $(p_{i1}P; Pp_{\sigma(i)n(\sigma(i))})$ is an i-pair for any i. This fact shows that each $(e_iF; Fe_{\sigma(i)})$ is an i-pair; whence each e_iF_F is injective.   □

From now on, we assume that F is a basic QF-ring with a Nakayama permutation

$$\begin{pmatrix} e_1 & \cdots & e_y \\ e_{\sigma(1)} & \cdots & e_{\sigma(y)} \end{pmatrix},$$

and we take the block extension $P = F(k(1), \ldots, k(y))$ of F.   Let $i \in \{1, \ldots, y\}$ and consider the i-pair $(e_iF; Fe_{\sigma(i)})$.   Put $S(A_{ij}) = S(_{Q_i}A_{ij}) = S(A_{ijQ_j})$. We define an upper staircase left $Q(i)$- right $Q(\sigma(i))$-subbimodule $S(i, \sigma(i))$ of $P(i, \sigma(i))$ with tiles $S(A_{ij})$ as follows:

(I)   Suppose that $i = \sigma(i)$:   Then, by Lemma 2.1.1 (2), we see that $S(A_{ij})$ is simple as both a left and a right ideal of $Q_i = A_{ii}$. Put $Q = Q_i$, $J = J(Q_i)$ and $S = S(Q_i)$. Then, in the $k(i) \times k(i)$ matrix ring

$$Q(i) = P(i, i) = \begin{pmatrix} Q & \cdots & \cdots & Q \\ J & \ddots & & \vdots \\ \vdots & \ddots & \ddots & \vdots \\ J & \cdots & J & Q \end{pmatrix},$$

we define an upper staircase left $Q(i)$- right $Q(i)$-subbimodule $S(i, i) = S(i, \sigma(i))$ of $Q(i)$ as follows:

$$S(i, i) = \qquad \qquad \qquad (\, (1,1)\text{-position} = 0 \,),$$

where, for the form of $S(i, i)$, we assume that, when Q is a division ring (namely, $Q = S$),

$$S(i, i) = .$$

Then, since S is an ideal of Q, we see that $S(i,i) = S(i,\sigma(i))$ is an ideal of $Q(i)$.

We let $\overline{Q(i)} = \overline{P(i,\sigma(i))} = P(i,\sigma(i))/S(i,\sigma(i))$ for the subbimodule $S(i,\sigma(i))$. In $Q(i)$, we replace Q or J of the (p,q)-position by $\overline{Q} = Q/S$ or $\overline{J} = J/S$, respectively, when the (p,q)-position of $S(i,i)$ is S. Then we may represent $\overline{Q(i)}$ with the matrix ring which is made of these replacements.

For example,

$$(1)\quad \overline{Q(i)} = \begin{pmatrix} Q & \overline{Q} & \overline{Q} & \overline{Q} & \overline{Q} & \overline{Q} \\ J & \overline{Q} & \overline{Q} & \overline{Q} & \overline{Q} & \overline{Q} \\ J & \overline{J} & \overline{Q} & \overline{Q} & \overline{Q} & \overline{Q} \\ J & J & J & Q & \overline{Q} & \overline{Q} \\ J & J & J & J & \overline{Q} & \overline{Q} \\ J & J & J & J & \overline{J} & \overline{Q} \end{pmatrix} = \begin{pmatrix} Q & Q & Q & Q & Q & Q \\ J & Q & Q & Q & Q & Q \\ J & J & Q & Q & Q & Q \\ J & J & J & Q & Q & Q \\ J & J & J & J & Q & Q \\ J & J & J & J & J & Q \end{pmatrix} \Big/ \begin{pmatrix} 0 & S & S & S & S & S \\ 0 & S & S & S & S & S \\ 0 & S & S & S & S & S \\ 0 & 0 & 0 & 0 & S & S \\ 0 & 0 & 0 & 0 & S & S \\ 0 & 0 & 0 & 0 & S & S \end{pmatrix},$$

$$(2)\quad \overline{Q(i)} = \begin{pmatrix} Q \cdots Q & \raisebox{0pt}{\rule{0pt}{8pt}}\!\!\lfloor & & \overline{Q} \\ & \ddots & & \\ & & \ddots\, Q & \raisebox{0pt}{\rule{0pt}{8pt}}\!\!\lfloor \\ & & & \ddots \\ 0 & & & Q \end{pmatrix} = \begin{pmatrix} Q \cdots \cdots \cdots Q \\ \ddots\quad \vdots \\ \ddots\ \vdots \\ \ddots\ \vdots \\ 0\qquad\qquad Q \end{pmatrix} \Big/ \begin{pmatrix} 0 \cdots 0 & \lfloor & & S \\ & \ddots & & \\ & & & \lfloor \\ & & & \ddots \\ 0 & & & 0 \end{pmatrix}.$$

(II) Now suppose that $i \neq \sigma(i)$: Put $S = S_{i\sigma(i)} = S(_{Q_i}A_{i\sigma(i)}) = S(A_{i\sigma(i)Q_{\sigma(i)}})$. Then S is a left Q_i- right $Q_{\sigma(i)}$-subbimodule of $A = A_{i\sigma(i)}$. In a left $Q(i)$- right $Q(\sigma(i))$-bimodule

$$P(i,\sigma(i)) = \begin{pmatrix} A & \cdots & A \\ & \cdots & \\ A & \cdots & A \end{pmatrix} \qquad (\,(k(i),k(\sigma(i)))\text{-matrix}\,),$$

we define an upper staircase subbimodule $S(i,\sigma(i))$ of $P(i,\sigma(i))$ with tiles S of $P(i,\sigma(i))$ as follows:

$$S(i,\sigma(i)) = \begin{pmatrix} 0 \cdots 0 & \lfloor & & S \\ & & & \\ & & & \lfloor \\ 0 & & & \rceil \end{pmatrix} \qquad (\,(1,1)\text{-position} = 0\,)$$

and put $\overline{P(i,\sigma(i))} = P(i,\sigma(i))/S(i,\sigma(i))$. We may represent $\overline{P(i,\sigma)}$ as

$$\overline{P(i,\sigma(i))} = \begin{pmatrix} A \cdots A & \overline{A} \\ & & \\ & & \\ A & \end{pmatrix} \qquad (\,(1,1)\text{-position} = A\,).$$

Next we make a subset X of $P = F(k(1), \dots, k(y))$ by

$$X = \begin{pmatrix} X(1,1) & X(1,2) & \cdots & X(1,y) \\ X(2,1) & X(2,2) & \cdots & X(2,y) \\ \cdots & \cdots & \cdots & \cdots \\ X(y,1) & X(y,2) & \cdots & X(y,y) \end{pmatrix},$$

where $X(i,j)\ (\subseteq Q_i)$ and $X(i,j)\ (\subseteq P(i,j))$ are defined by

$$X(i,i) = \begin{cases} 0 & \text{if } i \neq \sigma(i), \\ S(i,i) & \text{if } i = \sigma(i), \end{cases}$$

$$X(i,j) = \begin{cases} 0 & \text{if } j \neq \sigma(i), \\ S(i,j) & \text{if } j = \sigma(i). \end{cases}$$

Then X is an ideal of $P = F(k(1), \dots, k(y))$. The factor ring $F(k(1), \dots, k(y))/X$ is then called an *upper staircase factor ring* of $P = F(k(1), \dots, k(y))$. In the representation

$$P = F(k(1), \dots, k(y)) = \begin{pmatrix} P(1,1) & P(1,2) & \cdots & P(1,y) \\ P(2,1) & P(2,2) & \cdots & P(2,y) \\ \cdots & \cdots & \cdots & \cdots \\ P(y,1) & P(y,2) & \cdots & P(y,y) \end{pmatrix},$$

we replace $P(i,\sigma(i))$ with $\overline{P(i,\sigma(i))}$ and put $\overline{P} = F(k(1), \dots, k(y))/X$. We represent $\overline{P}$ with

$$\overline{P} = \begin{pmatrix} P(1,1) & \cdots & \overline{P(1,\sigma(1))} & \cdots & \cdots & \cdots & P(1,y) \\ \cdots & \cdots & \cdots & \cdots & \cdots & \cdots & \cdots \\ \cdots & \cdots & \cdots & \cdots & \cdots & \cdots & \cdots \\ P(i,1) & \cdots & \cdots & \overline{P(i,\sigma(i))} & \cdots & \cdots & \cdots \\ \cdots & \cdots & \cdots & \cdots & \cdots & \cdots & \cdots \\ \cdots & \cdots & \cdots & \cdots & \cdots & \cdots & \cdots \\ \cdots & \cdots & \cdots & \cdots & \cdots & \cdots & \cdots \\ P(y,1) & \cdots & \cdots & \cdots & \overline{P(y,\sigma(y))} & \cdots & P(y,y) \end{pmatrix}.$$

From the form of $\overline{P}$ together with $(1,1)$-position $= 0$ in the matrices $S(i,i)$ and $S(i,\sigma(i))$ in (**I**), (**II**) above, we see that $\overline{P} = F(k(1),\ldots,k(y))/X$ is a basic left H-ring.

Thus we obtain the following theorem

Theorem 4.2.3. *For a given basic QF-ring F, every upper staircase factor ring P/X of a block extension $P = F(k(1),\ldots,k(y))$ is a basic left H-ring.*

Let us confirm this theorem in two examples.

Example 4.2.4. As F, we take a local QF-ring Q and consider the block extension $Q(4)$:

$$Q(4) = \begin{pmatrix} Q & Q & Q & Q \\ J & Q & Q & Q \\ J & J & Q & Q \\ J & J & J & Q \end{pmatrix}$$

and let

$$X = \begin{pmatrix} 0 & 0 & S & S \\ 0 & 0 & S & S \\ 0 & 0 & S & S \\ 0 & 0 & 0 & S \end{pmatrix} \quad (S = S(Q)).$$

Then

$$Q(4)/X = \begin{pmatrix} Q & Q & \overline{Q} & \overline{Q} \\ J & Q & \overline{Q} & \overline{Q} \\ J & J & \overline{Q} & \overline{Q} \\ J & J & J & \overline{Q} \end{pmatrix}.$$

Example 4.2.5. As F, we take a basic QF-ring R with a $Pi(R) = \{e, f\}$. Assume that

$$\begin{pmatrix} e & f \\ f & e \end{pmatrix}$$

is a Nakayama permutation of R. Put

$$R = \begin{pmatrix} Q & A \\ B & W \end{pmatrix} = \begin{pmatrix} eRe & eRf \\ fRe & fRf \end{pmatrix}.$$

Consider a block extension $R(3,3)$ of R:

$$R(3,3) = \begin{pmatrix} Q & Q & Q & A & A & A \\ J & Q & Q & A & A & A \\ J & J & Q & A & A & A \\ B & B & B & W & W & W \\ B & B & B & K & W & W \\ B & B & B & K & K & W \end{pmatrix}$$

and let

$$X = \begin{pmatrix} 0 & 0 & 0 & 0 & 0 & S(A) \\ 0 & 0 & 0 & 0 & 0 & S(A) \\ 0 & 0 & 0 & 0 & 0 & 0 \\ 0 & 0 & S(B) & 0 & 0 & 0 \\ 0 & 0 & 0 & 0 & 0 & 0 \\ 0 & 0 & 0 & 0 & 0 & 0 \end{pmatrix}.$$

Then

$$R(3,3)/X = \begin{pmatrix} Q & Q & Q & A & A & \overline{A} \\ J & Q & Q & A & A & \overline{A} \\ J & J & Q & A & A & A \\ B & B & \overline{B} & W & W & W \\ B & B & B & K & W & W \\ B & B & B & K & K & W \end{pmatrix}.$$

4.3   The Representation of Left Harada Rings as Upper Staircase Factor Rings of Block Extensions of QF-Rings

Now, as our first stage of the matrix representation of left H-rings, we show that any basic left H-ring can be represented as an upper staircase factor ring of a block extension of a suitable semiperfect subring.

Let R be a basic left H-ring with a left well-indexed set $Pi(R) = \{e_{ij}\}_{i=1,j=1}^{m,\ n(i)}$. For each $i, j \in \{1, \ldots, m\}$, put

$$A_{ij} = e_{i1}Re_{j1}$$

and, in particular,

$$Q_i = A_{ii}.$$

We further put

$$
T = \begin{pmatrix} A_{11} & A_{12} & \cdots & A_{1m} \\ A_{21} & A_{22} & \cdots & A_{2m} \\ \vdots & \vdots & & \vdots \\ A_{m1} & A_{m2} & \cdots & A_{mm} \end{pmatrix} = \begin{pmatrix} Q_1 & A_{12} & \cdots & A_{1m} \\ A_{21} & Q_2 & \cdots & A_{2m} \\ \vdots & \vdots & & \vdots \\ A_{m1} & A_{m2} & \cdots & Q_m \end{pmatrix}.
$$

As in Section 4.2, define $P_{ij,st}$, $P(i,s)$ and a block extension $P = T(n(1), n(2), \ldots, n(m))$ with $Pi^*(P) = \{p_{ij}\}_{i=1,j=1}^{m \ \ n(i)}$, i.e., for each $i, s \in \{1, \ldots, m\}$, $j \in \{1, \ldots, n(i)\}$ and $t \in \{1, \ldots, n(s)\}$,

$$
P_{ij,st} = \begin{cases} Q_i & \text{if } i = s, \ j \le t, \\ J(Q_i) & \text{if } i = s, \ j > t, \\ A_{is} & \text{if } i \ne s, \end{cases}
$$

$$
P(i,s) = \begin{pmatrix} P_{i1,s1} & P_{i1,s2} & \cdots & P_{i1,sn(s)} \\ P_{i2,s1} & P_{i2,s2} & \cdots & P_{i2,sn(s)} \\ \cdots & \cdots & \cdots & \cdots \\ P_{in(i),s1} & P_{in(i),s2} & \cdots & P_{in(i),sn(s)} \end{pmatrix}
$$

$$
= \begin{cases} \begin{pmatrix} Q_i & \cdots & \cdots & Q_i \\ J(Q_i) & Q_i & & \vdots \\ \vdots & \ddots & \ddots & \vdots \\ J(Q_i) & \cdots & J(Q_i) & Q_i \end{pmatrix} & \text{if } i = s, \\[6pt] \begin{pmatrix} P_{is} & \cdots & P_{is} \\ \cdots & \cdots & \cdots \\ P_{is} & \cdots & P_{is} \end{pmatrix} & \text{if } i \ne s \end{cases}
$$

and

$$
P = \begin{pmatrix} P(1,1) & \cdots & P(1,m) \\ \cdots & \cdots & \cdots \\ P(m,1) & \cdots & P(m,m) \end{pmatrix} \qquad ((\textstyle\sum_{i=1}^{m} n(i), \sum_{i=1}^{m} n(i))\text{-matrix}).
$$

Here, instead of the $X(i,s) \subseteq P(i,s)$ used in Section 4.2, we define a simpler $K(i,s)$ as follows:

$$K(i,s) = \begin{cases} \begin{pmatrix} 0 \cdots \cdots 0 \\ \vdots \qquad \vdots \\ 0 \cdots \cdots 0 \end{pmatrix} & \text{if } s \neq \sigma(i), \\[2em] \begin{pmatrix} & & \overset{\rho(i)}{\vee} & \\ 0 \cdots \cdots 0 & S_{is} \cdots & S_{is} \\ \vdots \qquad \qquad \vdots \quad \vdots \qquad \vdots \\ 0 \cdots \cdots 0 & S_{is} \cdots & S_{is} \end{pmatrix} & \text{if } s = \sigma(i) \end{cases}$$

where $S_{is} = S(A_{is\ Q_s})$ and define K instead of X by

$$K = \begin{pmatrix} K(1,1) & K(1,2) & \cdots & K(1,m) \\ K(2,1) & K(2,2) & \cdots & K(2,n) \\ \cdots & \cdots & \cdots & \cdots \\ K(m,1) & K(m,2) & \cdots & K(m,m) \end{pmatrix}.$$

Put

$$\overline{P(i,\sigma(i))} = P(i,\sigma(i))/S(i,\sigma(i))$$

and

$$\overline{P} = \begin{pmatrix} P(1,1) & \cdots & \overline{P(1,\sigma(1))} & \cdots & & \cdots & \cdots & P(1,m) \\ \cdots & \cdots & \cdots & \cdots & \cdots & \cdots & \cdots \\ \cdots & \cdots & \cdots & \cdots & \cdots & \cdots & \cdots \\ P(i,1) & \cdots & \cdots & \overline{P(i,\sigma(i))} & \cdots & \cdots & \cdots \\ \cdots & \cdots & \cdots & \cdots & \cdots & \cdots & \cdots \\ \cdots & \cdots & \cdots & \cdots & \cdots & \cdots & \cdots \\ \cdots & \cdots & \cdots & \cdots & \cdots & \cdots & \cdots \\ P(m,1) & \cdots & \cdots & \cdots & \overline{P(m,\sigma(m))} & \cdots & P(m,m) \end{pmatrix}.$$

Under this preparation, we show the first representation theorem of left H-rings.

Theorem 4.3.1. *Let R be a basic left H-ring with a left well-indexed set $Pi(R) = \{e_{ij}\}_{i=1,j=1}^{m,\ n(i)}$. Let P be the ring defined above. Then there exists a surjective matrix ring homomorphism $\tau : P \to R$ such that $\operatorname{Ker}\tau = K$ and $\tau(p_{ij}) = e_{ij}$ for any $i = 1,\ldots,m$, $j = 1,\ldots,n(i)$, i.e., $R \cong \overline{P}$.*

Proof. (1). We construct a surjective ring homomorphism $\tau : P \to R$ such that $\mathrm{Ker}\,\tau = K$ and $\tau(p_{ij}) = e_{ij}$ for each $i = 1, \ldots, m$, $j = 1, \ldots, n(i)$. It will then follow that τ is a surjective matrix ring homomorphism.

Since R is a basic left H-ring and $Pi(R) = \{e_{ij}\}_{i=1,j=1}^{m\ \ n(i)}$ is a left well-indexed set of R, we have an isomorphism $\theta_{ij} : e_{ij}R \to e_{i1}J^{j-1}$ for any $i = 1, \ldots, m$ and $j = 1, \ldots, n(i)$.

Claim 1. *We can define an abelian group homomorphism*

$$\tau_{ij,st} : P_{ij,st} \to e_{ij}Re_{st}$$

by $x \mapsto (\theta_{ij})^{-1}(x\theta_{st}(e_{st}))$ for each i, $s \in \{1, \ldots, m\}$, $j \in \{1, \ldots, n(i)\}$ and $t \in \{1, \ldots, n(s)\}$.

Proof of Claim 1. For any $x \in P_{ij,st}$, we must show that $x\theta_{st}(e_{st}) \in e_{i1}J^{j-1} = \mathrm{Im}(\theta_{ij})$.

If $i = s$ and $j \le t$, then $x\theta_{st}(e_{st}) \in e_{i1}J^{t-1} \subseteq e_{i1}J^{j-1} = \mathrm{Im}(\theta_{ij})$.

If $i = s$ and $j > t$, then $x \in e_{i1}Je_{i1}$. Hence $xJ \subseteq e_{i1}J^{n(i)}$ by the definition of $\{e_{ij}\}_{i=1,j=1}^{m\ \ n(i)}$. Thus $x\theta_{st}(e_{st}) \in e_{i1}J^{n(i)} \subseteq e_{i1}J^{j-1} = \mathrm{Im}(\theta_{ij})$.

If $i \ne s$, $P_{ij,st} = e_{i1}Re_{s1}$ and $e_{i1}Re_{s1} \subseteq e_{i1}J^{n(i)}$ holds by the definition of $\{e_{ij}\}_{i=1,j=1}^{m\ \ n(i)}$. Therefore $x\theta_{st}(e_{st}) \subseteq e_{i1}Re_{st} \subseteq e_{i1}J^{n(i)} \subseteq e_{i1}J^t$.

Claim 2. $\tau_{ij,st} : P_{ij,st} \to e_{ij}Re_{st}$ *is an epimorphism for any i, $s \in \{1, \ldots, m\}$, $j \in \{1, \ldots, n(i)\}$ and $t \in \{1, \ldots, n(s)\}$.*

Proof of Claim 2. For any $y \in e_{ij}Re_{st}$, we consider the right R-homomorphism $\theta_{ij}(y)_L : e_{st}R \to e_{i1}R$. Then there exists $x \in e_{i1}Re_{s1}$ such that $(x)_L\theta_{st} = \theta_{ij}(y)_L$ since $e_{i1}R$ is an injective right R-module and θ_{ij} is an isomorphism. Thus $x\theta_{st}(e_{st}) = \theta_{ij}(y)$. Further we see that $x\theta_{st}(e_{st}) \in \mathrm{Im}(\theta_{ij})$ by the same argument as in the proof of Claim 1. This implies that Hence $\tau_{ij,st}(x) = (\theta_{ij})^{-1}(x\theta_{st}(e_{st})) = y$, and hence $\tau_{ij,st}$ is surjective.

Claim 3.

(1) *If $s \ne \sigma(i)$, then $\tau_{ij,st}$ is an isomorphism.*

(2) *If $s = \sigma(i)$ and $t \le \rho(i)$, then $\tau_{ij,st}$ is an isomorphism.*

(3) *If $s = \sigma(i)$ and $t > \rho(i)$, then the epimorphism $\tau_{ij,st}$ is not monic. In particular, $\mathrm{Ker}\,\tau_{ij,\sigma(i)t} = S_{(e_{i1}Re_{i1}}e_{i1}Re_{\sigma(i)1}) =$*

$S(e_{i1}Re_{\sigma(i)1}{}_{e_{\sigma(i)1}Re_{\sigma(i)1}})$ and this is simple both as a left $e_{i1}Re_{i1}$-module and as a right $e_{\sigma(i)1}Re_{\sigma(i)1}$-module.

Proof of Claim 3. $(e_{i1}R; Re_{\sigma(i)\rho(i)})$ is an i-pair by the definition of σ and ρ. Hence we have a right R-epimorphism $\varphi : e_{\sigma(i)\rho(i)}R \to S(e_{i1}R_R)$.

(1). We must show that $\tau_{ij,st}$ is monic. Assume, on the contrary, that $0 \neq x \in \operatorname{Ker} \tau_{ij,st}$, i.e., $(\theta_{ij})^{-1}(x\theta_{st}(e_{st})) = 0$. Then $x\theta_{st}(e_{st}) = 0$, i.e., $xJ^{t-1} = 0$.

Now consider the epimorphism $(x)_L : e_{s1}R \to xR$. Then $xR \supseteq S(e_{i1}R_R) = \operatorname{Im} \varphi$, and hence there exists a right R-homomorphism $\tilde{\varphi} : e_{\sigma(i)\rho(i)}R \to e_{s1}R$ with $(x)_L\tilde{\varphi} = \varphi$ since $e_{\sigma(i)\rho(i)}R$ is a projective right R-module. Put $y = \tilde{\varphi}(e_{\sigma(i)\rho(i)})$. Then $y \in e_{s1}Re_{\sigma(i)\rho(i)} \subseteq e_{s1}J^{n(s)}$ by the assumption that $s \neq \sigma(i)$ and the definition of $\{e_{ij}\}_{i=1,j=1}^{m,\ n(i)}$. Hence $xy \in xJ^{n(s)} \subseteq xJ^{t-1} = 0$, a contradiction.

(2). This follows by the same argument as used for (1) by using the assumption that $s = \sigma(i)$ and $t \leq \rho(i)$.

(3). First note that

$$\operatorname{Ker} \tau_{ij,\sigma(i)t} = \{\, x \in P_{ij,\sigma(i)t} \mid (\theta_{ij})^{-1}(x\theta_{\sigma(i)t}(e_{\sigma(i)t})) = 0 \,\}$$
$$= \{\, x \in P_{ij,\sigma(i)t} \mid x\theta_{\sigma(i)t}(e_{\sigma(i)t}) = 0 \,\}$$
$$= \{\, x \in P_{ij,\sigma(i)t} \mid xe_{\sigma(i)1}J^{t-1} = 0 \,\}$$

since $\theta_{\sigma(i)t}(e_{\sigma(i)t})R = \operatorname{Im} \theta_{\sigma(i)t} = e_{\sigma(i)1}J^{t-1}$.

Subclaim 1. $\operatorname{Ker} \tau_{ij,\sigma(i)t} \neq 0$.

Proof of Subclaim 1. Since $(e_{i1}R; Re_{\sigma(i)\rho(i)})$ is an i-pair, we have an epimorphism $\varphi' : e_{\sigma(i)1}J^{\rho(i)-1} \to S(e_{i1}R_R)$. Let $\tilde{\varphi}' : e_{\sigma(i)1}R \to e_{i1}R$ be an extension homomorphism of φ'. Then $\operatorname{Ker} \tilde{\varphi}' = \operatorname{Ker} \varphi'$ since $e_{\sigma(i)1}R/e_{\sigma(i)1}J_R^{\rho(i)}$ is uniserial. Moreover, $\operatorname{Ker} \varphi' = e_{\sigma(i)1}J^{\rho(i)} \supseteq e_{\sigma(i)1}J^{t-1}$ by the assumption that $t > \rho(i)$. Thus $\operatorname{Ker} \tilde{\varphi}' \supseteq e_{\sigma(i)1}J^{t-1}$. Therefore, if $x = \tilde{\varphi}'(e_{\sigma(i)1})$, then $0 \neq x \in \operatorname{Ker} \tau_{ij,\sigma(i)t}$ since $\operatorname{Ker} \tau_{ij,\sigma(i)t} = \{\, x \in P_{ij,\sigma(i)t} \mid xe_{\sigma(i)1}J^{t-1} = 0 \,\}$, as required.

Next, given any $x \in P_{ij,\sigma(i)t}$, we consider $(x)_L \in \operatorname{Hom}_R(e_{\sigma(i)1}R, e_{i1}R)$ and set $X = \{\, x \in P_{ij,\sigma(i)t} \mid \operatorname{Ker}(x)_L = e_{\sigma(i)1}J^{\rho(i)} \,\} \cup \{0\}$.

Subclaim 2. $\operatorname{Ker} \tau_{ij,\sigma(i)t} = X$. (We also see that $X = \{ x \in P_{ij,\sigma(i)t} \mid \operatorname{Ker}(x)_L \supseteq e_{\sigma(i)1}J^{\rho(i)} \}$.)

Proof of Subclaim 2. ($\supseteq$) We have $e_{\sigma(i)1}J^{t-1} \subseteq e_{\sigma(i)1}J^{\rho(i)}$ by the assumption that $t > \rho(i)$. Hence $\operatorname{Ker} \tau_{ij,\sigma(i)t} \supseteq X$.

($\subseteq$) For any $0 \neq x \in \operatorname{Ker} \tau_{ij,\sigma(i)t}$, there exists $u \leq t - 1$ with $\operatorname{Ker}(x)_L = e_{\sigma(i)1}J^u$ because $e_{\sigma(i)1}R/e_{\sigma(i)1}J_R^{t-1}$ is uniserial. Then $(x)_L$ induces $e_{\sigma(i)1}J^{u-1}/e_{\sigma(i)1}J^u \cong S(e_{i1}R_R)$. Moreover, $S(e_{i1}R_R) \cong T(e_{\sigma(i)\rho(i)}R_R)$ because $(e_{i1}R; Re_{\sigma(i)\rho(i)})$ is an i-pair. Hence $e_{\sigma(i)1}J^{u-1}/e_{\sigma(i)1}J^u \cong T(e_{\sigma(i)\rho(i)}R_R)$. Now $e_{\sigma(i)1}J^{u-1}$ is projective since $u \leq t - 1 \leq n(\sigma(i))$. Therefore $e_{\sigma(i)1}J^{u-1} \cong e_{\sigma(i)\rho(i)}R$ ($\cong e_{\sigma(i)1}J^{\rho(i)-1}$), i.e., $u = \rho(i)$. Hence $\operatorname{Ker} \tau_{ij,\sigma(i)t} \subseteq X$.

The parenthetical remark is also clear by the above proof.

Consider a map $\psi = (\theta_{\sigma(i)\rho(i)}(e_{\sigma(i)\rho(i)}))_R : e_{i1}Re_{\sigma(i)1} \to e_{i1}Re_{\sigma(i)\rho(i)}$. Then ψ is a left $e_{i1}Re_{i1}$-isomorphism since $\psi = \tau_{i1,\sigma(i)\rho(i)}$ and $\tau_{i1,\sigma(i)\rho(i)}$ is an isomorphism by (2) shown above.

Now $S(_{e_{i1}Re_{i1}}e_{i1}Re_{\sigma(i)\rho(i)}) = S(e_{i1}Re_{\sigma(i)\rho(i)}{}_{e_{\sigma(i)\rho(i)}Re_{\sigma(i)\rho(i)}})$ and this is simple both as a left $e_{i1}Re_{i1}$-module and as a right $e_{\sigma(i)\rho(i)}Re_{\sigma(i)\rho(i)}$-module because $(e_{i1}R; Re_{\sigma(i)\rho(i)})$ is an i-pair. Put $Y = S(_{e_{i1}Re_{i1}}e_{i1}Re_{\sigma(i)\rho(i)}) = S(e_{i1}Re_{\sigma(i)\rho(i)}{}_{e_{\sigma(i)\rho(i)}Re_{\sigma(i)\rho(i)}})$.

Subclaim 3. $\psi(X) = Y$.

Proof of Subclaim 3. Let $x \in e_{i1}Re_{\sigma(i)1}$ and put $y = \psi(x)$. Then $yR = \psi(x)R = x\theta_{\sigma(i)\rho(i)}(e_{\sigma(i)\rho(i)})R = x\theta_{\sigma(i)\rho(i)}(e_{\sigma(i)\rho(i)}R) = xe_{\sigma(i)1}J^{\rho(i)-1}$. Therefore, if $0 \neq x \in X$, then $yR = S(e_{i1}R_R)$ by Subclaim 2. Now $S(e_{i1}R_R) = S(e_{i1}Re_{\sigma(i)\rho(i)}{}_{e_{\sigma(i)\rho(i)}Re_{\sigma(i)\rho(i)}}) = Y$ since R is a basic ring. Hence $y \in Y$. On the other hand, if $0 \neq y \in Y$, then $x \in X$ by Subclaim 2.

Now $\tau_{\sigma(i)\rho(i),\sigma(i)\rho(i)} : e_{\sigma(i)1}Re_{\sigma(i)1} \to e_{\sigma(i)\rho(i)}Re_{\sigma(i)\rho(i)}$ is an abelian group isomorphism by (1), (2) above. Further it is clear that $\tau_{\sigma(i)\rho(i),\sigma(i)\rho(i)}$ is a ring isomorphism.

Subclaim 4. Consider $e_{i1}Re_{\sigma(i)\rho(i)}$ as a right $e_{\sigma(i)1}Re_{\sigma(i)1}$-module induced by $\tau_{\sigma(i)\rho(i),\sigma(i)\rho(i)}$. Then ψ is a (left $e_{i1}Re_{i1}$-) right $e_{\sigma(i)1}Re_{\sigma(i)1}$-isomorphism.

Proof of Subclaim 4. For any $x \in e_{i1}Re_{\sigma(i)\rho(i)}$ and $r \in e_{i1}Re_{i1}$,

$$\psi(xr) = xr\theta_{\sigma(i)\rho(i)}(e_{\sigma(i)\rho(i)}) = x\theta_{\sigma(i)\rho(i)}(\theta_{\sigma(i)\rho(i)}^{-1}(r\theta_{\sigma(i)\rho(i)}(e_{\sigma(i)\rho(i)}))) =$$
$$x\theta_{\sigma(i)\rho(i)}(e_{\sigma(i)\rho(i)})\,(\theta_{\sigma(i)\rho(i)}^{-1}\,(r\theta_{\sigma(i)\rho(i)}(e_{\sigma(i)\rho(i)}))) = \psi(x)\tau_{\sigma(i)\rho(i),\sigma(i)\rho(i)}(r).$$

By Subclaims 3 and 4, we see that X is simple and the essential socle of $e_{i1}Re_{\sigma(i)\rho(i)}$ both as a left $e_{i1}Re_{i1}$-module and as a right $e_{\sigma(i)\rho(i)}Re_{\sigma(i)\rho(i)}$-module since Y is simple essential. This completes the proof of Claim 3.

Claim 4. We can define a surjective ring homomorphism

$$\tau' : P \to R$$

by $e_{i'j'}\tau'([x_{ij,st}]_{ij,st})e_{s't'} = \tau_{i'j',s't'}(x_{i'j',s't'})$ for any $i', s' \in \{1, \ldots, m\}$, $j' \in \{1, \ldots, n(i')\}$ and $t' \in \{1, \ldots, n(s')\}$, where $x_{ij,st}$ is the (ij,st)-component of an element of the matrix ring P.

Proof of Claim 4. For any $x \in P_{ij,pq}$ and $y \in P_{pq,st}$,

$$\begin{aligned}
\tau_{ij,pq}(x)\tau_{pq,st}(y) &= (\theta_{ij})^{-1}(x\theta_{pq}(e_{pq})(\theta_{pq})^{-1}(y\theta_{st}(e_{st}))) \\
&= (\theta_{ij})^{-1}(x\theta_{pq}((\theta_{pq})^{-1}(y\theta_{st}(e_{st})))) \\
&= (\theta_{ij})^{-1}((xy)\theta_{st}(e_{st})) \\
&= \tau_{ij,st}(xy).
\end{aligned}$$

Therefore, for any $[x_{ij,st}]_{ij,st}, [y_{ij,st}]_{ij,st} \in P$, $i', s' \in \{1, \ldots, m\}$, $j' \in \{1, \ldots, n(i')\}$ and $t' \in \{1, \ldots, n(s')\}$, we have

$$\begin{aligned}
&e_{i'j'}\tau'([x_{ij,st}]_{ij,st}[y_{ij,st}]_{ij,st})e_{s't'} \\
&= \tau_{i'j',s't'}\Big(\sum_{1\leq p\leq m,1\leq q\leq n(p)} x_{i'j',pq}y_{pq,s't'} \Big) \\
&= \sum_{1\leq p\leq m,1\leq q\leq n(p)} \tau_{i'j',s't'}(x_{i'j',pq}y_{pq,s't'}) \\
&= \sum_{1\leq p\leq m,1\leq q\leq n(p)} \tau_{i'j',pq}(x_{i'j',pq})\tau_{pq,s't'}(y_{pq,s't'}) \\
&= \sum_{1\leq p\leq m,1\leq q\leq n(p)} (e_{i'j'}\tau'([x_{ij,st}]_{ij,st})e_{pq})(e_{pq}(\tau'([y_{ij,st}]_{ij,st}))e_{s't'}) \\
&= e_{i'j'}(\tau'([x_{ij,st}]_{ij,st})\tau'([y_{ij,st}]_{ij,st}))e_{s't'},
\end{aligned}$$

i.e., $\tau'([x_{ij,st}]_{ij,st}[y_{ij,st}]_{ij,st}) = \tau'([x_{ij,st}]_{ij,st})\tau'([y_{ij,st}]_{ij,st})$. Hence τ is a surjective ring homomorphism.

By Claim 3, $\operatorname{Ker}\tau = K$. Hence the matrix map

$$\tau = \begin{pmatrix} \tau_{11,11} & \cdots & \tau_{11,1n(1)} & \cdots & \tau_{11,m1} & \cdots & \tau_{11,mn(m)} \\ \vdots & & \vdots & & \vdots & & \vdots \\ \tau_{1n(1),11} & \cdots & \tau_{1n(1),1n(1)} & \cdots & \tau_{1n(1),m1} & \cdots & \tau_{1n(1),mn(m)} \\ \vdots & & \vdots & & \vdots & & \vdots \\ \tau_{m1,11} & \cdots & \tau_{m1,1n(1)} & \cdots & \tau_{m1,m1} & \cdots & \tau_{m1,mn(m)} \\ \vdots & & \vdots & & \vdots & & \vdots \\ \tau_{m(n(m)),11} & \cdots & \tau_{mn(m),1n(1)} & \cdots & \tau_{mn(m),m1} & \cdots & \tau_{mn(m),mn(m)} \end{pmatrix}$$

is a surjective ring homomorphism with $\operatorname{Ker}\tau = K$. Accordingly, a proof of Theorem 4.3.1 is completed. □

Notation. We let $R^{(*)}$ denote the factor ring $\overline{P} = P/K$ and identify R with $R^{(*)}$ when we are considering basic left H-rings R.

By the structure of $R = R^{(*)}$, we obtain the following theorem.

Theorem 4.3.2. *Given the situation above, we obtain the following:*

 (1) *For any $i \in \{1, \ldots, m\}$, if $i = \sigma(i)$, then Q_i is a local QF-ring.*

 (2) $R^{(*)}$ *is type* (♯) *if and only if T is QF.*

Proof. (1) follows from the structure of $R = R^{(*)}$ and (2) follows from Theorem 4.2.2. □

When $R^{(*)}$ is left H-ring of type (♯), we say that the QF-subring T is the *frame QF-subring* of R (with respect to $Pi(R)$) and we denote it by $F(R)$. We note that R is represented as an upper staircase factor ring of $F(R)(n(1), \ldots, n(m))$. We will later define the frame QF-subring $F(R)$ for any basic left H-ring R.

We now illustrate Theorem 4.2.2 with two examples.

Example 4.3.3. Let R be a basic left H-ring with a left well-indexed set $\{e_{11}, e_{12}, e_{13}, e_{21}, e_{22}\}$. Put $Q_i = e_{i1}Re_{i1}$, $J_i = J(Q_i)$ for $i = 1, 2$, $A = e_{11}Re_{21}$ and $B = e_{21}Re_{11}$.

(1) Suppose that $(e_{11}R; Re_{21})$ and $(e_{21}R; Re_{12})$ are i-pairs, so $\sigma(1) = 2$, $\sigma(2) = 1, \rho(1) = 1$ and $\rho(2) = 2$. This gives

$$P(1,1) = \begin{pmatrix} Q_1 & Q_1 & Q_1 \\ J_1 & Q_1 & Q_1 \\ J_1 & J_1 & Q_1 \end{pmatrix}, \quad P(1,2) = \begin{pmatrix} A & A \\ A & A \\ A & A \end{pmatrix},$$

$$P(2,1) = \begin{pmatrix} B & B & B \\ B & B & B \end{pmatrix}, \quad P(2,2) = \begin{pmatrix} Q_2 & Q_2 \\ J_2 & Q_2 \end{pmatrix},$$

$$K(1,1) = \begin{pmatrix} 0 & 0 & 0 \\ 0 & 0 & 0 \\ 0 & 0 & 0 \end{pmatrix}, \quad K(1,2) = \begin{pmatrix} 0 & S(A_{Q_2}) \\ 0 & S(A_{Q_2}) \\ 0 & S(A_{Q_2}) \end{pmatrix},$$

$$K(2,1) = \begin{pmatrix} 0 & 0 & S(B_{Q_1}) \\ 0 & 0 & S(B_{Q_1}) \end{pmatrix}, \quad K(2,2) = \begin{pmatrix} 0 & 0 \\ 0 & 0 \end{pmatrix}.$$

Therefore

$$R \cong P/K = \begin{pmatrix} P(1,1) & P(1,2) \\ P(2,1) & P(2,2) \end{pmatrix} \Big/ \begin{pmatrix} K(1,1) & K(1,2) \\ K(2,1) & K(2,2) \end{pmatrix}$$

$$= \begin{pmatrix} Q_1 & Q_1 & Q_1 & A & A \\ J_1 & Q_1 & Q_1 & A & A \\ J_1 & J_1 & Q_1 & A & A \\ B & B & B & Q_2 & Q_2 \\ B & B & B & J_2 & Q_2 \end{pmatrix} \Big/ \begin{pmatrix} 0 & 0 & 0 & 0 & S(A) \\ 0 & 0 & 0 & 0 & S(A) \\ 0 & 0 & 0 & 0 & S(A) \\ 0 & 0 & S(B) & 0 & 0 \\ 0 & 0 & S(B) & 0 & 0 \end{pmatrix}.$$

We abbreviate this as

$$R \cong \begin{pmatrix} Q_1 & Q_1 & Q_1 & A & \overline{A} \\ J_1 & Q_1 & Q_1 & A & \overline{A} \\ J_1 & J_1 & Q_1 & A & \overline{A} \\ B & B & \overline{B} & Q_2 & Q_2 \\ B & B & \overline{B} & J_2 & Q_2 \end{pmatrix}.$$

(2) Suppose that $(e_{11}R; Re_{21})$ and $(e_{21}R; Re_{22})$ are i-pairs. Then

$$R \cong \begin{pmatrix} Q_1 & Q_1 & Q_1 & A & \overline{A} \\ J_1 & Q_1 & Q_1 & A & \overline{A} \\ J_1 & J_1 & Q_1 & A & \overline{A} \\ B & B & B & Q_2 & Q_2 \\ B & B & B & J_2 & Q_2 \end{pmatrix}.$$

Later we can see that R is represented more simply as follows:

$$R \cong \begin{pmatrix} Q_1 & Q_1 & Q_1 & Q_1 & \overline{Q_1} \\ J_1 & Q_1 & Q_1 & Q_1 & \overline{Q_1} \\ J_1 & J_1 & Q_1 & Q_1 & \overline{Q_1} \\ J_1 & J_1 & J_1 & Q_1 & Q_1 \\ J_1 & J_1 & J_1 & J_1 & Q_1 \end{pmatrix}.$$

Remark 4.3.4. Any basic left H-ring of type ($\sharp$) with m blocks is trivially an upper staircase factor ring of type ($*$) with m blocks.

Now we are in a position to show the following fundamental representation theorem of left H-rings.

Theorem 4.3.5. *Let R be a basic left H-ring with a left well-indexed set $Pi(R) = \{e_{ij}\}_{i=1,j=1}^{m,\ n(i)}$. Then the following hold:*

(a) *There exists a subset*

$$\{e_{i_1 1}, e_{i_2 1}, \ldots, e_{i_y 1}\} = \{e_{\sigma(i_1)1}, e_{\sigma(i_2)1}, \ldots, e_{\sigma(i_y)1}\}$$

of $\{e_{i1}\}_{i=1}^{m}$ such that, putting $e = e_{i_1 1} + \cdots + e_{i_y 1}$,

$$F(R) = eRe = \begin{pmatrix} e_{i_1 1}Re_{i_1 1} & \cdots & e_{i_1 1}Re_{i_y 1} \\ \cdots & \cdots & \cdots \\ e_{i_y 1}Re_{i_1 1} & \cdots & e_{i_y 1}Re_{i_y 1} \end{pmatrix}$$

is a QF-ring with the Nakayama permutation

$$\begin{pmatrix} e_{i_1 1} & \cdots & e_{i_y 1} \\ e_{\sigma(i_1)1} & \cdots & e_{\sigma(i_y)1} \end{pmatrix}.$$

(b) *There exist $k(1), \ldots, k(y) \in \mathbb{N}$ with $k(1) + k(2) + \cdots + k(y) = n(1) + n(2) + \cdots + n(m)$ for which R can be represented as an upper staircase factor ring of the block extension $F(R)(k(1), \ldots, k(y))$ with respect to $\{e_{i_1 1}, \ldots, e_{i_y 1}\}$.*

Proof. When R is of type ($\sharp$), the statement follows from Theorems 4.2.2 and 4.3.1. Suppose that R is not of type ($\sharp$). Then $m > 1$ and $\{\sigma(1), \ldots, \sigma(m)\} \subsetneq \{1, \ldots, m\}$. Therefore there exists a subset $\{i_1, i_2, \ldots, i_l\}$ of $\{1, 2, \ldots, m\}$ such that

(α) $\sigma(i_1) = \sigma(i_2) = \cdots = \sigma(i_l) \neq \sigma(t)$ for any $t \in \{1, \ldots, m\} -$
$\{i_1, i_2, \ldots, i_l\}$, and

(β) $\rho(i_1) < \rho(i_2) < \ldots < \rho(i_l)$.

Without generality, we may let $1 = i_1$, $2 = i_2$, $\ldots$, $l = i_l$. So

(α) $\sigma(1) = \sigma(2) = \cdots = \sigma(l) \neq \sigma(t)$ for any $l < t$, and

(β) $\rho(1) < \rho(2) < \ldots < \rho(l)$.

Hence the representation $\overline{P} = R$ has the following form:

$$\overline{P} = \begin{pmatrix} P(1,1) & \cdots & \overline{P(1,\sigma(1))} & \cdots & & & \cdots & P(1,m) \\ \cdots & \cdots & \cdots & \cdots & \cdots & \cdots & \cdots & \\ \cdots & \cdots & \cdots & \cdots & \cdots & \cdots & \cdots & \cdots \\ P(l,1) & \cdots & \overline{P(l,\sigma(l))} & \cdots & & \cdots & & \\ \cdots & \cdots & \cdots & \cdots & \cdots & \cdots & \cdots & \cdots \\ \cdots & \cdots & \cdots & \cdots & \cdots & \cdots & \cdots & \cdots \\ \cdots & \cdots & \cdots & \cdots & \cdots & \cdots & \cdots & \cdots \\ P(m,1) & \cdots & & \cdots & \overline{P(m,\sigma(m))} & \cdots & P(m,m) \end{pmatrix}$$

and

$$\begin{pmatrix} \overline{P(1,\sigma(1))} \\ \overline{P(2,\sigma(2))} \\ \vdots \\ \overline{P(1,\sigma(1))} \end{pmatrix}$$

$$= \begin{matrix} & \overset{\rho(1)}{\vee} & & & \overset{\rho(2)}{\vee} & \overset{\rho(l)}{\vee} & & & \\ \begin{pmatrix} A_{1,\sigma(1)} & \cdots & A_{1,\sigma(1)} & \overline{A_{1,\sigma(1)}} & \cdots & & \cdots & \cdots & \overline{A_{1,\sigma(1)}} \\ A_{2,\sigma(2)} & \cdots & A_{2,\sigma(2)} & \cdots & A_{2,\sigma(2)} & \overline{A_{2,\sigma(2)}} & \cdots & \cdots & \overline{A_{2,\sigma(2)}} \\ \cdots & \cdots & \cdots & \cdots & \cdots & \cdots & \cdots & \\ \cdots & \cdots & \cdots & \cdots & \cdots & \cdots & \cdots & \\ A_{l,\sigma(l)} & \cdots & A_{l,\sigma(l)} & \cdots & & \cdots & A_{l,\sigma(l)} & \overline{A_{l,\sigma(l)}} & \cdots & \overline{A_{l,\sigma(l)}} \end{pmatrix} \end{matrix}.$$

We consider the following cases:

(i) $\sigma(1) = \sigma(2) \neq 1, 2$,   (ii) $\sigma(1) = \sigma(2) = 1$,   (iii) $\sigma(1) = \sigma(2) = 2$.

In case (i), R can be represented as

$$R = \begin{pmatrix} & & & \overset{\overset{\rho(1)}{\vee}}{} & \overset{\overset{\rho(2)}{\vee}}{} & & \\ * & * & & \begin{matrix} A \cdots A\,\overline{A} \cdots \overline{A}\,\overline{A} \cdots \overline{A} \\ \cdots \quad\quad \cdots \quad\quad \cdots \\ A \cdots A\,\overline{A} \cdots \overline{A}\,\overline{A} \cdots \overline{A} \end{matrix} & & * \\ \hline * & * & & \begin{matrix} B \cdots B\,B \cdots B\,\overline{B} \cdots \overline{B} \\ \cdots \quad\quad \cdots \quad\quad \cdots \\ B \cdots B\,B \cdots B\,\overline{B} \cdots \overline{B} \end{matrix} & & * \\ \hline * & * & & * & & * \end{pmatrix},$$

where $A = A_{11,\sigma(1)1} = e_{11}Re_{\sigma(1)1}$, $B = A_{21,\sigma(2)1} = e_{21}Re_{\sigma(2)1}$, $\overline{A} = A/S(A)$ and $\overline{B} = B/S(B)$. In the first row block, we replace the leftmost $\begin{pmatrix} \overline{A} \cdots \overline{A} \\ \cdots \\ \overline{A} \cdots \overline{A} \end{pmatrix}$ by $\begin{pmatrix} A \cdots A \\ \cdots \\ A \cdots A \end{pmatrix}$ to get W as follows:

$$W = \begin{pmatrix} & & & & \overset{\overset{\rho(2)}{\vee}}{} & & \\ * & * & & \begin{matrix} A \cdots A\,A \cdots A\,\overline{A} \cdots \overline{A} \\ \cdots \quad\quad \cdots \quad\quad \cdots \\ A \cdots A\,A \cdots A\,\overline{A} \cdots \overline{A} \end{matrix} & & * \\ \hline * & * & & \begin{matrix} B \cdots B\,B \cdots B\,\overline{B} \cdots \overline{B} \\ \cdots \quad\quad \cdots \quad\quad \cdots \\ B \cdots B\,B \cdots B\,\overline{B} \cdots \overline{B} \end{matrix} & & * \\ \hline * & * & & * & & * \end{pmatrix}.$$

Then, by (α) and (β), we see that W is a basic ring and $R = R^{(*)}$ is a factor ring of W. Furthermore we claim that W is a basic left H-ring with $m - 1$ row blocks. Let f_{ij} be the element of W, where each (ij, ij) entry is the unity (of Q_i or $\overline{Q_i}$) and all other entries are zero. Then $\{f_{ij}\}_{i=1,j=1}^{m,\ n(i)}$ is a complete set of orthogonal primitive idempotents of W. This implies that W is a basic left artinian ring such that

(I) $J(f_{ij}W_W)_W \cong f_{i,j+1}W_W$ for any $i = 1, \ldots, m$, $j = 1, \ldots, m(i) - 1$ and

(II) $f_{i1}W_W$ is injective for any $i \neq 2$.

Since W is basic and $S(f_{11}W_W) \cong \cdots \cong S(f_{1n(1)}W_W) \cong S(f_{21}W_W) \cong \cdots \cong S(f_{2n(2)}W_W)$, we see that $f_{21}W_W \subsetneq J(f_{1n(1)}W_W)_W$. It is easy to see that both $J(f_{1n(1)}W_W)$ and $f_{21}W$ canonically become right R-modules by (β). Since $J(f_{1n(1)}W)_R$ is indecomposable and $f_{21}R_R$ is injective, we have that $J(f_{1n(1)}W_W)_W \cong f_{21}W_W$. Thus W is a basic left H-ring with $m-1$ row blocks.

Our key method for making W is to replace

$$\binom{P(1,\sigma(1))}{P(2,\sigma(2))} = \begin{pmatrix} A \cdots A \overline{A} \cdots \overline{A} \; \overline{A} \cdots \overline{A} \\ \cdots \qquad \cdots \qquad \cdots \\ A \cdots A \overline{A} \cdots \overline{A} \; \overline{A} \cdots \overline{A} \\ B \cdots B \, B \cdots B \; \overline{B} \cdots \overline{B} \\ \cdots \qquad \cdots \qquad \cdots \\ B \cdots B \, B \cdots B \; \overline{B} \cdots \overline{B} \end{pmatrix}$$

by

$$\binom{P(1,\sigma(1))}{P(2,\sigma(2))}^{\star} := \begin{pmatrix} A \cdots A \, A \cdots A \; \overline{A} \cdots \overline{A} \\ \cdots \qquad \cdots \qquad \cdots \\ A \cdots A \, A \cdots A \; \overline{A} \cdots \overline{A} \\ B \cdots B \, B \cdots B \; \overline{B} \cdots \overline{B} \\ \cdots \qquad \cdots \qquad \cdots \\ A \cdots B \, B \cdots B \; \overline{B} \cdots \overline{B} \end{pmatrix}.$$

When $\sigma(1) = \cdots = \sigma(l) \neq i$ for each $i = 1, \dots, l$, by similar arguments we used all in one for

$$\begin{pmatrix} \overline{P(1,\sigma(1))} \\ \overline{P(2,\sigma(2))} \\ \vdots \\ \overline{P(l,\sigma(l))} \end{pmatrix},$$

we can make

$$\begin{pmatrix} \overline{P(1,\sigma(1))} \\ \overline{P(2,\sigma(2))} \\ \vdots \\ \overline{P(l,\sigma(l))} \end{pmatrix}^{\star}$$

and, by their replacement in the representation:

$$\overline{P} = \begin{pmatrix} P(1,1) & \cdots & \overline{P(1,\sigma(1))} & \cdots & \cdots & P(1,m) \\ \cdots & \cdots & \cdots & \cdots & \cdots & \cdots \\ \cdots & \cdots & \cdots & \cdots & \cdots & \cdots \\ P(l,1) & \cdots & \overline{P(l,\sigma(l))} & \cdots & \cdots & \cdots \\ \cdots & \cdots & \cdots & \cdots & \cdots & \cdots \\ \cdots & \cdots & \cdots & \cdots & \cdots & \cdots \\ \cdots & \cdots & \cdots & \cdots & \cdots & \cdots \\ P(m,1) & \cdots & & \overline{P(m,\sigma(m))} & \cdots & P(m,m) \end{pmatrix},$$

we can make a new basic left H-ring with $m - l + 1$ row blocks.

A similar argument produces such left H-rings for the other cases (ii) and (iii) or, more generally, for the case: $\sigma(k) = k$ for some k; $1 \leq k \leq l$. Though we only considered the subset $\{e_{i_1 1}, \ldots, e_{i_l 1}\} \subseteq \{e_{11}, e_{21}, \ldots, e_{m1}\}$, we also consider all other such subsets and, by replacements all in one, we can make a basic left H-ring, where the number of row blocks is smaller than m. We denote it by $W(R)$. Note that R is a factor ring of $W(R)$.

Next, if $W(R)$ is not type ($\sharp$), then we can also make $W(W(R))$. Put $W^2(R) = W(W(R))$. Inductively, we make $W(R), W^2(R), \ldots,$ and finally reach $W^k(R)$ which is a basic left H-ring of type ($\sharp$). Here we adopt the frame QF-subring of $W^k(R)$ as the *frame QF-subring* of R and we denote it by $F(R)$.

Now, by the construction of $F(R)$, there exists a subset

$$\{e_{i_1 1}, e_{i_2 1}, \ldots, e_{i_y 1}\} = \{e_{\sigma(i_1)1}, e_{\sigma(i_2)1}, \ldots, e_{\sigma(i_y)1}\}$$

of $\{e_{11}, e_{21}, \ldots, e_{m1}\}$ for which, putting $e = e_{i_1 1} + \cdots + e_{i_y 1}$,

$$F(R) = eRe = \begin{pmatrix} e_{i_1 1}Re_{i_1 1} & \cdots & e_{i_1 1}Re_{i_y 1} \\ \cdots & \cdots & \cdots \\ e_{i_y 1}Re_{i_1 1} & \cdots & e_{i_y 1}Re_{i_y 1} \end{pmatrix}$$

with the Nakayama permutation

$$\begin{pmatrix} e_{i_1 1} & \cdots & e_{i_y 1} \\ e_{\sigma(i_1)1} & \cdots & e_{\sigma(i_y)1} \end{pmatrix}$$

and there exist $k(1), \ldots, k(y) \in \mathbb{N}$ with $k(1) + k(2) + \cdots + k(y) = n(1) + n(2) + \cdots + n(m)$ and R can be represented as an upper staircase factor ring of the block extension $F(R)(k(1), \ldots, k(y))$ of $F(R)$ with respect to $\{e_{i_11}, \ldots, e_{i_y1}\}$. □

As an immediate corollary of Theorem 4.3.5 together with Proposition 3.1.8, we obtain the following important fact.

Corollary 4.3.6. *Left H-rings are left and right artinian rings.*

Example 4.3.7. Let R be a left H-ring with a left well-indexed set $Pi(R) = \{e_{11}, e_{21}, e_{22}\}$ such that $(e_{11}R; Re_{21})$ and $(e_{21}R; Re_{22})$ are i-pairs, so $\sigma(1) = 2, \sigma(2) = 2, \rho(1) = 1$ and $\rho(2) = 2$. Then, by Theorems 4.3.1,

$$R \cong \begin{pmatrix} Q_1 & A & \overline{A} \\ B & Q_2 & Q_2 \\ B & J_2 & Q_2 \end{pmatrix},$$

$$W(R) = \begin{pmatrix} Q_1 & A & A \\ B & Q_2 & Q_2 \\ B & J_2 & Q_2 \end{pmatrix} \cong \begin{pmatrix} Q_1 & Q_1 & Q_1 \\ J_1 & Q_1 & Q_1 \\ J_1 & J_1 & Q_1 \end{pmatrix},$$

where $Q_1 = e_{11}Re_{11}$, $J_2 = J(Q_2)$, $A = e_1Re_2$ and $B = e_2Re_1$. Since $W(R)$ is of type ($\sharp$), we obtain that $F(R) = Q_1$.

Example 4.3.8. Let R be a basic left H-ring with a left well-indexed set $Pi(R) = \{e_{11}, e_{12}, e_{21}, e_{22}, e_{31}, e_{32}\}$ such that $(e_{11}R; Re_{31})$, $(e_{21}R; Re_{32})$ and $(e_{31}R; Re_{22})$ are i-pairs, so $\sigma(1) = \sigma(2) = 3$, $\sigma(3) = 2$, $\rho(1) = 1$, $\rho(2) = 2$ and $\rho(3) = 2$. Then R is not of type ($\sharp$). Put $e_1 = e_{11}$, $e_2 = $

e_{12}, $e_3 = e_{21}$, $e_4 = e_{22}$, $e_5 = e_{31}$ and $e_6 = e_{32}$. We represent R as

$$R = \begin{pmatrix} e_1Re_1 & e_1Re_2 & e_1Re_3 & e_1Re_4 & e_1Re_5 & e_1Re_6 \\ e_2Re_1 & e_2Re_2 & e_2Re_3 & e_2Re_4 & e_2Re_5 & e_2Re_6 \\ e_3Re_1 & e_3Re_2 & e_3Re_3 & e_3Re_4 & e_3Re_5 & e_3Re_6 \\ e_4Re_1 & e_4Re_2 & e_4Re_3 & e_4Re_4 & e_4Re_5 & e_4Re_6 \\ e_5Re_1 & e_5Re_2 & e_5Re_3 & e_5Re_4 & e_5Re_5 & e_5Re_6 \\ e_6Re_1 & e_6Re_2 & e_6Re_3 & e_6Re_4 & e_6Re_5 & e_6Re_6 \end{pmatrix}$$

$$= \begin{pmatrix} Q_1 & A_{12} & A_{13} & A_{14} & A_{15} & A_{16} \\ A_{21} & Q_2 & A_{23} & A_{24} & A_{25} & A_{26} \\ A_{31} & A_{32} & Q_3 & A_{34} & A_{35} & A_{36} \\ A_{41} & A_{42} & A_{43} & Q_4 & A_{45} & A_{46} \\ A_{51} & A_{52} & A_{53} & A_{54} & Q_5 & A_{56} \\ A_{61} & A_{62} & A_{63} & A_{64} & A_{65} & Q_6 \end{pmatrix}$$

$$\cong \begin{pmatrix} Q_1 & Q_1 & A_{13} & A_{13} & A_{15} & \overline{A_{15}} \\ J_1 & Q_1 & A_{13} & A_{13} & A_{15} & \overline{A_{15}} \\ A_{31} & A_{31} & Q_3 & Q_3 & A_{35} & A_{35} \\ A_{31} & A_{31} & J_3 & Q_3 & A_{35} & A_{35} \\ A_{51} & A_{51} & A_{53} & A_{53} & Q_5 & Q_5 \\ A_{51} & A_{51} & A_{53} & A_{53} & J_5 & Q_5 \end{pmatrix},$$

where $J_1 = J(Q_1)$, $J_3 = J(Q_3)$ and $J_5 = J(Q_5)$. Then

$$W(R) \cong \begin{pmatrix} Q_1 & Q_1 & Q_1 & Q_1 & A_{15} & A_{15} \\ J_1 & Q_1 & Q_1 & Q_1 & A_{15} & A_{15} \\ J_1 & J_1 & Q_1 & Q_1 & A_{15} & A_{15} \\ J_1 & J_1 & J_1 & Q_1 & A_{15} & A_{15} \\ A_{51} & A_{51} & A_{51} & A_{51} & Q_5 & Q_5 \\ A_{51} & A_{51} & A_{51} & A_{51} & J_5 & Q_5 \end{pmatrix}.$$

Since $W(R)$ is of type $(\sharp)$, we obtain

$$F(R) = \begin{pmatrix} Q_1 & A_{15} \\ A_{51} & Q_5 \end{pmatrix}.$$

Example 4.3.9. Let R be a basic left H-ring with a left well-indexed set $Pi(R) = \{e_{11}, e_{12}, e_{21}, e_{22}, e_{23}, e_{31}, e_{32}\}$ such that $(e_{11}R; Re_{21})$, $(e_{21}R; Re_{22})$ and $(e_{31}R; Re_{23})$ are i-pairs, so $\sigma(1) = \sigma(2) = \sigma(3) = 2$, $\rho(1) = 1$, $\rho(2) = 2$ and $\rho(3) = 2$. Then R is not of type $(\sharp)$. Put

$e_1 = e_{11}$, $e_2 = e_{12}$, $e_3 = e_{21}$, $e_4 = e_{22}$, $e_5 = e_{31}$, $e_6 = e_{32}$ and $e_7 = e_{32}$. We represent R as

$$R = \begin{pmatrix} e_1Re_1 & e_1Re_2 & e_1Re_3 & e_1Re_4 & e_1Re_5 & e_1Re_6 & e_1Re_7 \\ e_2Re_1 & e_2Re_2 & e_2Re_3 & e_2Re_4 & e_2Re_5 & e_2Re_6 & e_2Re_7 \\ e_3Re_1 & e_3Re_2 & e_3Re_3 & e_3Re_4 & e_3Re_5 & e_3Re_6 & e_3Re_7 \\ e_4Re_1 & e_4Re_2 & e_4Re_3 & e_4Re_4 & e_4Re_5 & e_4Re_6 & e_4Re_7 \\ e_5Re_1 & e_5Re_2 & e_5Re_3 & e_5Re_4 & e_5Re_5 & e_5Re_6 & e_5Re_7 \\ e_6Re_1 & e_6Re_2 & e_6Re_3 & e_6Re_4 & e_6Re_5 & e_6Re_6 & e_6Re_7 \\ e_7Re_1 & e_7Re_2 & e_5Re_3 & e_7Re_4 & e_7Re_5 & e_7Re_6 & e_7Re_7 \end{pmatrix}$$

$$= \begin{pmatrix} Q_1 & A_{12} & A_{13} & A_{14} & A_{15} & A_{16} & A_{17} \\ A_{21} & Q_2 & A_{23} & A_{24} & A_{25} & A_{26} & A_{27} \\ A_{31} & A_{32} & Q_3 & A_{34} & A_{35} & A_{36} & A_{37} \\ A_{41} & A_{42} & A_{43} & Q_4 & A_{45} & A_{46} & A_{47} \\ A_{51} & A_{52} & A_{53} & A_{54} & Q_5 & A_{56} & A_{57} \\ A_{61} & A_{62} & A_{63} & A_{64} & A_{65} & Q_6 & A_{67} \\ A_{71} & A_{72} & A_{73} & A_{74} & A_{75} & A_{76} & Q_7 \end{pmatrix}$$

$$\cong \begin{pmatrix} Q_1 & Q_1 & A_{13} & \overline{A_{13}} & \overline{A_{13}} & A_{16} & A_{16} \\ J_1 & Q_1 & A_{13} & \overline{A_{13}} & \overline{A_{13}} & A_{16} & A_{16} \\ A_{31} & A_{31} & Q_3 & Q_3 & \overline{Q_3} & A_{36} & A_{36} \\ A_{31} & A_{31} & J_3 & Q_3 & \overline{Q_3} & A_{36} & A_{36} \\ A_{31} & A_{31} & J_3 & Q_3 & \overline{Q_3} & A_{36} & A_{36} \\ A_{61} & A_{61} & A_{63} & A_{63} & A_{63} & Q_6 & Q_6 \\ A_{61} & A_{61} & A_{63} & A_{63} & A_{63} & J_6 & Q_6 \end{pmatrix},$$

where $J_1 = J(Q_1)$, $J_3 = J(Q_3)$ and $J_6 = J(Q_6)$. Then

$$W(R) = \begin{pmatrix} Q_1 & Q_1 & A_{13} & A_{13} & A_{13} & A_{16} & A_{16} \\ J_1 & Q_1 & A_{13} & A_{13} & A_{13} & A_{16} & A_{16} \\ A_{31} & A_{31} & Q_3 & Q_3 & Q_3 & A_{36} & A_{36} \\ A_{31} & A_{31} & J_3 & Q_3 & \mathbb{Q}_3 & A_{36} & A_{36} \\ A_{31} & A_{31} & J_3 & Q_3 & Q_3 & A_{36} & A_{36} \\ A_{61} & A_{61} & A_{63} & A_{63} & A_{63} & Q_6 & Q_6 \\ A_{61} & A_{61} & A_{63} & A_{63} & A_{53} & J_6 & Q_6 \end{pmatrix}.$$

Since $W(R)$ is of type ($\sharp$) with one row block, we see that $F(R) = Q_1$ and

R is represented as

$$R \cong \begin{pmatrix} Q_1 & Q_1 & Q_1 & \overline{Q_1} & \overline{Q_1} & \overline{Q_1} & \overline{Q_1} \\ J_1 & Q_1 & Q_1 & \overline{Q_1} & \overline{Q_1} & \overline{Q_1} & \overline{Q_1} \\ J_1 & J_1 & Q_1 & Q_1 & \overline{Q_1} & \overline{Q_1} & \overline{Q_1} \\ J_1 & J_1 & J_1 & Q_1 & \overline{Q_1} & \overline{Q_1} & \overline{Q_1} \\ J_1 & J_1 & J_1 & Q_1 & \overline{Q_1} & \overline{Q_1} & \overline{Q_1} \\ J_1 & J_1 & J_1 & J_1 & J_1 & \overline{Q_1} & \overline{Q_1} \\ J_1 & J_1 & J_1 & J_1 & J_1 & \overline{J_1} & \overline{Q_1} \end{pmatrix} .$$

Remark 4.3.10. Let G be a basic indecomposable QF-ring and R an upper staircase factor ring of a block extension of G. Then we note that, in general, $F(R)$ need not be G. For example, let G be a local QF-ring and consider the ring R:

$$R = \begin{pmatrix} G & \overline{G} \\ G & G \end{pmatrix} = \begin{pmatrix} G & G \\ G & G \end{pmatrix} \Big/ \begin{pmatrix} 0 & S(G) \\ 0 & 0 \end{pmatrix}$$

Then $F(R) = R \neq G$.

In the following theorem, for a given basic indecomposable left H-ring, we shall observe the relationships between its minimal faithful right ideal and its frame QF-subring.

Theorem 4.3.11. *Let R be a basic indecomposable left H-ring with a left well-indexed set $Pi(R) = \{e_{ij}\}_{i=1,j=1}^{m,\ n(i)}$. Put $f = e_{11} + e_{21} + \cdots + e_{m1}$. Then the following hold:*

(1) *$fR = e_{11}R + e_{21}R + \cdots + e_{m1}R$ is a minimal faithful right ideal of R.*

(2) *$fRf = \mathrm{End}(fR_R)$ is also a basic indecomposable left H-ring. Put $A(R) = fRf$. If $A(R)$ is not QF, we can also define $A(A(R))$. Put $A^1(R) = A(R)$ and $A^2(R) = A(A(R))$, and inductively we define $A^3(R), A^4(R), \ldots, A^k(R), \ldots$. Then there exists p such that $A^1(R), A^2(R), \ldots, A^{p-1}(R)$ are not QF but $A^p(R) \ (= A^{p+1}(R) = \cdots)$ is QF. Put $M(R) = A^p(R)$. Then $M(R)$ coincides with the frame QF-subring $F(R)$.*

Proof. (1) is obvious.

(2) When R is QF, there is nothing to prove. Suppose that R is not QF.

Claim. Take any $k \in \{1, \ldots, m\}$ with $n(k) > 1$ and $l \in \{2, \ldots, n(k)\}$ and put $W = (1 - e_{kl})R(1 - e_{kl})$. Then the following hold:

(a) $e_{ij}W_W$ is not injective for any $i \in \{1, \ldots m\}$ and $j \in \{2, \ldots, n(i)\}$ except $(i, j) = (k, l)$,

(b) W is a basic left H-ring,

(c) $F(W) = F(R)$.

Proof of Claim. (a) The statement is easily shown.

(b) First we consider the case that both $_RRe_{k,l-1}$ and $_RRe_{kl}$ are injective. We may assume that there exists $s \in \{1, \ldots, m-1\}$ such that $(e_{s1}R; Re_{k,l-1})$ and $(e_{s+1,1}R; Re_{kl})$ are i-pairs. Then we see that

(I) (i) $e_{i1}W_W$ is injective for any $i \in \{1, \ldots, s, s+2, \ldots, m\}$ and

   (ii) $e_{s+1,1}W_W$ is not injective with $e_{sn(s)}J(W)_W \cong e_{s+1,1}W_W$,

(II) (i) $e_{ij}J(W)_W \cong e_{i,j+1}W_W$ for any $i \in \{1, \ldots m\}$ and $j \in \{1, \ldots, n(i) - 1\}$ except $(i, j) = (k, l-1), (k, l)$,

   (ii) $e_{k,l-1}J(W)_W \cong e_{k,l+1}W_W$ if $l + 1 \le n(k)$.

Therefore W is a left H-ring.

Next we consider the remainder case. Then

(I) $e_{i1}W_W$ is injective for any $i \in \{1, \ldots, m\}$,

(II) (i) $e_{ij}J(W)_W \cong e_{i,j+1}W_W$ for any $i \in \{1, \ldots, m\}$ and $j \in \{1, \ldots, n(i) - 1\}$ except $(i, j) = (k, l-1), (k, l)$,

   (ii) $e_{k,l-1}J(W)_W \cong e_{k,l+1}W_W$ if $l + 1 \le n(k)$.

Therefore W is a left H-ring.

(c) By the way of making the frame QF-subrings $F(W)$ and $F(R)$, we can see that $F(W) = F(R)$.

Using the claim above finite times, we see that $A(R)$ is a basic indecomposable left H-ring with $F(A(R)) = F(R)$. Therefore clearly the statement holds. $\qquad\square$

Example 4.3.12. Let R be a basic left H-ring with the well-ordered indexed set $Pi(R) = \{e_{11}, e_{12}, e_{21}, e_{22}\}$ such that $(e_{11}R; Re_{21})$ and $(e_{21}R; Re_{11})$ are i-pairs, so $\sigma(1) = 2$, $\sigma(2) = 1$, $\rho(1) = 1$ and $\rho(2) = 1$. Put $e_1 = e_{11}$, $e_2 = e_{12}$, $e_3 = e_{21}$ and $e_4 = e_{22}$ and represent R as

$$R = \begin{pmatrix} e_1Re_1 & e_1Re_2 & e_1Re_3 & e_1Re_4 \\ e_2Re_1 & e_2Re_2 & e_2Re_3 & e_2Re_4 \\ e_3Re_1 & e_3Re_2 & e_3Re_3 & e_3Re_4 \\ e_4Re_1 & e_4Re_2 & e_4Re_3 & e_4Re_4 \end{pmatrix} = \begin{pmatrix} Q_1 & A_{12} & A_{13} & A_{14} \\ A_{21} & Q_2 & A_{23} & A_{24} \\ A_{31} & A_{32} & Q_3 & A_{34} \\ A_{41} & A_{42} & A_{43} & Q_4 \end{pmatrix}.$$

Then, since R is of type $(\sharp)$,

$$F(R) = \begin{pmatrix} Q_1 & A_{13} \\ A_{31} & Q_3 \end{pmatrix}.$$

On the other hand, a minimal faithful right ideal of R is $e_{11}R + e_{21}R$ and its endomorphism ring is

$$F(R) = \begin{pmatrix} Q_1 & A_{13} \\ A_{31} & Q_3 \end{pmatrix}.$$

Example 4.3.13. Let R be the left H-ring mentioned in Example 4.3.8, that is, R is a basic left H-ring with a left well-indexed set $Pi(R) = \{e_{11}, e_{12}, e_{21}, e_{22}, e_{31}, e_{32}\}$ such that $(e_{11}R; Re_{31})$, $(e_{21}R; Re_{32})$ and $(e_{31}R; Re_{22})$ are i-pairs and $\sigma(1) = \sigma(2) = 3$, $\sigma(3) = 2$, $\rho(1) = 1$, $\rho(2) = 2$ and $\rho(3) = 2$. Then

$$R = \begin{pmatrix} Q_1 & A_{12} & A_{13} & A_{14} & A_{15} & A_{16} \\ A_{21} & Q_2 & A_{23} & A_{24} & A_{25} & A_{26} \\ A_{31} & A_{32} & Q_3 & A_{34} & A_{35} & A_{36} \\ A_{41} & A_{42} & A_{43} & Q_4 & A_{45} & A_{46} \\ A_{51} & A_{52} & A_{53} & A_{54} & Q_5 & A_{56} \\ A_{61} & A_{62} & A_{63} & A_{64} & A_{65} & Q_6 \end{pmatrix}$$

$$\cong \begin{pmatrix} Q_1 & Q_1 & A_{13} & A_{13} & A_{15} & \overline{A_{15}} \\ J_1 & Q_1 & A_{13} & A_{13} & A_{15} & \overline{A_{15}} \\ A_{31} & A_{31} & Q_3 & Q_3 & A_{35} & A_{35} \\ A_{31} & A_{31} & J_3 & Q_3 & A_{35} & A_{35} \\ A_{51} & A_{51} & A_{53} & A_{53} & Q_5 & Q_5 \\ A_{51} & A_{51} & A_{53} & A_{53} & J_5 & Q_5 \end{pmatrix},$$

where $J_1 = J(Q_1)$, $J_3 = J(Q_3)$ and $J_5 = J(Q_5)$.

As we saw, R is not of type ($\sharp$) and

$$F(R) = \begin{pmatrix} Q_1 & A_{15} \\ A_{51} & Q_5 \end{pmatrix}.$$

On the other hand, the minimal faithful right ideal of R is $e_{11}R + e_{21}R + e_{31}R$ and hence

$$A(R) = \begin{pmatrix} Q_1 & A_{13} & A_{15} \\ A_{31} & Q_3 & A_{35} \\ A_{51} & A_{15} & Q_5 \end{pmatrix}.$$

Therefore $A(R)$ is not QF and

$$A(A(R)) = \begin{pmatrix} Q_1 & A_{15} \\ A_{51} & Q_5 \end{pmatrix}.$$

Example 4.3.14. Let R be a basic left H-ring with a left well-indexed set $Pi(R) = \{e_{11}, e_{12}, e_{21}, e_{22}, e_{31}, e_{32}\}$ such that $(e_{11}R; Re_{21})$, $(e_{21}R; Re_{22})$ and $(e_{31}R; Re_{11})$ are i-pairs, so $\sigma(1) = \sigma(2) = 2$ and $\sigma(3) = 1$. R is not of type ($\sharp$) and it has the representation:

$$R \cong \begin{pmatrix} Q_1 & Q_1 & A_{13} & \overline{A_{13}} & A_{15} & A_{15} \\ J_1 & Q_1 & A_{13} & \overline{A_{13}} & A_{15} & A_{15} \\ A_{31} & A_{31} & Q_3 & Q_3 & A_{35} & A_{35} \\ A_{31} & A_{31} & J_3 & Q_3 & A_{35} & A_{35} \\ A_{51} & \overline{A_{51}} & A_{53} & A_{53} & Q_5 & Q_5 \\ A_{51} & \overline{A_{51}} & A_{53} & A_{53} & J_5 & Q_5 \end{pmatrix},$$

where $J_1 = J(Q_1), J_3 = J(Q_3)$ and $J_5 = J(Q_5)$. Then

$$W(R) \cong \begin{pmatrix} Q_1 & Q_1 & Q_1 & Q_1 & A_{15} & A_{15} \\ J_1 & Q_1 & Q_1 & Q_1 & A_{15} & A_{15} \\ J_1 & J_1 & Q_1 & Q_1 & A_{15} & A_{15} \\ J_1 & J_1 & J_1 & Q_1 & A_{15} & A_{15} \\ A_{51} & \overline{A_{51}} & \overline{A_{51}} & \overline{A_{51}} & Q_5 & Q_5 \\ A_{51} & \overline{A_{51}} & \overline{A_{51}} & \overline{A_{51}} & J_5 & Q_5 \end{pmatrix} \quad (matrix\ isomorphism).$$

We note that $W(R)$ is not of type ($\sharp$) yet and

$$F(R) \neq \begin{pmatrix} Q_1 & A_{15} \\ A_{51} & Q_5 \end{pmatrix}.$$

We represent $W(R)$ as

$$W(R) \cong \begin{pmatrix} Q_5 & Q_5 & A_{51} & \overline{A_{51}} & \overline{A_{51}} & \overline{A_{51}} \\ J_5 & Q_5 & A_{51} & \overline{A_{51}} & \overline{A_{51}} & \overline{A_{51}} \\ A_{15} & A_{15} & Q_1 & Q_1 & Q_1 & Q_1 \\ A_{15} & A_{15} & J_1 & Q_1 & Q_1 & Q_1 \\ A_{15} & A_{15} & J_1 & J_1 & Q_1 & Q_1 \\ A_{15} & A_{15} & J_1 & J_1 & J_1 & Q_1 \end{pmatrix}.$$

In view of this representation, we reach the representation:

$$W(W(R)) \cong \begin{pmatrix} Q_5 & Q_5 & A_{51} & A_{51} & A_{51} & A_{51} \\ J_5 & Q_5 & A_{51} & A_{51} & A_{51} & A_{51} \\ A_{15} & A_{15} & Q_1 & Q_1 & Q_1 & Q_1 \\ A_{15} & A_{15} & J_1 & Q_1 & Q_1 & Q_1 \\ A_{15} & A_{15} & J_1 & J_1 & Q_1 & Q_1 \\ A_{15} & A_{15} & J_1 & J_1 & J_1 & Q_1 \end{pmatrix} \cong \begin{pmatrix} Q_5 & Q_5 & Q_5 & Q_5 & Q_5 & Q_5 \\ J_5 & Q_5 & Q_5 & Q_5 & Q_5 & Q_5 \\ J_5 & J_5 & Q_5 & Q_5 & Q_5 & Q_5 \\ J_5 & J_5 & J_5 & Q_5 & Q_5 & Q_5 \\ J_5 & J_5 & J_5 & J_5 & Q_5 & Q_5 \\ J_5 & J_5 & J_5 & J_5 & J_5 & Q_5 \end{pmatrix}.$$

Accordingly $F(R) = Q_5$ and

$$R \cong \begin{pmatrix} Q_5 & Q_5 & Q_5 & \overline{Q_5} & \overline{Q_5} & \overline{Q_5} \\ J_5 & Q_5 & Q_5 & \overline{Q_5} & \overline{Q_5} & \overline{Q_5} \\ J_5 & J_5 & Q_5 & Q_5 & Q_5 & \overline{Q_5} \\ J_5 & J_5 & J_5 & Q_5 & Q_5 & \overline{Q_5} \\ J_5 & J_5 & J_5 & J_5 & Q_5 & Q_5 \\ J_5 & J_5 & J_5 & J_5 & J_5 & Q_5 \end{pmatrix}.$$

Next, by using a minimal faithful right ideal of R, we show $F(R) = Q_5$. Put $e_1 = e_{11}$, $e_2 = e_{21}$ and $e_3 = e_{31}$. Then $e_1 R + e_2 R + e_3 R$ is a minimal faithful right ideal of R. We represent $A(R)$ as

$$A(R) = \begin{pmatrix} e_1 R e_1 & e_1 R e_2 & e_1 R e_3 \\ e_2 R e_1 & e_2 R e_2 & e_2 R e_3 \\ e_3 R e_1 & e_3 R e_2 & e_3 R e_3 \end{pmatrix} = \begin{pmatrix} Q_1 & A_{13} & A_{15} \\ A_{31} & Q_3 & A_{35} \\ A_{51} & A_{53} & Q_5 \end{pmatrix}.$$

Then

$$S(A(R)) = \begin{pmatrix} 0 & S(A_{13}) & 0 \\ 0 & S(Q_3) & 0 \\ S(A_{51}) & 0 & 0 \end{pmatrix}.$$

Therefore $A(R)$ is not QF and its minimal faithful right ideal is $e_1 A(R) + e_3 A(R)$. This implies that

$$A^2(R) = \begin{pmatrix} Q_1 & A_{15} \\ A_{51} & Q_5 \end{pmatrix}.$$

Further, since the first row can be embedded to the second row as right ideals, $A^2(R)$ is not QF and its minimal faithful right ideal is the second row $e_3 A^2(R)$. Hence we reach $A^3(R) = Q_5 = F(R)$.

COMMENTS

Most of the material in this chapter is taken from Oshiro [150]–[152]. In Chapter 5, using the relationships between left H-rings and their frame QH-subrings, we study the self-duality of left H-rings. In Chapter 7, the fact that any left H-ring can be represented as an upper staircase factor ring of its frame QF-subring plays an important role for developing the structure theory of Nakayama rings, and thereby we can give a complete classification of these classical artinian rings.

Chapter 5

Self-Duality of Left Harada Rings

In Chapters 3 and 4, we obtained fundamental theorems on left H-rings. One of these is Theorem 3.3.3, in which we showed that left H-rings and right co-H-rings are one and the same. This fact strongly urges us to study the following problem:

Problem A: Are left H-rings self-dual?

For the study of the self-duality of left H-rings, by the Azumaya-Morita Theorem 1.1.10, we should confirm that each left H-ring has a finitely generated left injective cogenerator. This is guaranteed. Actually, by Theorem 3.3.4, if R is a left H-ring, then $E(T(_R R))$ is finitely generated. Hence R has a Morita duality. Then, as we see later, the corresponding ring $D(R) = \mathrm{End}(E(T(_R R)))$ becomes right co-Harada and so is a left H-ring. Since the structure of $D(R)$ is also similar to R, by the Azumaya-Morita Theorem 1.1.10, Problem A is equivalent to the following problem:

Problem B: Is R isomorphic to $D(R)$?

In this chapter, by examining Problem B thoroughly, we show that Problem B can be translated into the following problem:

Problem C: Do QF-rings have Nakayama automorphisms?

However the answer to Problem C (and hence to Problems A and B) is negative in general as we see later by Example 5.3.2 due to Koike [98] which is deduced from an example in Kraemer [103]. Although, in general, left H-rings need not be self-dual, we show instead that left H-rings are almost self-dual in the sense of Simson [166].

There is another reason why we study Problem A. In the early 1980s, the self-duality of Nakayama rings was studied by Haack [56], Mano [110], Dischinger-Müller [37] and Waschbüsch [181], among others. However Waschbüsch pointed out that, in 1967, Amdal and Ringdal [4] had already proved the self-duality for this class of rings and he himself supplied a proof of this following ideas of Amdal and Ringdal.

In Chapter 7, we can see that Nakayama rings are H-rings. Consequently the problem as to whether Nakayama rings are self-dual can be translated into the problem of whether Nakayama automorphisms of Nakayama QF-rings exist or not. In this regard, although Haack [56] did not succeed in showing the self-duality of Nakayama rings, he did establish the existence of Nakayama automorphisms for Nakayama QF-rings. Therefore, in this respect, Problem A also arises naturally. In Chapter 7, we confirm the existence of Nakayama automorphisms for Nakayama QF-rings and hence, by this approach, the self-duality of Nakayama rings.

5.1   Nakayama Isomorphisms, Weakly Symmetric Left H-Rings and Almost Self-Duality

We first recall the concept of Nakayama automorphisms for QF-rings. Let R be a basic QF-ring with $Pi(R) = \{e_i\}_{i=1}^n$ and let

$$\begin{pmatrix} e_1 & e_2 & \cdots & e_n \\ e_{\sigma(1)} & e_{\sigma(2)} & \cdots & e_{\sigma(n)} \end{pmatrix}$$

be a Nakayama permutation of R. If there exists a ring automorphism φ of R satisfying $\varphi(e_i) = e_{\sigma(i)}$ for each $i = 1, \ldots, n$, then φ is called a *Nakayama automorphism* of R.

We now generalize a Nakayama automorphism to the concept of a Nakayama isomorphism for a basic left artinian ring with a finitely generated left injective cogenerator. Let R be a basic left artinian ring with

$Pi(R) = \{e_i\}_{i=1}^n$. We represent R as

$$R = \begin{pmatrix} (e_1, e_1) & \cdots & (e_n, e_n) \\ & \cdots & \\ (e_1, e_n) & \cdots & (e_n, e_n) \end{pmatrix},$$

where we put $(e_i, e_j) = \text{Hom}_R(e_j R, e_i R)$. In this expression, $e_i = \langle 1 \rangle_{ii}$, where 1 is the unity of (e_i, e_i).

Now suppose that the injective hull $E(T(_R R))$ is finitely generated. Put $G = E(T(_R R))$ and $T = \text{End}(_R G)$. Then R-$FMod$ is Morita dual to $FMod$-T (see Theorem 1.1.10). Put $G_i = E(T(_R R e_i))$ for each $i = 1, \ldots, n$. Then $G = \oplus_{i=1}^n G_i$. We represent T as

$$T = \begin{pmatrix} [G_1, G_1] & \cdots & [G_1, G_n] \\ & \cdots & \\ [G_n, G_1] & \cdots & [G_n, G_n] \end{pmatrix},$$

where $[G_i, G_j] = \text{Hom}_R(G_i, G_j)$. Put $f_i = \langle i \rangle_{ii}$ in this matrix representation. Then $\{f_i\}_{i=1}^n$ is a complete set of orthogonal primitive idempotents of T. Let φ be an isomorphism from R to T. Then the following are equivalent:

(1) φ is a matrix isomorphism.

(2) $\varphi(e_i) = f_i$ for each $i = 1, \ldots, n$.

(3) $\varphi(e_i R) = f_i T$ for each $i = 1, \ldots, n$.

If these equivalent conditions are satisfied, we say that φ is a *Nakayama isomorphism* with respect to $Pi(R)$. Of course, when R is a basic QF-ring, it is just a Nakayama automorphism of R.

Remark 5.1.1. (1) If a basic left artinian ring R has a Nakayama isomorphism and is Morita equivalent to W, then W has also a Nakayama isomorphism.

(2) If R is a basic left artinian ring and has a Nakayama isomorphism φ with respect to $Pi(R) = \{e_i\}_{i=1}^n$, then we claim that R has also a Nakayama isomorphism with respect to any other complete set of orthogonal primitive idempotents of R. Let T and $\{f_i\}_{i=1}^n$ be as above for $Pi(R)$. Consider

another complete set $Pi(R)' = \{g_i\}_{i=1}^n$ of orthogonal primitive idempotents, and put $T' = \mathrm{End}(E(T(_RRg_1)) \oplus \cdots \oplus E(T(_RRg_n)))$ and $f_i' = \langle 1 \rangle_{ii}$ in T'. Let η be an inner automorphism of R such that $\eta(\{e_i\}) = \{g_i\}$. We may assume that $\eta(e_i) = g_i$ for any i. Since $e_iR \cong g_iR$ for any i, there exists a canonical isomorphism θ from T to T'. Then $\theta\varphi\eta^{-1}$ is a Nakayama isomorphism of R for $Pi(R)'$. Thus Nakayama isomorphisms are uniquely determined up to inner automorphisms.

Let R be a basic left H-ring with a left well-indexed set $Pi(R) = \{e_{ij}\}_{i=1,j=1}^{m \ \ n(i)}$. Recall that the maps σ and $\rho : \{1,\ldots,m\} \to N$ are defined by $\sigma(i) = s$ and $\rho(i) = t$ when $(e_{i1}R; Re_{st})$ is an i-pair. Then we say that R is a *weakly symmetric* left H-ring if $\sigma(i) = i$ for all i. We note that R is weakly symmetric if and only if it is of type ($\sharp$) and its frame QF-subring $F(R)$ is a weakly symmetric QF-ring, i.e., the Nakayama permutation of $F(R)$ is identity.

Let $A = B_1$ and $B = B_n$ be rings, where $n > 1$. Following Simson [166], we say that A is *almost Morita dual* to B if there exist rings $B_2, \ldots, B_{n-1}$ such that B_{i+1} is almost dual to B_i for each $i = 1, \ldots, n-1$. In particular, we say that B is *almost self-dual* when $B = A$.

5.2 Self-Duality and Almost Self-Duality of Left Harada Rings

Let R be a basic left H-ring with a left well-indexed set $Pi(R) = \{e_{ij}\}_{i=1,j=1}^{m \ \ n(i)}$. For each $e_{i1}R$, by Theorem 3.3.4, we have

(1) $Re_{\sigma(i)\rho(i)}/S_{k-1}(_RRe_{\sigma(i)\rho(i)}) \cong E(T(_RRe_{ik}))$ for each $i = 1,\ldots,m$, $k = 1,\ldots,n(i)$,

(2) $\oplus_{i=1}^m \oplus_{k=1}^{n(i)} Re_{\sigma(i)\rho(i)}/S_{k-1}(_RRe_{\sigma(i)\rho(i)}) \cong E(T(_RR))$,

(3) $\oplus_{i=1}^m \oplus_{k=1}^{n(i)} Re_{\sigma(i)\rho(i)}/S_{k-1}(_RRe_{\sigma(i)\rho(i)})$ is a finitely generated injective cogenerator in R-FMod.

Put $g_i = e_{\sigma(i)\rho(i)}$, and denote the generator $g_i + S_{k-1}(_RRg_i)$ of $Rg_i/S_{k-1}(_RRg_i)$ by g_{ik} for each $i = 1, \ldots, m$, $k = 1, \ldots, n(i)$. Note that $g_{i1} = g_i = e_{\sigma(i)\rho(i)}$ for each $i = 1, \ldots, m$. Put

$$G = Rg_{11} \oplus \cdots \oplus Rg_{1n(1)} \oplus \cdots \oplus Rg_{m1} \oplus \cdots \oplus Rg_{mn(m)}.$$

Then, since $G \cong E(T(_RR))$, $_RG$ is a finitely generated injective cogenerator. Set $T = \text{End}(_RG)$. Then we say that T is the *dual ring* of R and also denote it by $D(R)$.

By the Azumaya-Morita Theorem 1.1.10, R-$FMod$ is Morita dual to $FMod$-T under the functor $\text{Hom}_R(-, _RG_T)$. Hence it follows that every indecomposable injectiveve right T-module are finitely generated. Since the class of all finitely generated injective left R-modules is closed under small covers, we see, from the duality theorem, that the class of all finitely generated projective right T-modules is closed under essential extensions. As a result, by Remark 3.1.13, T is a left H-ring.

To observe T in more detail, we express it as

$$T = \begin{pmatrix} [g_{11}, g_{11}] & \cdots & [g_{11}, g_{1n(1)}] & \cdots & [g_{11}, g_{m1}] & \cdots & [g_{11}, g_{mn(m)}] \\ \vdots & & \vdots & & \vdots & & \vdots \\ \vdots & & \vdots & & \vdots & & \vdots \\ [g_{mn(m)}, g_{11}] & \cdots & [g_{mn(m)}, g_{1n(1)}] & \cdots & [g_{mn(m)}, g_{m1}] & \cdots & [g_{mn(m)}, g_{mn(m)}] \end{pmatrix},$$

where $[g_{ij}, g_{kl}] = \text{Hom}_R(Rg_{ij}, Rg_{kl})$ for each $i, k \in \{1, \ldots, m\}$, $j \in \{1, \ldots, n(i)\}$ and $l \in \{1, \ldots, n(k)\}$. In this representation, put $h_{ij} = \langle 1 \rangle_{ij,ij}$, i.e., h_{ij} is the matrix such that the (ij, ij)-position is the identity of $[g_{ij}, g_{ij}]$ and all other entries are the zero maps. Then $Pi(T) = \{h_{ij}\}_{i=1,j=1}^{m,\ n(i)}$, i.e., this set is a complete set of orthogonal primitive idempotents of T. Again using the duality, together with (1) and (2) above, we shall show the following theorem.

Theorem 5.2.1. *Under the above assumptions, the following hold:*

(1) T *is a basic left H-ring for which*

    (a) $h_{i1}T_T$ *is injective for each $i = 1, \ldots, m$ and*

(b) $J(h_{i,k-1}T) \cong h_{ik}T$ for each $i = 1, \ldots, m,\ k = 2, \ldots, n(i)$.

(2) If $(e_{i1}R; Re_{kt})$ is an i-pair, then $(h_{i1}T; Th_{kt})$ is also an i-pair. In particular, the functions σ and ρ for $Pi(T)$ are as given for $Pi(R)$.

Proof. (1). (a). Since $h_{i1}T = \operatorname{Hom}_R(Rg_{i1}, G)$, $h_{i1}T_T$ is injective for each $i = 1, \ldots, m$.

(b). From the exact sequence

$$0 \to S_k({}_RRg_i)/S_{k-1}({}_RRg_i) \to Rg_i/S_{k-1}({}_RRg_i) \xrightarrow{\varphi} Rg_i/S_k({}_RRg_i) \to 0,$$

we can obtain the exact sequence

$$0 \to h_{ik}T = \operatorname{Hom}_R(Rg_{ik}, G) \xrightarrow{\varphi^*} h_{i,k-1}T = \operatorname{Hom}_R(Rg_{i,k-1}, G)$$
$$\to \operatorname{Hom}_R(S_k({}_RRg_i)/S_{k-1}({}_RRg_i),\, G) \to 0.$$

Since the last term of the above exact sequence is a simple right T-module, $\operatorname{Im}\varphi^*$ is a maximal submodule of $h_{i,k-1}T_T$. Hence (b) holds.

(2). We assume that $(e_{i1}R; Re_{kt})$ is an i-pair, i.e., $(k,t) = (\sigma(i), \rho(i))$. Then $g_{i1} = e_{kt}$. Since $S({}_RRg_{kt}) \cong T({}_RRe_{kt})$, there exists an epimorphism $\theta : Rg_{i1} \to S({}_RRg_{kt})$. Then we have an exact sequence

$$0 \to \operatorname{Hom}_R(S({}_RRg_{kt}), G) \xrightarrow{\theta^*} h_{i1}T = \operatorname{Hom}_R(Rg_{i1}, G)$$
$$\to \operatorname{Hom}_R(J(Rg_{i1}), G) \to 0.$$

Therefore $\operatorname{Im}\theta^* = S(h_{i1}T_T)$. On the other hand, from the exact sequence

$$0 \to S({}_RRg_{kt}) \to Rg_{kt},$$

we have the exact sequence

$$h_{kt}T = \operatorname{Hom}_R(Rg_{kt}, G) \to \operatorname{Hom}_R(S({}_RRg_{kt}), G) \to 0,$$

i.e., we have an epimorphism $h_{kt}T = \operatorname{Hom}_R(Rg_{kt}, G) \to S(h_{i1}T_T)$. Then $S({}_TTh_{kt}) = S(h_{i1}T_T)$ and $(h_{i1}T; Th_{kt})$ is also an i-pair. □

In view of Theorem 5.2.1, the structure of T is quite similar to that of R.

Now we consider the case when R is type ($\sharp$). Then the frame QF-subring $F(R)$ of R is

$$F(R) = \begin{pmatrix} (e_{11}, e_{11}) & (e_{21}, e_{11}) & \cdots & (e_{m1}, e_{11}) \\ (e_{11}, e_{21}) & (e_{21}, e_{21}) & \cdots & (e_{m1}, e_{21}) \\ \cdots & \cdots & \cdots & \cdots \\ (e_{11}, e_{m1}) & (e_{21}, e_{m1}) & \cdots & (e_{m1}, e_{m1}) \end{pmatrix}$$

$$= \begin{pmatrix} e_{11}Re_{11} & e_{11}Re_{21} & \cdots & e_{11}Re_{m1} \\ e_{21}Re_{11} & e_{21}Re_{21} & \cdots & e_{21}Re_{m1} \\ \cdots & \cdots & \cdots & \cdots \\ e_{m1}Re_{11} & e_{m1}Re_{21} & \cdots & e_{m1}Re_{m1} \end{pmatrix}$$

and R can be represented as an upper staircase factor ring of $F(R)(n(1), \ldots, n(m))$, which is of $type(\sharp)$.

Note that $g_{i1} = e_{\sigma(i)\rho(\sigma(i))}$ for each $i = 1, \ldots, m$. By Theorem 5.2.2, we see that the frame QF-subring of T is

$F(T)$

$$= \begin{pmatrix} h_{11}Th_{11} & h_{11}Th_{21} & \cdots & h_{11}Th_{m1} \\ h_{21}Th_{11} & h_{21}Th_{21} & \cdots & h_{21}Th_{m1} \\ \cdots & \cdots & \cdots & \cdots \\ h_{m1}Th_{11} & h_{m1}Th_{21} & \cdots & h_{m1}Th_{m1} \end{pmatrix}$$

$$= \begin{pmatrix} [g_{11}, g_{11}] & [g_{11}, g_{21}] & \cdots & [g_{11}, g_{m1}] \\ [g_{21}, g_{11}] & [g_{21}, g_{21}] & \cdots & [g_{21}, g_{m1}] \\ \cdots & \cdots & \cdots & \cdots \\ [g_{m1}, g_{11}] & [g_{m1}, g_{21}] & \cdots & [g_{m1}, g_{m1}] \end{pmatrix}$$

$$= \begin{pmatrix} [e_{\sigma(1)\rho(\sigma(1))}, e_{\sigma(1)\rho(\sigma(1))}] & [e_{\sigma(1)\rho(\sigma(1))}, e_{\sigma(2)\rho(\sigma(2))}] & \cdots & [e_{\sigma(1)\rho(\sigma(1))}, e_{\sigma(m)\rho(\sigma(m))}] \\ [e_{\sigma(2)\rho(\sigma(2))}, e_{\sigma(1)\rho(\sigma(1))}] & [e_{\sigma(2)\rho(\sigma(2))}, e_{\sigma(2)\rho(\sigma(2))}] & \cdots & [e_{\sigma(2)\rho(\sigma(2))}, e_{\sigma(m)\rho(\sigma(m))}] \\ \cdots & \cdots & \cdots \\ [e_{\sigma(m)\rho(\sigma(m))}, e_{\sigma(1)\rho(\sigma(1))}] & [e_{\sigma(m)\rho(\sigma(m))}, e_{\sigma(2)\rho(\sigma(2))}] & \cdots & [e_{\sigma(m)\rho(\sigma(m))}, e_{\sigma(m)\rho(\sigma(m))}] \end{pmatrix}$$

$$\cong \begin{pmatrix} e_{\sigma(1)\rho(\sigma(1))}Re_{\sigma(1)\rho(\sigma(1))} & e_{\sigma(1)\rho(\sigma(1))}Re_{\sigma(2)\rho(\sigma(2))} & \cdots & e_{\sigma(1)\rho(\sigma(1))}Re_{\sigma(m)\rho(\sigma(m))} \\ e_{\sigma(2)\rho(\sigma(2))}Re_{\sigma(1)\rho(\sigma(1))} & e_{\sigma(2)\rho(\sigma(2))}Re_{\sigma(2)\rho(\sigma(2))} & \cdots & e_{\sigma(2)\rho(\sigma(2))}Re_{\sigma(m)\rho(\sigma(m))} \\ \cdots & \cdots & \cdots \\ e_{\sigma(m)\rho(\sigma(m))}Re_{\sigma(1)\rho(\sigma(1))} & e_{\sigma(m)\rho(\sigma(m))}Re_{\sigma(2)\rho(\sigma(2))} & \cdots & e_{\sigma(m)\rho(\sigma(1))}Re_{\sigma(m)\rho(\sigma(m))} \end{pmatrix}$$

$$\cong \begin{pmatrix} e_{\sigma(1)1}Re_{\sigma(1)1} & e_{\sigma(1)1}Re_{\sigma(2)1} & \cdots & e_{\sigma(1)1}Re_{\sigma(m)1} \\ e_{\sigma(2)1}Re_{\sigma(1)1} & e_{\sigma(2)1}Re_{\sigma(2)1} & \cdots & e_{\sigma(2)1}Re_{\sigma(m)1} \\ \cdots & \cdots & \cdots & \cdots \\ e_{\sigma(m)1}Re_{\sigma(1)1} & e_{\sigma(m)1}Re_{\sigma(2)1} & \cdots & e_{\sigma(m)1}Re_{\sigma(m)1} \end{pmatrix} \quad \text{(matrix isomorphisms)} \cdots (*).$$

Therefore we can adopt $(*)$ as $F(T)$. Hence $F(T) = eRe$, where $e = \sum_{i=1}^{m} e_{\sigma(i)1}$. And T can be represented as an upper staircase factor ring of

$F(T)(n(1), \ldots, n(m))$ with respect to $\{e_{\sigma(i)1}\}_{i=1}^{m}$, which has the same form of $type(\sharp)$ as R.

Now, since $D(R)$ is also a basic left H-ring, we can obtain the dual ring $D(D(R))$ of $D(R)$. Let $T_1 = D(R)$ and $T_2 = D(T_1)$. Then, by the same argument as above, we can take $F(T_2)$ as follows:

$$F(T_2) = \begin{pmatrix} e_{\sigma^2(1)1}Re_{\sigma^2(1)1} & e_{\sigma^2(1)1}Re_{\sigma^2(2)1} & \cdots & e_{\sigma^2(1)1}Re_{\sigma^2(m)1} \\ e_{\sigma^2(2)1}Re_{\sigma^2(1)1} & e_{\sigma^2(2)1}Re_{\sigma^2(2)1} & \cdots & e_{\sigma^2(2)1}Re_{\sigma^2(m)1} \\ \cdots & \cdots & \cdots & \cdots \\ e_{\sigma^2(m)1}Re_{\sigma^2(1)1} & e_{\sigma^2(m)1}Re_{\sigma^2(2)1} & \cdots & e_{\sigma^2(m)1}Re_{\sigma^2(m)1} \end{pmatrix}$$

and T_2 can be represented as an upper staircase factor ring of $F(T_2)(n(1), \ldots, n(m))$ with respect to $\{e_{\sigma^2(i)1}\}_{i=1}^{m}$, which has the same form of $type(\sharp)$ as R. Proceeding inductively in this fashion, we can define $T_1, T_2, T_3, \ldots$ such that T_{i+1} is right Morita dual to T_i for each $i = 1, 2, \ldots$. However, since $\sigma^{n!} = 1$ (the identity permutation), we see that $T^{n!+1} = R$. Thus we obtain the following theorem.

Theorem 5.2.2. *If R is type $(\sharp)$, then the following hold:*

(1) *R is almost self-dual.*

(2) *$D(R)$ is also type $(\sharp)$.*

(3) $F(D(R)) = \begin{pmatrix} e_{\sigma(1)1}Re_{\sigma(1)1} & e_{\sigma(1)1}Re_{\sigma(2)1} & \cdots & e_{\sigma(1)1}Re_{\sigma(m)1} \\ e_{\sigma(2)1}Re_{\sigma(1)1} & e_{\sigma(2)1}Re_{\sigma(2)1} & \cdots & e_{\sigma(2)1}Re_{\sigma(m)1} \\ \cdots & \cdots & \cdots & \cdots \\ e_{\sigma(m)1}Re_{\sigma(1)1} & e_{\sigma(m)1}Re_{\sigma(2)1} & \cdots & e_{\sigma(m)1}Re_{\sigma(m)1} \end{pmatrix}.$

(4) *$D(R)$ can be represented as an upper staircase factor ring of $F(D(R))(n(1), n(2), \ldots, n(m))$ with respect to $\{e_{\sigma(i)1}\}_{i=1}^{m}$, which has the same form of type $(\sharp)$ as R.*

Noting that σ is the identity if and only if $D(R) = R$, we obtain the following theorem.

Theorem 5.2.3. *R is a weakly symmetric left H-ring if and only if $F(R)$ is a weakly symmetric QF-ring, i.e., the Nakayama permutation of $F(R)$ is the identity.*

Now we return to our general setting for R by dropping the assumption that R is of "type $(\sharp)$". We recall the representation of R as an upper staircase factor ring of a block extension of the frame QF-subring $F(R)$.

$F(R)$ is made by taking a subset $\{e_{i_j1}\}_{j=1}^{y} \subseteq \{e_{i1}\}_{i=1}^{m}$ with $\{e_{i_j1}\}_{j=1}^{y} = \{e_{\sigma(i_j)1}\}_{j=1}^{y}$ as follows:

$$
F(R) = \begin{pmatrix}
e_{i_11}Re_{i_11} & e_{i_11}Re_{i_21} & \cdots & e_{i_11}Re_{i_y1} \\
e_{i_21}Re_{i_11} & e_{i_21}Re_{i_21} & \cdots & e_{i_21}Re_{i_y1} \\
\cdots & \cdots & \cdots & \cdots \\
\cdots & \cdots & \cdots & \cdots \\
e_{i_y1}Re_{i_11} & e_{i_y1}Re_{i_21} & \cdots & e_{i_y1}Re_{i_y1}
\end{pmatrix}.
$$

By Theorem 5.2.2 and the process above, we can take $F(T)$ as

$F(T)$

$$
= \begin{pmatrix}
e_{\sigma(i_1)\rho(i_1)}Re_{\sigma(i_1)\rho(i_1)} & e_{\sigma(i_1)\rho(i_1)}Re_{\sigma(i_2)\rho(i_2)} & \cdots & e_{\sigma(i_1)\rho(i_1)}Re_{\sigma(i_y)\rho(i_y)} \\
e_{\sigma(i_2)\rho(i_2)}Re_{\sigma(i_1)\rho(i_1)} & e_{\sigma(i_2)\rho(i_2)}Re_{\sigma(i_2)\rho(i_2)} & \cdots & e_{\sigma(i_2)\rho(i_2)}Re_{\sigma(i_y)\rho(i_y)} \\
\cdots & \cdots & \cdots & \cdots \\
\cdots & \cdots & \cdots & \cdots \\
e_{\sigma(i_y)\rho(i_y)}Re_{\sigma(i_1)\rho(i_1)} & e_{\sigma(i_y)\rho(i_y)}Re_{\sigma(i_2)\rho(i_2)} & \cdots & e_{\sigma(i_y)\rho(i_y)}Re_{\sigma(i_y)\rho(i_y)}
\end{pmatrix}
$$

$$
\cong \begin{pmatrix}
e_{\sigma(i_1)1}Re_{\sigma(i_1)1} & e_{\sigma(i_1)1}Re_{\sigma(i_2)1} & \cdots & e_{\sigma(i_1)1}Re_{\sigma(i_y)1} \\
e_{\sigma(i_2)1}Re_{\sigma(i_1)1} & e_{\sigma(i_2)1}Re_{\sigma(i_2)1} & \cdots & e_{\sigma(i_2)1}Re_{\sigma(i_y)1} \\
\cdots & \cdots & \cdots & \cdots \\
\cdots & \cdots & \cdots & \cdots \\
e_{\sigma(i_y)1}Re_{\sigma(i_1)1} & e_{\sigma(i_y)1}Re_{\sigma(i_2)1} & \cdots & e_{\sigma(i_y)1}Re_{\sigma(i_y)1}
\end{pmatrix}
$$

($\cong$: matrix isomorphism).

Thus, by taking $k(1), \ldots, k(y) \in \mathbb{N}$ such that $\sum_{i=1}^{y} k(i) = \sum_{i=1}^{m} n(i)$, we can represent R as an upper staircase factor ring of $F(R)(k(1), \ldots, k(y))$ with respect to $\{e_{i_i1}\}_{i=1}^{y}$. In view of the features shared in common by R and T given in Theorem 5.2.2, we see that $T = D(R)$ can also be represented as an upper staircase factor ring of the block extension $F(T)(k(1), \ldots, k(y))$ with respect to $\{e_{\sigma(i_j)1}\}_{j=1}^{y}$, where the upper staircase representation forms of R and T are the same. Since $D(R)$ is a basic left H-ring, by using the same argument to $D(D(R))$, we see that its frame QF-subring is

$$
F(D(D(R))) = \begin{pmatrix}
e_{\sigma^2(i_1)1}Re_{\sigma^2(i_1)1} & e_{\sigma^2(i_1)1}Re_{\sigma^2(i_2)1} & \cdots & e_{\sigma^2(i_1)1}Re_{\sigma^2(i_y)1} \\
e_{\sigma^2(i_2)1}Re_{\sigma^2(i_1)1} & e_{\sigma^2(i_2)1}Re_{\sigma^2(i_2)1} & \cdots & e_{\sigma^2(i_2)1}Re_{\sigma^2(i_y)1} \\
\cdots & \cdots & \cdots & \cdots \\
\cdots & \cdots & \cdots & \cdots \\
e_{\sigma^2(i_y)1}Re_{\sigma^2(i_1)1} & e_{\sigma^2(i_y)1}Re_{\sigma^2(i_2)1} & \cdots & e_{\sigma^2(i_y)1}Re_{\sigma^2(i_y)1}
\end{pmatrix}
$$

and $D(D(R))$ can be represented as an upper staircase factor ring of the block extension $F(D(D(R)))(k(1), \ldots, k(y))$ with respect to $\{e_{\sigma^2(i_j)1}\}_{j=1}^{y}$, which is of the same form as R. As earlier, let $T_1 = D(R)$ and $T_2 = D(T_1) = D(D(R))$, and proceed inductively to form T_k for any $k \in \mathbb{N}$. Then

$$F(T_k) = \begin{pmatrix} e_{\sigma^k(i_1)1}Re_{\sigma^k(i_1)1} & e_{\sigma^k(i_1)1}Re_{\sigma^k(i_2)1} & \cdots & e_{\sigma^k(i_1)1}Re_{\sigma^k(i_y)1} \\ e_{\sigma^k(i_2)1}Re_{\sigma^k(i_1)1} & e_{\sigma^k(i_2)1}Re_{\sigma^k(i_2)1} & \cdots & e_{\sigma^k(i_2)1}Re_{\sigma^k(i_y)1} \\ \cdots & \cdots & \cdots & \cdots \\ \cdots & \cdots & \cdots & \cdots \\ e_{\sigma^k(i_y)1}Re_{\sigma^k(i_1)1} & e_{\sigma^k(i_y)1}Re_{\sigma^k(i_2)1} & \cdots & e_{\sigma^k(i_y)1}Re_{\sigma^k(i_y)1} \end{pmatrix}$$

and $T_k = F(T_k)(k(1), \ldots, k(y))$ with respect to $\{e_{\sigma^k(i_j)1}\}_{j=1}^{y}$.

Since T_{i+1} is almost dual to T_i for each $i = 2, 3, \ldots$, and $T_{n!+1} = T_1 = R$, R is almost self-dual. Accordingly we can improve Theorem 5.2.2 for the general case as follows.

Theorem 5.2.4. *Let R be a basic left H-ring with a left well-indexed set $Pi(R) = \{e_{ij}\}_{i=1,j=1}^{m,\ n(i)}$. Then the following hold:*

(1) *There exist a subset $\{e_{i_j 1}\}_{j=1}^{y} \subseteq \{e_{i1}\}_{i=1}^{m}$ and $k(1), k(2), \ldots, k(y) \in \mathbb{N}$ such that*

  (a) $\sum_{i=1}^{y} k(i) = \sum_{i=1}^{m} n(i)$,
  (b) $\{e_{i_j 1}\}_{j=1}^{y} = \{e_{\sigma(i_j)1}\}_{j=1}^{y}$,

  (c) $F(R) = \begin{pmatrix} e_{i_1 1}Re_{i_1 1} & e_{i_1 1}Re_{i_2 1} & \cdots & e_{i_1 1}Re_{i_y 1} \\ e_{i_2 1}Re_{i_1 1} & e_{i_2 1}Re_{i_2 1} & \cdots & e_{i_2 1}Re_{i_y 1} \\ \cdots & \cdots & \cdots & \cdots \\ \cdots & \cdots & \cdots & \cdots \\ e_{i_y 1}Re_{i_1 1} & e_{i_y 1}Re_{i_2 1} & \cdots & e_{i_y 1}Re_{i_y 1} \end{pmatrix}$ *and*

  (d) $R = F(R)(k(1), k(2), \ldots, k(y))$ *with respect to* $\{e_{\sigma(i)1}\}_{i=1}^{m}$.

(2) $F(D(R)) = \begin{pmatrix} e_{\sigma(i_1)1}Re_{\sigma(i_1)1} & e_{\sigma(i_1)1}Re_{\sigma(i_2)1} & \cdots & e_{\sigma(i_1)1}Re_{\sigma(i_y)1} \\ e_{\sigma(i_2)1}Re_{\sigma(i_1)1} & e_{\sigma(i_2)1}Re_{\sigma(i_2)1} & \cdots & e_{\sigma(i_2)1}Re_{\sigma(i_y)1} \\ \cdots & \cdots & \cdots & \cdots \\ \cdots & \cdots & \cdots & \cdots \\ e_{\sigma(i_y)1}Re_{\sigma(i_1)1} & e_{\sigma(i_y)1}Re_{\sigma(i_2)1} & \cdots & e_{\sigma(i_y)1}Re_{\sigma(i_y)1} \end{pmatrix}$

and $D(R)$ can be represented as an upper staircase factor ring of $F(D(R))(k(1), k(2), \ldots, k(y))$ with respect to $\{e_{\sigma(i)1}\}_{i=1}^{m}$, which is the same form as R.

Theorem 5.2.5. *Every left H-ring is almost self-dual. Hence every left H-ring has a finitely generated right injective cogenerator.*

As an immediate consequence of Theorem 5.2.4, we obtain the following theorem.

Theorem 5.2.6. *Let R be a left H-ring represented as an upper staircase factor ring $\overline{P}$ of a block extension P of $F(R)$. Then the following are equivalent:*

(1) R *has a Nakayama isomorphism.*

(2) $F(R)$ *has a Nakayama automorphism.*

(3) P *has a Nakayama isomorphism.*

(4) *Every upper staircase factor ring of a block extension of $F(R)$ has a Nakayama isomorphism or, equivalently, any basic left H-ring with a frame QF-subring isomorphic to $F(R)$ has a Nakayama isomorphism.*

The following theorem is the main theorem in this section.

Theorem 5.2.7. *Let F be a basic QF-ring. The following are equivalent:*

(1) F *has a Nakayama automorphism.*

(2) *Every left H-ring R with $F(R)$ isomorphic to F has a Nakayama isomorphism.*

(3) *Every left H-ring R with $F(R)$ isomorphic to F is self-dual.*

Proof. (1) $\Leftrightarrow$ (2). This follows from Theorem 5.2.4.

(2) $\Rightarrow$ (3) is obvious.

(3) $\Rightarrow$ (1). Let $Pi(F) = \{e_i\}_{i=1}^m$. Consider a block extension $R = F(n(1), n(2), \ldots, n(m))$, where

$$n(1) < n(2) < \cdots < n(m) \quad \cdots\cdots \quad (**).$$

Let $\{h_{ij}\}_{i=1,j=1}^{m \ \ n(i)}$ be a complete set of orthogonal primitive idempotents of the dual ring $D(R)$ (see Theorem 5.2.2). We shall prove that there exists an isomorphism $\Phi : R \to D(R)$ such that $\Phi(e_{ij}) = h_{ij}$ for all e_{ij}. Since R has a self-duality, there exists a ring isomorphism $\Phi_1 : R \to D(R)$. Put $h'_{ij} = \Phi_1(e_{ij})$ for each $i = 1, \ldots, m$ and $j = 1, \ldots, n(i)$. Then $\{h'_{ij}\}_{i=1,j=1}^{m \ \ n(i)}$ is also a complete set of orthogonal primitive idempotents of the dual ring $D(R)$ such that

(i) $h'_{i1}D(R)_{D(R)}$ is injective for each $i = 1, \ldots, m$ and

(ii) $J(h'_{i,k-1}D(R)) \cong h'_{ik}D(R)$ for any $i = 1, \ldots, m$ and $k = 2, \ldots, n(i)$.

Then there exists an inner automorphism $\Phi_2 : D(R) \to D(R)$ such that $\Phi_2(h'_{ij}) = h_{ij}$ for all h'_{ij}. Considering the injective modules $h'_{i1}D(R)_{D(R)}$ and $h_{i1}D(R)_{D(R)}$ and the condition (**), we see that $\Phi_2(h'_{ij}) = h_{ij}$. Accordingly we have a desired ring isomorphism $\Phi = \Phi_2\Phi_1$ such that $\Phi(e_{ij}) = h_{ij}$ for all e_{ij}. Hence R has a Nakayama isomorphism and, by Theorem 5.2.6, F has a Nakayama automorphism. $\qquad\square$

5.3 Koike's Example of a QF-Ring without a Nakayama Automorphism

For Problem C, Koike [98] pointed out that an example of Kraemer [103] is a QF-ring without a weakly symmetric self-duality and this QF-ring does not have a Nakayama automorphism, Thus Problem C was solved negatively. We shall introduce this example following Koike's presentation.

Let A, A', B and B' be rings, let $\alpha : A \to A'$ and $\beta : B \to B'$ be ring homomorphisms, and let $_AM_B$ and $_{A'}M'_{B'}$ be bimodules. An additive homomorphism $\Phi : M \to M'$ is said to be (α, β)-*semilinear* if $\Phi(amb) = \alpha(a)\Phi(m)\beta(b)$ holds for any $a \in A$, $b \in B$ and $m \in M$.

In the following, given $m \in \mathbb{N}$, we let $[i]$ denote the least positive residue of the integer i modulo m.

Proposition 5.3.1. *Let* $A_1, A_2, \ldots, A_m$ $(m > 1)$ *be basic artinian rings and let* $_{A_1}U_{1A_2}, {}_{A_2}U_{2A_3}, \cdots, {}_{A_m}U_{mA_1}$ *be bimodules, each of which defines a Morita duality. Let*

$$R = \begin{pmatrix} A_1 & U_1 & 0 & \cdots & 0 & 0 \\ 0 & A_2 & U_2 & 0 & 0 & 0 \\ \vdots & 0 & A_3 & U_3 & \ddots & \vdots \\ \vdots & \vdots & \ddots & \ddots & \ddots & 0 \\ 0 & 0 & 0 & \cdots & A_{m-1} & U_{m-1} \\ U_m & 0 & 0 & \cdots & 0 & A_m \end{pmatrix}$$

and define a ring structure on R by the usual matrix operation and the relations $U_iU_{[i+1]} = 0$ for $1 \le i \le m$. Then the following hold:

(1) *R is a QF-ring.*

(2) *The following two conditions are equivalent:*

    (a) *R has a Nakayama automorphism.*

(b)   For each $i = 1, \ldots, m$, there exist a ring isomorphism τ_i : $A_i \to A_{[i+1]}$ and a $(\tau_i, \tau_{[i+1]})$-semilinear isomorphism Φ_i : $U_i \to U_{[i+1]}$ such that $T(_{A_i}A_i e) \cong S(_{A_i}U_i \tau(e))$ for any $e \in Pi(A_i)$.

Proof. (1).   By assumption, all A_i have complete sets of orthogonal primitive idempotents consisting of the same number (n, say). Thus, for each $i = 1, \ldots, m$, there exists a complete set $\{e_{ij}\}_{j=1}^{n}$ of orthogonal primitive idempotents of A_i. Then, since $_{A_i}U_{iA_{i+1}}$ defines a Morita duality, there exists a permutation π_i on $\{1, 2, \ldots, n\}$ such that $S(e_{ij}U_{iA_{[i+1]}}) \cong T(e_{[i+1], \pi_i(j)}A_{[i+1]}{}_{A_{[i+1]}})$ and $S(_{A_i}U_i e_{[i+1], \pi_i(j)}) \cong T(_{A_i}A_i e_{ij})$ for each $j = 1, \ldots, n$. For each $i = 1, \ldots, m$ and $j = 1, \ldots, n$, let f_{ij} be the idempotent of R in which the (i, j)-entry is e_{ij} and all other entries are zero and let $f_i = \sum_{j=1}^{n} f_{ij}$ be the idempotent of R in which the (i, i)-entry is 1_{A_i} and all other entries are zero. Then $\{f_{ij}\}_{i=1,j=1}^{m,\ n}$ is a complete set of orthogonal primitive idempotents of R. By the definition of R, we can easily check that $S(f_{ij}R_R) \cong T(f_{[i+1], \pi_i(j)}R_R)$ and $S(_RRf_{[i+1], \pi_i(j)}) \cong T(_RRf_{ij})$ for any $i = 1, \ldots, m$ and $j = 1, \ldots, n$. Thus each $(f_{ij}R; Rf_{[i+1], \pi_i(j)})$ becomes an i-pair, and hence R is a QF-ring.

(2).   (a) $\Rightarrow$ (b).   For the setting above, a Nakayama permutation of R is given by $\sigma : f_{ij} \mapsto f_{[i+1], \pi_i(j)}$. Assume that R has a Nakayama automorphism τ. Then $\tau(f_{ij}) = f_{[i+1], \pi_i(j)}$ for any $i = 1, \ldots, m$ and $j = 1, \ldots, n$. Since $\tau(f_i) = \tau(\sum_{j=1}^{n} f_{ij}) = \sum_{j=1}^{n} \tau(f_{ij}) = \sum_{j=1}^{n} f_{[i+1], \pi_i(j)} = f_{[i+1]}$, it follows, from $A_i \cong f_i R f_i$ and $U_i \cong f_i R f_{[i+1]}$, that τ induces ring isomorphisms $\tau_i : A_i \to A_{[i+1]}$ and additive isomorphisms $\Phi : U_i \to U_{[i+1]}$.

(b) $\Rightarrow$ (a).   Let τ be the automorphism of R defined by τ_i and Φ_i. It is clear that $\tau_i(e_{ij}) = e_{[i+1], \pi_i(j)}$. From this, it follows that $T(_{A_i}A_i e_{ij}) \cong S(_{A_i}U_i e_{[i+1], \pi_i(j)}) = S(_{A_i}U_i \tau_i(e_{ij}))$. Thus $T(_{A_i}A_i e) \cong S(_{A_i}U_i \tau_i(e))$ for any $e \in Pi(R)$. Therefore τ is a Nakayama automorphism of R. $\square$

Let C and D be division rings and $_CM_D$ a bimodule. Set $M_1 = {}_CM_D$ and, for each $i = 2, 3, \ldots$, define inductively

$$M_i = \begin{cases} _C\mathrm{Hom}_C(_DM_{i-1C}, \ _CC_C)_D & \text{if } i \text{ is odd,} \\ _D\mathrm{Hom}_D(_CM_{i-1D}, \ _DD_D)_C & \text{if } i \text{ is even.} \end{cases}$$

Example 5.3.2. (Koike [98], "Kraemer's Formulation")   There exists an extension $C \supset D$ of division rings satisfying the following conditions (see [103, Theorems 6.1; 6.2]):

(1) $\dim ({}_D C) = 2$ and $\dim (C_D) = 3$.

(2)   There exist ring isomorphisms $\lambda : D \to C$ and $\mu : C \to D$.

(3)   There exists a (λ, μ)-semilinear isomorphism: $\varphi : {}_D C_6{}_C \to {}_C C_1{}_D$.

(4) $(a_1, a_2, a_3, a_4, a_5) = (3, 1, 2, 2, 1)$ and $(b_1, b_2, b_3, b_4, b_5) = (1, 3, 1, 2, 2)$, where a_i and b_i are the right and left dimensions of C_i, respectively.

Then, by [103, Lemma 6.3],

(5)   the map $\psi : {}_C C_7{}_D \to {}_D C_2{}_C$ defined by $\psi(c_7)(c_1) = \mu(c_7(\varphi^{-1}(c_1)))$ (for each $c_7 \in C_7$, $c_1 \in C_1$) is a (μ, λ)-semilinear isomorphism.

Let

$$
A_i = \begin{cases} \begin{pmatrix} C & C_i \\ 0 & D \end{pmatrix} & \text{if } i \text{ is odd,} \\[2mm] \begin{pmatrix} D & C_i \\ 0 & C \end{pmatrix} & \text{if } i \text{ is even} \end{cases}
$$

and

$$
U_i = \begin{cases} \begin{pmatrix} D & 0 \\ C_i & C \end{pmatrix} & \text{if } i \text{ is odd,} \\[2mm] \begin{pmatrix} C & 0 \\ C_i & D \end{pmatrix} & \text{if } i \text{ is even.} \end{cases}
$$

Then, by [184, Corollary 10.3], U_{i+1} becomes an (A_{i+2}, A_i)-bimodule which defines a Morita duality. By using λ, μ, ψ and φ, we can see that $A_6 \cong A_1$ and $A_7 \cong A_2$ as rings. Therefore we may regard U_5 as an (A_1, A_4)-bimodule that defines a Morita duality. Similarly we regard U_1 as an (A_2, A_5)-bimodule that defines a Morita duality.

Now let us consider the QF-ring

$$
R = \begin{pmatrix} A_5 & U_4 & 0 & 0 & 0 \\ 0 & A_3 & U_2 & 0 & 0 \\ 0 & 0 & A_1 & U_5 & 0 \\ 0 & 0 & 0 & A_4 & U_3 \\ U_1 & 0 & 0 & 0 & A_2 \end{pmatrix}.
$$

Using the dimensions a_i and b_i of C_i, we see that all A_i ($i = 1, 2, 3, 4, 5$) are pairwise non-isomorphic. Therefore, by Proposition 5.3.1, R is a basic QF-ring that does not have a Nakayama automorphism.

The ring R was given in Kraemer [103] as an example of a QF-ring without weakly symmetric self-duality.

5.4 Factor Rings of QF-Rings with a Nakayama Automorphism

By Theorem 5.2.7, if R is a basic QF-ring with a Nakayama automorphism, then any upper staircase factor ring of a block extension of R has a Nakayama isomorphism and hence it has a self-duality. However, for a basic QF-ring R without a Nakayama automorphism, we can make a block extension of R without a self-duality. This difference is a remarkable fact concerning Nakayama automorphisms.

In this section, we provide the following additional result.

Theorem 5.4.1. *Let R be a basic indecomposable QF-ring R with $Pi(R) = \{e_i\}_{i=1}^n$ and let*

$$\begin{pmatrix} e_1 & \cdots & e_n \\ e_{\sigma(1)} & \cdots & e_{\sigma(n)} \end{pmatrix}$$

be a Nakayama permutation. Put $\overline{R} = R/S(R)$. Then the following hold:

(1) *Je_i is injective as a left $\overline{R}$-module and $E(T(_RRe_i)) \cong J(R)e_{\sigma(i)}$ as a left $\overline{R}$-module for each $i = 1, \ldots, n$.*

(2) *$Je_1 \oplus \cdots \oplus Je_n$ is an injective co-generator in $\overline{R}$-FMod.*

(3) *$\overline{R}$ has a self-duality.*

(4) *If R has a Nakayama automorphism, then $\overline{R}$ has a Nakayama isomorphism.*

Proof. Let $e \in Pi(R)$. Assume that $_RRe$ is simple. Then eR_R is also simple and $eRf = fRe = 0$ for any $f \in Pi(R)$ with $f \neq e$. Since R is indecomposable as a ring, this implies that $R = eRe$ and R is a division ring. Therefore we may assume that all Re_i are not simple, i.e., all Je_i are non-zero modules.

(1). Since $S(R)Je_i = 0$, each Je_i is a left $\overline{R}$-module. To show that Je_i is injective as a left $\overline{R}$-module, let $\overline{I}$ be a left ideal of $\overline{R}$ and take any $0 \neq \psi \in \mathrm{Hom}_{\overline{R}}(\overline{I}, Je_i)$. Since $_RRe_i$ is injective, we have a homomorphism $\psi^* : _R\overline{R} \to _RRe_i$ which is an extension of ψ. Setting $x = \psi(\overline{1})$, we must show that $Rx \subseteq Je_i$. If $Rx = Re_i$, then $0 = S(R)x = S(R)e_i = S(Re_i)$, whence $Je_i = 0$, a contradiction. Therefore $Rx \subseteq Je_i$. Since $S(_RJe_{\sigma(i)}) = S(_RRe_{\sigma(i)}) \cong T(_RRe_i)$, we see that $Je_{\sigma(i)} \cong E(T(_RRe_i))$.

(2). Since $Je_i \not\cong Je_j$ for any distinct $i, j \in \{1, \ldots, n\}$, $Je_1 \oplus \cdots \oplus Je_n$ is an injective co-generator in $\overline{R}$-FMod.

(3). Put $g_i = e_{\sigma(i)}$ for each $i = 1, \ldots, n$ and $W = Je_1 \oplus \cdots \oplus Je_n = Je_{\sigma(1)} \oplus \cdots \oplus Je_{\sigma(n)}$. We represent R and $\mathrm{End}(_RW)$ as follows:

$$R = \begin{pmatrix} [e_1, e_1] & \cdots & [e_1, e_{\sigma(1)}] & \cdots & \cdots & [e_1, e_n] \\ & \cdots & & & \cdots & \\ & \cdots & & & & \\ [e_n, e_1] & \cdots & & [e_n, e_{\sigma(n)}] & \cdots & [e_n, e_n] \end{pmatrix},$$

$$\mathrm{End}(W) = \begin{pmatrix} [Je_1, Je_1] & \cdots & [Je_1, Je_{\sigma(1)}] & \cdots & \cdots & [Je_1, Je_n] \\ & \cdots & & & \cdots & \\ & \cdots & & & \cdots & \\ [Je_n, Je_1] & \cdots & & [Je_n, Je_{\sigma(n)}] & \cdots & [Je_n, Je_n] \end{pmatrix},$$

where $[e_i, e_j] = \mathrm{Hom}_R(Re_i, Re_j)$ and $[Je_i, Je_j] = \mathrm{Hom}_R(Je_i, Je_j)$. We shall define a surjective matrix ring homomorphism $\Phi = (\varphi_{ij})_{1 \leq i, j \leq n} : R \to \mathrm{End}(_RW)$. First we define

$$\varphi_{ij} : [Re_i, Re_j] \to [Je_i, Je_j]$$

by $\alpha \mapsto \alpha|_{Je_i}$ for any i, j and put

$$\Phi = \begin{pmatrix} \varphi_{11} & \cdots & \varphi_{1n} \\ & \cdots & \\ \varphi_{n1} & \cdots & \varphi_{nn} \end{pmatrix}.$$

Then $\Phi : R \to \mathrm{End}(_RW)$ is a matrix ring homomorphism.

We claim that each φ_{ij} is epimorphic. Let $\beta \in [Je_i, Je_j]$ and consider the diagram

$$0 \to Je_i \to Re_i$$
$$\downarrow \beta$$
$$Je_j$$

Since $_RRe_j$ is injective, there exists $\alpha \in \mathrm{Hom}_R(Re_i, Je_j)$ such that $\alpha|_{Je_i} = \beta$. Thus φ_{ij} is epimorphic.

Further we claim that

$$\mathrm{Ker}\,\varphi_{ij} = \begin{cases} 0 & \text{if } j \neq \sigma(i), \\ \mathrm{Hom}_R(Re_i, S(_RRe_j)) & \text{if } j = \sigma(i). \end{cases}$$

Suppose that $j \neq \sigma(i)$ and there exists $(0 \neq)\, \alpha \in \mathrm{Ker}\,\varphi_{ij}$, i.e., $\varphi_{ij}(\alpha) = \alpha|_{Je_i} = 0$. Then $T(_RRe_i) \cong S(_RRe_j)$. Hence $(e_iR; Re_j)$ is an i-pair, and hence $j = \sigma(i)$, a contradiction. Hence $\mathrm{Ker}\,\varphi_{ij} = 0$ if $j \neq \sigma(i)$. Next assume that $j = \sigma(i)$. Then it is clear that $\mathrm{Ker}\,\varphi_{ij} = \mathrm{Ker}\,\varphi_{i\sigma(i)} = \mathrm{Hom}_R(Re_i, S(_RRe_j))$.

We put $X_{i\sigma(i)} = \text{Ker}\,\varphi_{i\sigma(i)} = \text{Hom}_R(Re_i, S(_RRe_j))$ for each $i = 1, \ldots, n$. Then

$$\text{Ker}\,\Phi = \begin{pmatrix} 0 \cdots & X_{1\sigma(1)} & 0 & \cdots 0 \\ & \cdots & & \cdots \\ & \cdots & & \cdots \\ 0 \cdots & 0 & X_{n\sigma(n)} & \cdots 0 \end{pmatrix}.$$

Thus, by Lemma 2.1.1, $\text{Ker}\,\Phi = S(R)$. Hence Φ induces an isomorphism $\overline{\Phi} : \overline{R} \cong \text{End}(_RW)$, which shows the self-duality of $\overline{R}$.

(4). Since R has a Nakayama automorphism, there exists a matrix isomorphism $\Theta = (\theta_{ij})_{1 \le i, j \le n}$ from

$$R = \begin{pmatrix} [e_1, e_1] & \cdots & [e_1, e_{\sigma(1)}] & \cdots & \cdots & [e_1, e_n] \\ & \cdots & & & \cdots & \\ & \cdots & & & \cdots & \\ [e_n, e_1] & \cdots & & [e_n, e_{\sigma(n)}] & \cdots & [e_n, e_n] \end{pmatrix}$$

to

$$D(R) = \begin{pmatrix} [g_1, g_1] & \cdots & [g_1, g_{\sigma(1)}] & \cdots & \cdots & [g_1, g_n] \\ & \cdots & & & \cdots & \\ & \cdots & & & \cdots & \\ [g_n, g_1] & \cdots & & [g_n, g_{\sigma(n)}] & \cdots & [g_n, g_n] \end{pmatrix}.$$

Let

$$\overline{\Theta} : R/S(R) \to D(R)/S(D(R))$$

be the induced isomorphism. We put

$$X = \begin{pmatrix} [Jg_1, Jg_1] & \cdots & [Jg_1, Jg_{\sigma(1)}] & \cdots & \cdots & [Jg_1, Jg_n] \\ & \cdots & & & \cdots & \\ & \cdots & & & \cdots & \\ [Jg_n, Jg_1] & \cdots & & [Jg_n, Jg_{\sigma(n)}] & \cdots & [Jg_n, Jg_n] \end{pmatrix}.$$

Applying the argument in the proof of (3) to $D(R)$, we obtain a matrix isomorphism

$$\overline{\Psi} = (\psi_{ij})_{1 \le i, j \le n} : D(R)/S(D(R)) \to X.$$

Further, by using (1), there exists a matrix isomorphism

$\Omega = (\omega_{ij}) : X \to \text{End}(E(T(_RR))) = \text{End}(E(T(_RRe_1)) \oplus \cdots \oplus E(T(_RRe_n)))$.

Consequently $\Omega\,\overline{\Psi}\,\overline{\Theta} : \overline{R} \to \text{End}(E(T(_RR)))$ is a matrix isomorphism, which completes the proof. $\qquad \square$

As an immediate consequence, we obtain the following result.

Corollary 5.4.2. *If R is a weakly symmetric QF-ring, then $R/S(R)$ is self-dual.*

COMMENTS

Most of the material in this chapter is based on Kado-Oshiro [89] and Koike [98], [99]. Koike gave Example 5.3.2 by using Kraemer's example in [103]. This important example shows that in general left H-rings need not be self-dual though QF-rings and Nakayama rings are self-dual. This is an important fact about the structure of left H-rings. The almost self-duality of left H-rings is shown in Koike [99] but the proof presented here is different from Koike's original one. For recent developments on artinian rings with Morita duality, the reader is referred to Koike [101].

Chapter 6

Skew Matrix Rings

In this chapter, we introduce skew matrix rings which generalize the usual matrix rings. These rings were first introduced by Kupisch [104] in 1975 under the name VPE-rings for his study on Nakayama rings and independently by Oshiro [149] in 1987 for the matrix representation of left H-rings and its applications to Nakayama rings. In this book, we use the definition of these rings given in the latter article. The main purpose of this chapter is to describe fundamental properties of these rings and apply them to QF-rings. One such application produces a QF- ring whose Nakayama permutation corresponds to any given permutation. In Chapter 7, we give a complete classification of Nakayama rings by using skew matrix rings over local Nakayama rings.

6.1 Definition of a Skew Matrix Ring

Let Q be a ring and let $c \in Q$ and $\sigma \in \text{End}(Q)$ with $\sigma(c) = c$ and $\sigma(q)c = cq$ for any $q \in Q$. Let R denote the set of all $n \times n$ matrices over Q:

$$R = \begin{pmatrix} Q \cdots Q \\ \cdots \\ Q \cdots Q \end{pmatrix}.$$

We define a multiplication on R with respect to (σ, c) as follows:

For any (x_{ik}), $(y_{ik}) \in R$, the multiplication of these matrices is defined by

$$(z_{ik}) = (x_{ik})(y_{ik}),$$

where z_{ik} is given by the following:

$$
z_{ik} = \begin{cases} \sum_{j<i} x_{ij}\sigma(y_{jk})c + \sum_{i\leq j\leq k} x_{ij}y_{jk} + \sum_{k<j} x_{ij}y_{jk}c & \text{if } i \leq k, \\[2mm] \sum_{j\leq k} x_{ij}\sigma(y_{jk}) + \sum_{k<j<i} x_{ij}\sigma(y_{jk})c + \sum_{i\leq j} x_{ij}y_{jk} & \text{if } k < i. \end{cases}
$$

We may write this multiplication as follows:

$$\langle a \rangle_{ij} \langle b \rangle_{kl} = 0 \quad \text{if } j \neq k,$$

$$
\langle a \rangle_{ij} \langle b \rangle_{jk} = \begin{cases} \langle a\sigma(b) \rangle_{ik} & \text{if } j \leq k < i, \\ \langle a\sigma(b)c \rangle_{ik} & \text{if } k < j < i \text{ or } j < i \leq k, \\ \langle ab \rangle_{ik} & \text{if } i = j, \\ \langle abc \rangle_{ik} & \text{if } i \leq k < j, \\ \langle ab \rangle_{ik} & \text{if } k < i < j \text{ or } i < j \leq k. \end{cases}
$$

It is straightforward to check that this multiplication satisfies the associative law:

$$(\langle x \rangle_{ij} \langle y \rangle_{jk})\langle z \rangle_{kl} = \langle x \rangle_{ij}(\langle y \rangle_{jk}\langle z \rangle_{kl})$$

Furthermore R becomes a ring under this multiplication and the usual addition of matrices. We call this ring R the *skew matrix ring* over Q with respect to (σ, c, n) and denote it by

$$
R = \begin{pmatrix} Q & \cdots & Q \\ & \cdots & \\ Q & \cdots & Q \end{pmatrix}_{\sigma,c,n}
$$

or, simply, by $(Q)_{\sigma,c,n}$. Note that, for the identity map id_Q of Q and the identity $1 \in Q$, $(Q)_{id_Q,1,n}$ is the usual matrix ring over Q and $(Q)_{\sigma,c,1}$ is Q itself since σ and c do not impact.

We may consider Q as a subring of R by the map:

$$
q \longmapsto \begin{pmatrix} q & 0 & \cdots & 0 \\ 0 & \ddots & \ddots & \vdots \\ \vdots & \ddots & \ddots & 0 \\ 0 & \cdots & 0 & q \end{pmatrix}.
$$

Put $\alpha_{ij} = \langle 1 \rangle_{ij}$ for each $i, j \in \{1, \ldots, n\}$. We call $\{\alpha_{ij}\}_{i,j\in\{1,\ldots,n\}}$ the *skew matrix units* of R. For these matrix units, we obtain the following relations:

$$(*) \cdots \begin{cases} \text{When} \quad j \neq k, \quad \alpha_{ij}\alpha_{kl} = 0. \\[2mm] \text{When} \quad i > j, \quad \begin{cases} \sigma(q)\alpha_{ij} = \alpha_{ij}q \ \text{ for any } \ q \in Q, \\[1mm] \alpha_{ij}\alpha_{jk} = \begin{cases} \alpha_{ik} & \text{if } i > k \geq j, \\ \alpha_{ik}c & \text{if } k \geq i \text{ or } j > k. \end{cases} \end{cases} \\[4mm] \text{When} \quad i = j, \quad \begin{cases} q\alpha_{ij} = \alpha_{ij}q \ \text{ for all } \ q \in Q, \\[1mm] \alpha_{ij}\alpha_{jk} = \alpha_{ik}. \end{cases} \\[4mm] \text{When} \quad i < j, \quad \begin{cases} q\alpha_{ij} = \alpha_{ij}q \ \text{ for all } \ q \in Q, \\[1mm] \alpha_{ij}\alpha_{jk} = \begin{cases} \alpha_{ik}c & \text{if } i \leq k < j, \\ \alpha_{ik} & \text{if } k < i \text{ or } j \leq k. \end{cases} \end{cases} \end{cases}$$

Consider the set R' defined by

$$R' = \begin{pmatrix} Q\alpha_{11} & \cdots & Q\alpha_{1n} \\ & \cdots & \\ Q\alpha_{n1} & \cdots & Q\alpha_{nn} \end{pmatrix}.$$

Then R' becomes a ring by the relations $(*)$ above. And there exists a canonical matrix ring isomorphism between R and R' by the map

$$\begin{pmatrix} q_{11} & \cdots & q_{1n} \\ & \cdots & \\ q_{n1} & \cdots & q_{nn} \end{pmatrix} \longleftrightarrow \begin{pmatrix} q_{11}\alpha_{11} & \cdots & q_{1n}\alpha_{1n} \\ & \cdots & \\ q_{n1}\alpha_{n1} & \cdots & q_{nn}\alpha_{n,n} \end{pmatrix}.$$

Under this isomorphism, we often identify R and R' and represent R as

$$R = \begin{pmatrix} Q\alpha_{11} & Q\alpha_{12} & \cdots & Q\alpha_{1n} \\ Q\alpha_{21} & Q\alpha_{22} & \cdots & Q\alpha_{2n} \\ & \cdots & \cdots & \\ Q\alpha_{n1} & Q\alpha_{n2} & \cdots & Q\alpha_{nn} \end{pmatrix}_{\sigma,c,n}.$$

Furthermore using the canonical ring isomorphisms $Q\alpha_{ii} \cong Q$, we often represent R as

$$R = \begin{pmatrix} Q & Q\alpha_{12} & \cdots & Q\alpha_{1n} \\ Q\alpha_{21} & Q & \ddots & \vdots \\ \vdots & \ddots & \ddots & Q\alpha_{n-1,n} \\ Q\alpha_{n1} & \cdots & Q\alpha_{n,n-1} & Q \end{pmatrix}_{\sigma,c,n}.$$

In order to make calculations easier, we put $\beta_{ji} = \alpha_{ji}$ for each $j > i$ in these representations. Moreover, instead of $\{\alpha_{ij}\}_{i,j\in\{1,\dots,n\}}$, we take the sets $\{\alpha_{ij}\}_{1\le i\le j\le n} \cup \{\beta_{ji}\}_{1\le i<j\le n}$ as the skew matrix units of R. Then

$$R = \begin{pmatrix} Q\alpha_{11} & Q\alpha_{12} & \cdots & & Q\alpha_{1n} \\ Q\beta_{21} & Q\alpha_{22} & \ddots & & \vdots \\ \vdots & \ddots & \ddots & & Q\alpha_{n-1,n} \\ Q\beta_{n1} & \cdots & Q\beta_{n,n-1} & & Q\alpha_{nn} \end{pmatrix}_{\sigma,c,n} = \begin{pmatrix} Q & Q\alpha_{12} & \cdots & & Q\alpha_{1n} \\ Q\beta_{21} & Q & \ddots & & \vdots \\ \vdots & \ddots & \ddots & & Q\alpha_{n-1,n} \\ Q\beta_{n1} & \cdots & Q\beta_{n,n-1} & & Q \end{pmatrix}_{\sigma,c,n}$$

with relations:

$$\begin{cases} \alpha_{ij}q = q\alpha_{ij} & \text{for any } q \in Q, \\ \beta_{ij}q = \sigma(q)\beta_{ij} & \text{for any } q \in Q, \\ \alpha_{ij}\alpha_{jk} = \alpha_{ik} & \text{if } i < j < k, \\ \beta_{ij}\beta_{jk} = \beta_{ik}c & \text{if } k < j < i, \\ \alpha_{ij}\beta_{jk} = \begin{cases} \alpha_{ik}c & \text{if } i \le k < j, \\ \beta_{ik} & \text{if } k < i < j, \end{cases} \\ \beta_{ij}\alpha_{jk} = \begin{cases} \alpha_{ik}c & \text{if } j < i \le k, \\ \beta_{ik} & \text{if } j < k < i. \end{cases} \end{cases}$$

Remark 6.1.1. We look $(Q)_{\sigma,c.n}$ for some particular cases.

(1) If $n = 2$, then the multiplication is as follows:
$$\begin{pmatrix} x_1 & x_2 \\ x_3 & x_4 \end{pmatrix}\begin{pmatrix} y_1 & y_2 \\ y_3 & y_4 \end{pmatrix} = \begin{pmatrix} x_1y_1 + x_2y_3c & x_1y_2 + x_2y_4 \\ x_3\sigma(y_1) + x_4y_3 & x_3\sigma(y_2)c + x_4y_4 \end{pmatrix}.$$

(2) If Q is a local ring, then R is a semiperfect ring since $e_iRe_i \cong Q$ for each $i = 1, \dots, n$, where $e_i = \langle 1 \rangle_{ii}$.

(3) If Q is a right (left) artinian ring, then so is R.

(4) If Q is a right (left) noetherian ring, then so is R.

The following theorem is an important result on a skew matrix ring.

Theorem 6.1.2. *The matrix map* $\tau = (\tau_{ij}) : R \longrightarrow R$ *given by*

$$\begin{pmatrix} x_{11} & x_{12} & \cdots & x_{1n} \\ x_{21} & x_{22} & \cdots & x_{2n} \\ \cdots & \cdots & \cdots & \cdots \\ x_{n1} & x_{n2} & \cdots & x_{nn} \end{pmatrix} \longmapsto \begin{pmatrix} x_{nn} & x_{n1} & \cdots & x_{n,n-1} \\ \sigma(x_{1n}) & \sigma(x_{11}) & \cdots & \sigma(x_{1,n-1}) \\ \cdots & \cdots & \cdots & \cdots \\ \sigma(x_{n-1,n}) & \sigma(x_{n-1,1}) & \cdots & \sigma(x_{n-1,n-1}) \end{pmatrix}$$

is a ring homomorphism; in particular, if $\sigma \in \mathrm{Aut}(Q)$, *then* $\tau \in \mathrm{Aut}(R)$.

Proof.   This is straightforward.   □

Now, for each $i = 1, \ldots, n$, let

$$W_i = \langle Q \rangle_{i1} + \cdots + \langle Q \rangle_{i,i-1} + \langle Qc \rangle_{ii} + \langle Q \rangle_{i,i+1} + \cdots + \langle Q \rangle_{in}$$

$$= \begin{pmatrix} & & & 0 & & & \\ Q & \cdots & Q & Qc & Q & \cdots & Q \\ & & & 0 & & & \end{pmatrix} < i.$$

Then W_i is a submodule of $e_i R_R$. For $i = 2, \ldots, n$, let $\theta_i : e_i R \longrightarrow W_{i-1}$ be the map given by

$$\begin{pmatrix} & & & & 0 & & & \\ x_1 & \cdots & x_{i-2} & x_{i-1} & x_i & \cdots & x_n \\ & & & & 0 & & & \end{pmatrix} < i \;\; \longmapsto \;\; \begin{pmatrix} & & & & 0 & & & \\ x_1 & \cdots & x_{i-2} & x_{i-1}c & x_i & \cdots & x_n \\ & & & & 0 & & & \end{pmatrix} < i-1$$

and let $\theta_1 : e_1 R \longrightarrow W_n$ be the map given by

$$\begin{pmatrix} x_1 & \cdots\cdots & & x_n \\ 0 & \cdots & \cdots & 0 \\ & \cdots & \cdots & \\ 0 & \cdots\cdots & & 0 \end{pmatrix} \longmapsto \begin{pmatrix} 0 & \cdots & \cdots & 0 \\ & \cdots & \cdots & \\ 0 & \cdots & \cdots & 0 \\ \sigma(x_1) & \cdots & \sigma(x_{n-1}) & \sigma(x_n)c \end{pmatrix}.$$

Then, for each $i = 2, \ldots, n$, θ_i is realized by the left multiplication by $\langle 1 \rangle_{i-1,i}$ and θ_1 is realized by the left multiplication by $\langle 1 \rangle_{n1}$. Hence we have the following result.

Proposition 6.1.3. *Each θ_i is an R-homomorphism. Moreover, if σ is an onto homomorphism, then each θ_i is an onto homomorphism and*

$$\mathrm{Ker}\,\theta_1 = \begin{pmatrix} 0 & \cdots & 0 & l_Q(c) \\ 0 & \cdots & 0 & 0 \\ \cdots & \cdots & \cdots & \cdots \\ 0 & \cdots\cdots & & 0 \end{pmatrix},$$

$$\mathrm{Ker}\,\theta_i = \langle l_Q(c) \rangle_{i,i-1} = \begin{pmatrix} & & \overset{i-1}{\overset{\vee}{0}} & & \\ 0 & l_Q(c) & 0 \\ & & 0 & & \end{pmatrix} < i \quad \text{for } i = 2, \ldots, n.$$

By the above observations, we show the following fundamental result.

Theorem 6.1.4. *Let Q be a local QF-ring, $\sigma \in \mathrm{Aut}(Q)$ and $c \in J(Q)$ with $\sigma(c) = c$ and $\sigma(q)c = cq$ for any $q \in Q$. Set $R = (Q)_{\sigma,c,n}$. Then the following hold:*

(1)　R *is a basic indecomposable QF-ring with the Nakayama permutation:*

$$\begin{pmatrix} e_1 \ e_2 \ \cdots \ e_n \\ e_n \ e_1 \ \cdots \ e_{n-1} \end{pmatrix},$$

　　　where $e_i = \langle 1 \rangle_{ii}$ for $i = 1, \ldots, n$. Namely R is a basic indecomposable QF-ring with a cyclic Nakayama permutation.

(2)　*For any idempotent e of R, eRe is represented as a skew matrix ring over Q with respect to $(\sigma, c, k \ (\leq n))$, where $k = \#Pi(eRe)$. Thus eRe is a basic indecomposable QF-ring with a cyclic Nakayama permutation.*

(3)　R *has a Nakayama automorphism.*

Proof. (1). Put $X = S(Q_Q) \ (= S(\,_Q Q))$. Noting that $cX = Xc = 0$, we can easily see that

$$S(e_1 R_R) = \begin{pmatrix} 0 & \cdots & 0 & X \\ 0 & \cdots & 0 & 0 \\ & \cdots & & \\ 0 & \cdots & 0 & 0 \end{pmatrix} = S(Re_n) \quad \text{and}$$

$$S(e_i R_R) = \begin{pmatrix} 0 & & & & & \\ 0 & \cdot & & \mathbf{0} & & \\ & \cdot & \cdot & & & \\ & & 0 & 0 & & \\ & & & X & 0 & \\ & & & & 0 & \cdot \\ \mathbf{0} & & & & & \cdot \cdot \\ & & & & & 0 \ 0 \end{pmatrix} = S(_R Re_{i-1}) \quad \text{for } i = 2, \ldots, n.$$

Hence we infer that $(e_1 R; Re_n)$, $(e_2 R; Re_1)$, $\ldots$, $(e_n R; Re_{n-1})$ are i-pairs, whence R is a basic indecomposable QF-ring.

　　(2). For any subset $\{f_1, \ldots, f_k\} \subseteq Pi(R)$, clearly fRf is represented as a skew matrix ring over Q with respect to (σ, c, k), where $f = f_1 + \cdots + f_k$, whence eRe is represented as a skew matrix ring for any idempotent e of R.

　　(3). This follows from Theorem 6.1.2.　　　　　　　　　　　　□

6.2 Nakayama Permutations vs Given Permutations

The purpose of this section is to provide several examples of basic indecomposable QF-rings by using skew-matrix rings.

Let k be a field. We consider $Q = k[x]/(x^4)$ and put $c = x + (x^4)$. Then Q is a local ring and its only ideals are Q, Qc, Qc^2, Qc^3, 0. Using Q, for any given permutation, we shall construct an example of a QF-ring with a Nakayama permutation which corresponds to the given permutation.

Let R be the skew matrix ring $(Q)_{id_Q,c,w}$ with skew matrix units $\{\alpha_{ij}\} \cup \{\beta_{ji}\}$. We take $\{n(1), n(2), \ldots, n(m)\} \subset \{1, 2, \ldots, w\}$ such that

$$1 < n(1) \leq n(2) \leq \cdots \leq n(m) \quad \text{and} \quad n(1) + n(2) + \cdots + n(m) = w$$

and put

$$(1,1) = 1, \ (1,2) = 2, \ \ldots, \ (1, n(1)) = n(1),$$
$$(2,1) = n(1)+1, \ (2,2) = n(1)+2, \ \ldots, \ (2, n(2)) = n(1)+n(2), \ \ldots$$
$$\vdots$$
$$(m,1) = n(1) + n(2) + \cdots + n(m-1) + 1, \ \ldots, \ (m, n(m)) = w.$$

For the sake of convenience, we use kl instead of (k, l). Then

$$\{1, 2, \ldots, w\} = \{11, \ 12, \ldots, \ 1n(1), \ 21, \ldots, \ 2n(2), \ \ldots, \ m1, \ldots, \ mn(m)\}.$$

We make the following partition of $\{1, 2, \ldots, w\}$:

$$\{1 = 11, \ 12, \ldots, \ 1n(1))\} \cup \{21, \ 22, \ldots, \ 2n(2)\} \cup \cdots$$
$$\cdots \cup \{(m-1)1, \ldots, (m-1)n(m-1)\} \cup \{m1, \ m2, \ldots, \ mn(m)\}.$$

For each $i, j \in \{1, \ldots, m\}$, we construct blocks R_{ij} of R as follows:

$$R_{ii} = \begin{pmatrix} Q\alpha_{i1,i1} & Q\alpha_{i1,i1+1} & \cdots & & Q\alpha_{i1,in(i)} \\ Q\beta_{i1+1,i1} & Q\alpha_{i1+1,i1+1} & \cdots & & \cdots \\ \cdots & \cdots & & \cdots & Q\alpha_{in(i)-1,in(i)} \\ Q\beta_{in(i),i1} & \cdots & Q\beta_{in(i),in(i)-1} & & Q\alpha_{in(i),in(i)} \end{pmatrix},$$

$$R_{ij} = \begin{pmatrix} Q\alpha_{i1,j1} & Q\alpha_{i1,j1+1} & \cdots & Q\alpha_{i1,jn(j)} \\ Q\alpha_{i1+1,j1} & Q\alpha_{i1+1,j1+1} & \cdots & Q\alpha_{i1+1,jn(j)} \\ \cdots & \cdots & \cdots & \cdots \\ Q\alpha_{in(i),j1} & Q\alpha_{in(i),j1+1} & \cdots & Q\alpha_{in(i),jn(j)} \end{pmatrix} \quad \text{for } i < j,$$

$$R_{ji} = \begin{pmatrix} Q\beta_{j1,i1} & Q\beta_{j1,i1+1} & \cdots & Q\beta_{j1,in(i)} \\ Q\beta_{j1+1,i1} & Q\beta_{j1+1,i1+1} & \cdots & Q\beta_{j1+1,in(i)} \\ \cdots & \cdots & \cdots & \cdots \\ Q\beta_{jn(j),i1} & Q\beta_{jn(j),i1+1} & \cdots & Q\beta_{jn(j),in(i)} \end{pmatrix} \quad \text{for } i < j.$$

Then R is represented as

$$R = \begin{pmatrix} R_{11} & \cdots & R_{1m} \\ & \cdots & \\ R_{m1} & \cdots & R_{mm} \end{pmatrix}.$$

For $i < j$, we consider the following subsets $T_{ij} \subsetneqq R_{ij}$ and $T_{ji} \subsetneqq R_{ji}$:

$$T_{ij} = \begin{pmatrix} Qc\alpha_{i1,j1} & \cdots & & & \cdots & & \cdots & Qc\alpha_{i1,jn(j)} \\ Qc^2\alpha_{i1+1,j1} & \ddots & & & & & & \vdots \\ \vdots & & \ddots & & \ddots & & & \vdots \\ Qc^2\alpha_{in(i),j1} & \cdots & Qc^2\alpha_{in(i),jx} & Qc\alpha_{in(i),jx+1} & \cdots & Qc\alpha_{in(i),jn(j)} \end{pmatrix},$$

where $x = n(i) - 1$,

$$T_{ji} = \begin{pmatrix} Qc\beta_{j1,i1} & \cdots & & \cdots & & Qc\beta_{j1,in(i)} \\ Qc^2\beta_{j1+1,i1} & \ddots & & & & Qc\beta_{j1+1,in(i)} \\ \vdots & & \ddots & & & \vdots \\ \vdots & & & Qc^2\beta_{js,in(i)-1} & Qc\beta_{js,in(i)} \\ \vdots & & & & \vdots & \vdots \\ Qc^2\beta_{jn(j),i1} & \cdots & Qc^2\beta_{jn(j),in(i)-1} & Qc\beta_{jn(j),in(i)} \end{pmatrix},$$

where $s = n(i) - 1$. Put

$$T = \begin{pmatrix} R_{11} & T_{12} & \cdots & & \cdots & T_{1m} \\ T_{21} & \ddots & \ddots & & & \vdots \\ \vdots & \ddots & \ddots & \ddots & & \vdots \\ \vdots & & \ddots & \ddots & T_{m-1,m} \\ T_{m1} & \cdots & \cdots & T_{m,m-1} & R_{mm} \end{pmatrix}.$$

Then T is a basic indecomposable artinian subring of R. In T, we put $f_k = \langle 1 \rangle_{kk}$ for each $k = 1, \ldots, w$. Then $\{f_1, \ldots, f_w\}$ is a complete set of orthogonal primitive idempotents of T; note that $f_i T f_i \cong Q$.

For each $i < j$, we define $I_{ij} \subsetneq T_{ij}$ and $I_{ji} \subsetneq T_{ji}$ as follows:

$$
I_{ij} = \begin{pmatrix}
Qc^2\alpha_{i1,j1} & \cdots & & \cdots & & \cdots & Qc^2\alpha_{i1,jn(j)} \\
Qc^3\alpha_{i1+1,j1} & \ddots & & & & & \vdots \\
\vdots & \ddots & \ddots & & & & \vdots \\
Qc^3\alpha_{in(i),j1} & \cdots & Qc^3\alpha_{in(i),x} & Qc^2\alpha_{in(i),x+1} & \cdots & Qc^2\alpha_{in(i),jn(j)}
\end{pmatrix},
$$

$$
I_{ji} = \begin{pmatrix}
Qc^2\beta_{j1,i1} & \cdots & & \cdots & & Qc^2\beta_{j1,in(i)} \\
Qc^3\beta_{j1+1,i1} & \ddots & & & & \vdots \\
\vdots & \ddots & & \ddots & & \vdots \\
\vdots & & & Qc^3\beta_{s,in(i)-1} & Qc^2\beta_{s,in(i)} & \\
\vdots & & & \vdots & \vdots & \\
Qc^3\beta_{jn(j),i1} & \cdots & & Qc^3\beta_{jn(j),in(i)-1} & Qc^2\beta_{jn(j),in(i)}
\end{pmatrix}.
$$

Put

$$
I = \begin{pmatrix}
0 & I_{12} & \cdots & & \cdots & I_{1m} \\
I_{21} & 0 & \ddots & & & \vdots \\
\vdots & \ddots & \ddots & \ddots & & \vdots \\
\vdots & & \ddots & \ddots & I_{m-1,m} \\
I_{m1} & \cdots & & \cdots & I_{m,m-1} & 0
\end{pmatrix}.
$$

Then I is an ideal of T. Here we put $G = T/I$ and $g_i = f_i + I$ for all i. We show that G is a basic indecomposable QF-ring. Actually, by the description of G and I together with the structure of R (cf. Theorem 6.1.4), we can easily see that the following pairs are i-pairs:

$$(g_{11}G; Gg_{1n(1)}),\ (g_{11+1}G; Gg_{11}),\ (g_{11+2}G; Gg_{11+1}),\ \ldots,$$
$$(g_{1n(1)-1}G; Gg_{1n(1)-2}),\ (g_{1n(1)}G; Gg_{11}),\ \ldots,$$
$$(g_{m1}G; Gg_{mn(m)})),\ (g_{m1+1,m1}G; Gg_{m1}),\ (g_{m1+2}G; Gg_{m1+1}),\ \ldots,$$
$$(g_{mn(m)-1}G; Gg_{mn(m)-2}),\ (g_{mn(m)}G; Gg_{mn(m)-1}).$$

Hence G is a QF-ring whose Nakayama permutation is the product of the following cyclic permutations:

$$
\begin{pmatrix}
g_{11} & g_{11+1} & g_{11+2} & \cdots & g_{1n(1)} \\
g_{1n(1)} & g_{11} & g_{11+1} & \cdots & g_{1n(1)-1}
\end{pmatrix},
$$

$$\begin{pmatrix} g_{21} & g_{21+1} & g_{21+2} & \cdots & g_{2n(2)} \\ g_{2n(2)} & g_{21} & g_{21+1} & \cdots & g_{2n(2)-1} \end{pmatrix},$$

$$\cdots \qquad \cdots \quad \cdots$$

$$\begin{pmatrix} g_{m1} & g_{m1+1} & g_{m1+2} & \cdots & g_{mn(m)} \\ g_{mn(m)} & g_{m1} & g_{m1+1} & \cdots & g_{mn(m)-1} \end{pmatrix}.$$

Next, under this QF-ring G, we shall make another type of QF-rings. In order to make it, put $Q^\star = k[x]/(x^5)$ and $c = x + (x^5)$ in $Q^\star$. Though we already used $c = x + (x^4)$ in Q, no confusions occur in below arguments. We note that $Q^\star/S(Q^\star) \cong Q$ and each $Q\alpha_{ij,kl}$ and $Q\beta_{ij,kl}$ are $(Q^\star, Q^\star)$-bimodule.

Let $t > 0$. In addition to the above ring

$$T = \begin{pmatrix} T_{11} & \cdots & T_{1m} \\ & \cdots & \\ T_{m1} & \cdots & T_{mm} \end{pmatrix},$$

we make more blocks $t \times t$-matrix T_{00}, $t \times n(i)$-matrix T_{0i} and $n(j) \times t$-matrix T_{j0} as follows: Put

$$T_{00} = \begin{pmatrix} Q^\star & Qc^2\alpha_{11} & \cdots & Qc^2\alpha_{11} \\ Qc^2\alpha_{11} & Q^\star & \cdots & Qc^2\alpha_{11} \\ \cdots & \cdots & \cdots & \cdots \\ \cdots & \cdots & \cdots & \cdots \\ Qc^2\alpha_{11} & \cdots & Qc^2\alpha_{11} & Q^\star \end{pmatrix} \quad (\, t \times t\text{-matrix}\,)$$

$$= \begin{pmatrix} Q^\star & Qc^2\alpha_{01,02} & \cdots & Qc^2\alpha_{01,0t} \\ Qc^2\alpha_{02,01} & Q^\star & \cdots & Qc^2\alpha_{02,0t} \\ \cdots & \cdots & \cdots & \cdots \\ \cdots & \cdots & \cdots & \cdots \\ Qc^2\alpha_{0t,01} \cdots & \cdots & Qc^2\alpha_{0t,0(t-1)} & Q^\star \end{pmatrix},$$

where, of course, $\alpha_{oi,ok} = \alpha_{i0,k0} = \alpha_{11}$ for any $1 \le i, k \le t$. In T_{00}, for $1 \le i \le t$, we can define multiplications: $\langle c^2 \rangle_{1i} \langle c^2 \rangle_{i1} = \langle c^4 \rangle_{11}$, so $\langle c^3 \rangle_{1i} \langle c^2 \rangle_{i1} = \langle c^5 \rangle_{11} = 0$ and $\langle c^2 \rangle_{1i} \langle c^3 \rangle_{i1} = \langle c^5 \rangle_{11} = 0$. Therefore T_{00} canonically becomes a ring.

Put

$$T_{01} = \begin{pmatrix} Qc^2\alpha_{11,11} & Qc\alpha_{11,11+1} & Qc\alpha_{11,11+2} & \cdots & Qc\alpha_{11,1n(1)} \\ \vdots & \vdots & \vdots & & \vdots \\ \vdots & \vdots & \vdots & & \vdots \\ Qc^2\alpha_{11,11} & Qc\alpha_{11,11+1} & Qc\alpha_{11,11+2} & \cdots & Qc\alpha_{11,1n(1)} \end{pmatrix}$$

$$= \begin{pmatrix} Qc^2 & Qc\alpha_{12} & Qc\alpha_{13} & \cdots & Qc\alpha_{1n(1)} \\ \vdots & \vdots & \vdots & & \vdots \\ \vdots & \vdots & \vdots & & \vdots \\ Qc^2 & Qc\alpha_{12} & Qc\alpha_{13} & \cdots & Qc\alpha_{1n(1)} \end{pmatrix} \quad (t \times n(1)\text{-matrix}),$$

$$T_{10} = \begin{pmatrix} Qc^2\alpha_{11,11} & \cdots & Qc^2\alpha_{11,11} \\ Qc\beta_{11+1,11} & \cdots & Qc\beta_{11+1,11} \\ Qc^2\beta_{11+2,11} & \cdots & Qc^2\beta_{11+2,11} \\ Qc^2\beta_{11+3,11} & \cdots & Qc^2\beta_{11+3,11} \\ \cdots \\ Qc^2\beta_{1n(1),11} & \cdots & Qc^2\beta_{1n(1),11} \end{pmatrix}$$

$$= \begin{pmatrix} Qc^2 & \cdots & Qc^2 \\ Qc\beta_{21} & \cdots & Qc\beta_{21} \\ Qc^2\beta_{31} & \cdots & Qc^2\beta_{31} \\ Qc^2\beta_{41} & \cdots & Qc^2\beta_{41} \\ \cdots \\ Qc^2\beta_{n(1)1} & \cdots & Qc^2\beta_{n(1)1} \end{pmatrix} \quad (n(1) \times t\text{-matrix}).$$

For $1 < i$, put

$$T_{0i} = \begin{pmatrix} Qc\alpha_{11,i1} & Qc\alpha_{11,i1+1} & \cdots & Qc\alpha_{11,in(i)} \\ & \cdots & \cdots & \\ Qc\alpha_{11,i1} & Qc\alpha_{11,i1+1} & \cdots & Qc\alpha_{11,in(i)} \end{pmatrix} \quad (t \times n(i)\text{-matrix}).$$

For $1 < j$, put

$$T_{j0} = \begin{pmatrix} Qc\beta_{j1,11} & \cdots & Qc\beta_{j1,11} \\ Qc^2\beta_{j1+1,11} & \cdots & Qc^2\beta_{j1+1,11} \\ Qc^2\beta_{j1+2,11} & \cdots & Qc^2\beta_{j1+2,11} \\ \cdots \\ Qc^2\beta_{jn(j),11} & \cdots & Qc^2\beta_{jn(j),11} \end{pmatrix} \quad (n(j) \times t\text{-matrix}).$$

And put

$$U = \begin{pmatrix} T_{00} & T_{01} & \cdots & T_{0m} \\ T_{10} & T_{11} & \cdots & T_{1m} \\ & \cdots & \cdots & \\ T_{m0} & T_{m1} & \cdots & T_{mm} \end{pmatrix}.$$

Then we can canonically define a multiplication: $(T_{0k}, T_{k0}) \to T_{00}$. And hence we can define a multiplication on U. By this multiplicative operation, U becomes a (basic indecomposable artinian) ring.

Next we make $I_{00} \subsetneq T_{00}$, $I_{ij} \subsetneq T_{ij}$ and $I_{ji} \subsetneq T_{ji}$ for any $0 \le i < j$ as follows:

$$I_{00} = \begin{pmatrix} 0 & Qc^3 & \cdots & Qc^3 \\ Qc^3 & \ddots & \ddots & \vdots \\ \vdots & \ddots & \ddots & Qc^3 \\ Qc^3 & \cdots & Qc^3 & 0 \end{pmatrix} \quad (t \times t\text{-matrix}),$$

$$I_{01} = \begin{pmatrix} Qc^3 & Qc^2\alpha_{12} & \cdots & Qc^2\alpha_{1n(1)} \\ \vdots & \vdots & & \vdots \\ \vdots & \vdots & & \vdots \\ Qc^3 & Qc^2\alpha_{12} & \cdots & Qc^2\alpha_{1n(1)} \end{pmatrix} \quad (t \times n(1)\text{-matrix}),$$

$$I_{10} = \begin{pmatrix} Qc^3 & \cdots & Qc^3 \\ Qc^2\beta_{21} & \cdots & Qc^2\beta_{21} \\ Qc^3\beta_{31} & \cdots & Qc^3\beta_{31} \\ Qc^3\beta_{41} & \cdots & Qc^3\beta_{41} \\ & \cdots & \\ Qc^3\beta_{n(1)1} & \cdots & Qc^3\beta_{n(1)1} \end{pmatrix} \quad (n(1) \times t\text{-matrix}).$$

For $1 < i$,

$$I_{0i} = \begin{pmatrix} Qc^2\alpha_{11,i1} & \cdots & Qc^2\alpha_{11,in(i)} \\ & \cdots & \\ Qc^2\alpha_{11,i1} & \cdots & Qc^2\alpha_{11,in(i)} \end{pmatrix} \quad (t \times n(i)\text{-matrix}).$$

For $1 < j$,

$$I_{j0} = \begin{pmatrix} Qc^2\beta_{j1,11} & \cdots & Qc^2\beta_{j1,11} \\ Qc^3\beta_{j1+1,11} & \cdots & Qc^3\beta_{j1+1,11} \\ & \cdots & \\ Qc^3\beta_{jn(j),11} & \cdots & Qc^3\beta_{jn(j),11} \end{pmatrix} \quad (n(j) \times t\text{-matrix}).$$

For $1 \leq i < j$, we use I_{ij} and I_{ji} mentioned above:

$$I_{ij} = \begin{pmatrix} Qc^2\alpha_{i1,j1} & \cdots & & \cdots & & \cdots & Qc^2\alpha_{i1,jn(j)} \\ Qc^3\alpha_{i1+1,j1} & \ddots & & & & & \vdots \\ \vdots & & \ddots & \ddots & & & \vdots \\ Qc^3\alpha_{in(i),j1} & \cdots & Qc^3\alpha_{in(i),x} & Qc^2\alpha_{in(i),x+1} & \cdots & Qc^2\alpha_{in(i),jn(j)} \end{pmatrix},$$

$$I_{ji} = \begin{pmatrix} Qc^2\beta_{j1,i1} & \cdots & & \cdots & & Qc^2\beta_{j1,in(i)} \\ Qc^3\beta_{j1,ii} & \ddots & & & & \vdots \\ \vdots & & \ddots & \ddots & & \vdots \\ \vdots & & & Qc^3\beta_{s,in(i)-1} & Qc^2\beta_{s,in(i)} & \\ \vdots & & & & \vdots & \vdots \\ Qc^3\beta_{jn(j),i1} & \cdots & Qc^3\beta_{jn(j),in(i)-1} & Qc^2\beta_{jn(j),in(i)} \end{pmatrix}.$$

Put

$$I = \begin{pmatrix} I_{00} & I_{01} & I_{02} & \cdots & & \cdots & I_{0m} \\ I_{10} & 0 & I_{12} & \cdots & & \cdots & I_{1m} \\ I_{20} & I_{21} & 0 & I_{23} & & \cdots & I_{2m} \\ \vdots & & \ddots & \ddots & & & \vdots \\ \vdots & & & \ddots & \ddots & & I_{m-1,m} \\ I_{m0} & I_{m1} & \cdots & \cdots & & I_{m,m-1} & 0 \end{pmatrix}.$$

Then I is an ideal of U. And put $H = U/I$, and H is a basic indecomposable artinian ring.

In H, we use $0i$-th row ($i0$-th column) to mean i-th row (i-th column) for any $1 \leq i \leq t$ and use p-th row (column) to denote $(t+p)$-th row $((t+p)$-th column) for any $p \in \{1 = 11, 11+1, \ldots, 1n(1)\} \cup \cdots \cup \{m(1), m(1)+1, \ldots, mn(m) = w\}$. Put

$$h_{0i} = \langle 1 \rangle_{i0,0i} \text{ for any } i = 1, \ldots, t \text{ and}$$
$$h_p = \langle 1 \rangle_{t+p,t+p} \text{ for any } p \in \{1 = 11, 11+1, \ldots, 1n(1)\} \cup \cdots \cup \{m(1), m(1)+1, \ldots, mn(m) = w\}.$$

Then $Pi(H) = \{h_{0i}\}_{1 \leq i \leq t} \cup \{h_{1i}\}_{1 \leq i \leq n(1)} \cdots \cup \{h_{mi}\}_{1 \leq i \leq n(m)}$ and we can easily see that the following pairs are i-pairs:

$$(h_{01}H; Hh_{01}), \ldots, (h_{0t}H; Hh_{0t}),$$
$$(h_{11}H; Hh_{1n(1)}), \ (h_{11+1}H; Hh_{11}), \ (h_{11+2}H; Hh_{11+1}), \ldots$$
$$\ldots, (h_{1n(1)-1}H; Hh_{1n(1)-2}), \ (h_{1n(1)}H; Hh_{11}), \ldots$$
$$(h_{m1}H; Hh_{mn(m)}), \ (h_{m1+1}H; Hh_{m1}), \ (h_{m1+2}H; Hh_{m1+1}), \ldots$$
$$(h_{mn(m)-1}H; Hh_{mn(m)-2}), \ (h_{mn(m)}H; Hh_{mn(m)-1}).$$

Hence H is a basic indecomposable QF-ring and its Nakayama permutation is the product of the permutations:

$$\begin{pmatrix} h_{01} & \cdots & h_{0t} \\ h_{01} & \cdots & h_{0t} \end{pmatrix}, \ \begin{pmatrix} h_{11} & h_{11+1} & \cdots & h_{1n(1)} \\ h_{1n(1)} & h_{11} & \cdots & h_{1,n(1)-1} \end{pmatrix}, \ldots\ldots$$

$$\ldots\ldots, \begin{pmatrix} h_{m1} & h_{m1+1} & \cdots & h_{mn(m)} \\ h_{mn(m)} & h_{m1} & \cdots & h_{m,n(m)-1} \end{pmatrix}.$$

Here we note that, for $e = h_{01} + \cdots + h_{0t}$, eHe is a basic indecomposable QF-ring with the identity Nakayama permutation:

$$\begin{pmatrix} h_{01} & \cdots & h_{0t} \\ h_{01} & \cdots & h_{0t} \end{pmatrix}$$

Thus we obtain the following result.

Theorem 6.2.1. *For a given permutation*

$$\begin{pmatrix} 1 & \cdots & n \\ \rho(1) & \cdots & \rho(n) \end{pmatrix},$$

we can construct a basic indecomposable QF-ring R with $Pi(R) = \{e_1, \ldots, e_n\}$ with a Nakayama permutation

$$\begin{pmatrix} e_1 & \cdots & e_n \\ e_{\rho(1)} & \cdots & e_{\rho(n)} \end{pmatrix}.$$

Example 6.2.2.

(1) For $w = 4$ and the partition $\{1, 2, 3, 4\} = \{1, 2\} \cup \{3, 4\}$,

$$G = \begin{pmatrix} Q & Q\alpha_{12} & Qc\alpha_{13} & Qc\alpha_{14} \\ Q\beta_{21} & Q & Qc^2\alpha_{23} & Qc\alpha_{24} \\ Qc\beta_{31} & Qc\beta_{32} & Q & Q\alpha_{34} \\ Qc^2\beta_{41} & Qc\beta_{42} & Q\beta_{43} & Q \end{pmatrix} \Big/ \begin{pmatrix} 0 & 0 & Qc^2\alpha_{13} & Qc^2\alpha_{14} \\ 0 & 0 & Qc^3\alpha_{23} & Qc^2\alpha_{24} \\ Qc^2\beta_{31} & Qc^2\beta_{32} & 0 & 0 \\ Qc^3\beta_{41} & Qc^2\beta_{42} & 0 & 0 \end{pmatrix}.$$

(2) For $t = 2$ and $w = 2$,

$$H = \begin{pmatrix} Q^\star & Qc^2 & Qc^2 & Qc\alpha_{12} \\ Qc^2 & Q^\star & Qc^2 & Qc\alpha_{12} \\ Qc^2 & Qc^2 & Q & Q\alpha_{12} \\ Qc\beta_{21} & Qc\beta_{21} & Q\beta_{21} & Q \end{pmatrix} \Big/ \begin{pmatrix} 0 & Qc^3 & Qc^3 & Qc^2\alpha_{12} \\ Qc^3 & 0 & Qc^3 & Qc^2\alpha_{12} \\ Qc^3 & Qc^3 & 0 & 0 \\ Qc^2\beta_{21} & Qc^2\beta_{21} & 0 & 0 \end{pmatrix}.$$

6.3   *QF*-Rings with a Cyclic Nakayama Permutation

In this section, we show the following theorem (cf. Theorem 6.1.4).

Theorem 6.3.1. *Let R be a basic indecomposable QF-ring such that, for any idempotent e of R, eRe is a QF-ring with a cyclic Nakayama permutation. Then there exist a local QF-ring Q, an element $c \in J(Q)$ and $\sigma \in \mathrm{Aut}(Q)$ satisfying $\sigma(c) = c$ and $\sigma(q)c = cq$ for any $q \in Q$ and R is represented as the skew matrix ring:*

$$R \cong \begin{pmatrix} Q & \cdots & Q \\ & \cdots & \\ Q & \cdots & Q \end{pmatrix}_{\sigma,c,n} \quad ,$$

where $n = \sharp Pi(R)$.

Proof. For any $e, f \in Pi(R)$, let (e, f) denote $\mathrm{Hom}_R(eR, fR)$.
We first consider the case $\sharp Pi(R) = 2$. Letting $Pi(R) = \{e, f\}$, we have

$$R = \begin{pmatrix} Q & A \\ B & T \end{pmatrix},$$

where $Q = (e, e)$, $A = (f, e)$, $B = (e, f)$ and $T = (f, f)$. By our assumption, Q is a local *QF*-ring. Since

$$\begin{pmatrix} e & f \\ f & e \end{pmatrix}$$

is a Nakayama permutation, we see that

$$S(eR_R) = S(_RRf) = \begin{pmatrix} 0 & S(A) \\ 0 & 0 \end{pmatrix} \quad \text{and} \quad S(fR_R) = S(_RRe) = \begin{pmatrix} 0 & 0 \\ S(B) & 0 \end{pmatrix}.$$

Noting these facts, we can easily prove the following:

Lemma A.

(1) $\{ a \in A \mid aB = 0 \} = \{ a \in A \mid Ba = 0 \}$,
(2) $\{ b \in B \mid bA = 0 \} = \{ b \in B \mid Ab = 0 \}$.

We denote the sets in (1) and (2) by A^* and B^*, respectively. Note that these are submodules of $_QA_T$ and $_TB_Q$, respectively, and

$$\begin{pmatrix} 0 & A^* \\ 0 & 0 \end{pmatrix} \quad \text{and} \quad \begin{pmatrix} 0 & 0 \\ B^* & 0 \end{pmatrix}$$

are ideals of R.

Now, for the sake of convenience, set

$$\overline{R} = \begin{pmatrix} Q & A \\ B & T \end{pmatrix} \Big/ \begin{pmatrix} 0 & 0 \\ B^* & 0 \end{pmatrix} = \begin{pmatrix} Q & A \\ \overline{B} & T \end{pmatrix}$$

and

$$\overline{r} = r + \begin{pmatrix} 0 & 0 \\ B^* & 0 \end{pmatrix}$$

for each $r \in R$. Then $\{\overline{e}, \overline{f}\}$ is a complete set of orthogonal primitive idempotents of $\overline{R}$ and

$$S(\overline{fR}_R) = \begin{pmatrix} 0 & 0 \\ 0 & S(T) \end{pmatrix}.$$

Since eR_R is injective and $S(\overline{fR}_R)$ is simple, we see

$$\begin{pmatrix} Q & A \\ 0 & 0 \end{pmatrix} \supseteq_{\sim} \begin{pmatrix} 0 & 0 \\ \overline{B} & T \end{pmatrix}$$

as R (and as $\overline{R}$)-modules. Since $S(A_T)$ is simple, it follows that

$$A_T \cong T_T.$$

Hence $\alpha T = A$ for some $\alpha \in A$. If $Q\alpha \subsetneq {}_Q A$, then $S(Q)\alpha = S(Q)Q\alpha = 0$, whence $S(Q)A = 0$, which is a contradiction. Hence $Q\alpha = \alpha T = A$. If $q \in Q$, then there exists $t \in T$ with $q\alpha = \alpha t$. Then the map $\psi : Q \to T$ given by $\psi(q) = t$ is a ring isomorphism. We may therefore replace T by Q to get

$$R = \begin{pmatrix} Q & A \\ B & Q \end{pmatrix}$$

with $q\alpha = \alpha q$ for any $q \in Q$.

Next, similarly, considering the factor ring

$$\begin{pmatrix} Q & A \\ B & Q \end{pmatrix} \Big/ \begin{pmatrix} 0 & A^* \\ 0 & 0 \end{pmatrix} = \begin{pmatrix} Q & \overline{A} \\ B & Q \end{pmatrix},$$

we can obtain $\beta \in B$ and $\sigma \in \mathrm{Aut}(Q)$ such that

$$B = Q\beta = \beta Q \quad \text{and} \quad \beta q = \sigma(q)\beta \text{ for any } q \in Q.$$

Let $c = \alpha\beta$. Noting that $\langle \beta \rangle_{21}(\langle \alpha \rangle_{12}\langle \beta \rangle_{21}) = (\langle \beta \rangle_{21}\langle \alpha \rangle_{12})\langle \beta \rangle_{21}$, we see that

$$\beta(\alpha\beta) = (\beta\alpha)\beta.$$

Suppose that $\alpha\beta - \beta\alpha \neq 0$. Then $(\alpha\beta - \beta\alpha)A \neq 0$ and hence $0 \neq (\alpha\beta - \beta\alpha)\alpha = \alpha\beta\alpha - \beta\alpha\alpha = \alpha\beta\alpha - \alpha\beta\alpha = 0$, a contradiction. Thus $\alpha\beta = \beta\alpha$. Hence

$$\sigma(c) = c.$$

Moreover it is easy to see that $c \in J(Q)$ and $\sigma(q)c = cq$ for any $q \in Q$.

Now, for

$$X = \begin{pmatrix} x_1 & x_2\alpha \\ x_3\beta & x_4 \end{pmatrix}, \; Y = \begin{pmatrix} y_1 & y_2\alpha \\ y_3\beta & y_4 \end{pmatrix} \in R = \begin{pmatrix} Q_1 & Q\alpha \\ Q\beta & Q \end{pmatrix},$$

we have

$$XY = \begin{pmatrix} x_1y_1 + x_2y_3c & (x_1y_2 + x_2y_4)\alpha \\ (x_3\sigma(y_1) + x_4y_3)\beta & x_3\sigma(y_2)c + x_4y_4 \end{pmatrix}.$$

Hence

$$R \cong \begin{pmatrix} Q & Q \\ Q & Q \end{pmatrix}_{\sigma,c}$$

under the map

$$\begin{pmatrix} x_1 & x_2\alpha \\ x_3\beta & x_4 \end{pmatrix} \longmapsto \begin{pmatrix} x_1 & x_2 \\ x_3 & x_4 \end{pmatrix}.$$

We note that the maps

$$\begin{pmatrix} 0 & 0 \\ x\beta & y \end{pmatrix} \longmapsto \begin{pmatrix} xc & y\alpha \\ 0 & 0 \end{pmatrix}, \quad \begin{pmatrix} x & y\alpha \\ 0 & 0 \end{pmatrix} \longmapsto \begin{pmatrix} 0 & 0 \\ \sigma(x)\beta & \sigma(y) \end{pmatrix}$$

determine right R-epimorphisms:

$$\begin{pmatrix} 0 & 0 \\ B & Q \end{pmatrix} \longmapsto \begin{pmatrix} Qc & A \\ 0 & 0 \end{pmatrix}, \quad \begin{pmatrix} Q & A \\ 0 & 0 \end{pmatrix} \longmapsto \begin{pmatrix} 0 & 0 \\ B & Qc \end{pmatrix}$$

with kernels

$$\begin{pmatrix} 0 & 0 \\ B^* & 0 \end{pmatrix}, \quad \begin{pmatrix} 0 & A^* \\ 0 & 0 \end{pmatrix},$$

respectively. Hence it follows that

$$Qc_Q \cong \overline{A}_Q \cong \overline{B}_Q.$$

Next we consider the case $\#Pi(R) = 3$, say $Pi(R) = \{e_1, e_2, e_3\}$. We may assume that

$$\begin{pmatrix} e_1 & e_2 & e_3 \\ e_3 & e_1 & e_2 \end{pmatrix}$$

is the Nakayama permutation. Then we can represent R as

$$R = \begin{pmatrix} (e_1, e_1) & (e_2, e_1) & (e_3, e_1) \\ (e_1, e_2) & (e_2, e_2) & (e_3, e_2) \\ (e_1, e_3) & (e_2, e_3) & (e_3, e_3) \end{pmatrix} = \begin{pmatrix} Q_1 & A_{12} & A_{13} \\ A_{21} & Q_2 & A_{23} \\ A_{31} & A_{32} & Q_3 \end{pmatrix}.$$

Put $Q = Q_1$. Considering

$$\begin{pmatrix} Q_1 & A_{12} \\ A_{21} & Q_2 \end{pmatrix}, \quad \begin{pmatrix} Q_1 & A_{13} \\ A_{31} & Q_3 \end{pmatrix}, \quad \begin{pmatrix} Q_2 & A_{23} \\ A_{32} & Q_3 \end{pmatrix},$$

we can assume that $Q = Q_2 = Q_3$ by the same argument as above; hence

$$R = \begin{pmatrix} Q & A_{12} & A_{13} \\ A_{21} & Q & A_{23} \\ A_{31} & A_{32} & Q \end{pmatrix},$$

and then $A_{ij\,Q} \cong Q_Q$ for each distinct $i, j \in \{1, 2, 3\}$.

Noting that

$$S(e_1 R_R) = S(_R R e_3) = \begin{pmatrix} 0 & 0 & S(A_{13}) \\ 0 & 0 & 0 \\ 0 & 0 & 0 \end{pmatrix},$$

$$S(e_2 R_R) = S(_R R e_1) = \begin{pmatrix} 0 & 0 & 0 \\ S(A_{21}) & 0 & 0 \\ 0 & 0 & 0 \end{pmatrix},$$

$$S(e_3 R_R) = S(_R R e_2) = \begin{pmatrix} 0 & 0 & 0 \\ 0 & 0 & 0 \\ 0 & S(A_{32}) & 0 \end{pmatrix},$$

we prove the following:

Lemma B.

(1) $\{ x \in A_{32} \mid x A_{23} = 0 \} = \{ x \in A_{32} \mid A_{23} x = 0 \}$
$$= \{ x \in A_{32} \mid x A_{21} = 0 \}$$
$$= \{ x \in A_{32} \mid A_{13} x = 0 \}.$$

(2) $\{ x \in A_{21} \mid x A_{12} = 0 \} = \{ x \in A_{21} \mid x A_{13} = 0 \}$
$$= \{ x \in A_{21} \mid A_{12} x = 0 \}$$
$$= \{ x \in A_{21} \mid A_{32} x = 0 \}.$$

(3) $\{x \in A_{13} \mid xA_{31} = 0\} = \{x \in A_{13} \mid A_{31}x = 0\}$
$$= \{x \in A_{13} \mid xA_{32} = 0\}$$
$$= \{x \in A_{13} \mid A_{21}x = 0\}.$$

Proof of Lemma B. (1). By Lemma A, $\{x \in A_{32} \mid xA_{23} = 0\} = \{x \in A_{32} \mid A_{23}x = 0\}$.

Now let $x \in A_{32}$ such that $xA_{23} = 0$. If $xA_{21} \neq 0$, then $A_{23}xA_{21} \neq 0$, whence $A_{23}x \neq 0$, a contradiction. If $A_{13}x \neq 0$, then $A_{13}xA_{23} \neq 0$, whence $xA_{23} \neq 0$, a contradiction. Thus $\{x \in A_{32} \mid xA_{23} = 0\} \subseteq \{x \in A_{32} \mid xA_{21} = 0\}$ and $\{x \in A_{32} \mid xA_{23} = 0\} \subseteq \{x \in A_{32} \mid A_{13}x = 0\}$. Let $x \in A_{32}$ such that $xA_{21} = 0$. If $xA_{23} \neq 0$, then it follows from $_QQ \cong {_Q}A_{31}$ that $xA_{23}A_{31} \neq 0$, so $xA_{21} \neq 0$, a contradiction. Thus $\{x \in A_{32} \mid xA_{23} = 0\} = \{x \in A_{32} \mid xA_2 = 0\}$.

Now let $x \in A_{32}$ be such that $A_{13}x = 0$. If $xA_{23} \neq 0$, then $A_{13}xA_{23} \neq 0$, so $A_{13}x \neq 0$, a contradiction. Thus $\{x \in A_{32} \mid xA_{23} = 0\} = \{x \in A_{32} \mid A_{13}x = 0\}$.

Similarly we can prove (2) and (3).

We denote the sets in (1), (2) and (3) of Lemma B by A_{32}^*, A_{21}^* and A_{13}^*, respectively. It is easy to see that $_QA_{32Q}^*$, $_QA_{21Q}^*$ and $_QA_{13Q}^*$ are submodules of $_QA_{32Q}$, $_QA_{21Q}$ and $_QA_{13Q}$, respectively. Put

$$X_{13} = \begin{pmatrix} 0 & 0 & A_{13}^* \\ 0 & 0 & 0 \\ 0 & 0 & 0 \end{pmatrix}, \quad X_{21} = \begin{pmatrix} 0 & 0 & 0 \\ A_{21}^* & 0 & 0 \\ 0 & 0 & 0 \end{pmatrix}, \quad X_{32} = \begin{pmatrix} 0 & 0 & 0 \\ 0 & 0 & 0 \\ 0 & A_{32}^* & 0 \end{pmatrix}.$$

These are ideals of R. Put $X = X_{13} + X_{21} + X_{32}$. We consider the factor ring $\overline{R} = R/X$ and put $\overline{r} = r + X$ for any $r \in R$. Then we represent $\overline{R}$ as

$$\overline{R} = \begin{pmatrix} Q_1 & A_{12} & \overline{A_{13}} \\ \overline{A_{21}} & Q_2 & A_{23} \\ A_{31} & \overline{A_{32}} & Q_3 \end{pmatrix} = \begin{pmatrix} Q_1 & A_{12} & A_{13} \\ A_{21} & Q_2 & A_{23} \\ A_{31} & A_{32} & Q_3 \end{pmatrix} \Big/ \begin{pmatrix} 0 & 0 & X_{13} \\ X_{21} & 0 & 0 \\ 0 & X_{32} & 0 \end{pmatrix}.$$

It is easy to see that

$$S(\overline{e_1}\overline{R}_{\overline{R}}) = S(\overline{e_1}\overline{R}_R) = \begin{pmatrix} 0 & S(A_{12}) & 0 \\ 0 & 0 & 0 \\ 0 & 0 & 0 \end{pmatrix},$$

$$S(\overline{e_2}\overline{R}_{\overline{R}}) = S(\overline{e_2}\overline{R}_R) = \begin{pmatrix} 0 & 0 & 0 \\ 0 & 0 & S(A_{23}) \\ 0 & 0 & 0 \end{pmatrix},$$

$$S(\overline{e_3 R_{\overline{R}}}) = S(\overline{e_3 R_R}) = \begin{pmatrix} 0 & 0 & 0 \\ 0 & 0 & 0 \\ S(A_{31}) & 0 & 0 \end{pmatrix}.$$

Therefore, as each $e_i R_R$ is injective, there are left multiplications $\langle \theta_{23} \rangle_{23} : \overline{e_3 R_R} \to e_2 R_R$, $\langle \theta_{12} \rangle_{12} : \overline{e_2 R_R} \to e_1 R_R$ and $\langle \theta_{31} \rangle_{31} : \overline{e_1 R_R} \to e_3 R_R$ which are monomorphisms.

We put $\gamma_1 = \langle \theta_{31} \rangle_{31} \eta_1$, $\gamma_2 = \langle \theta_{12} \rangle_{12} \eta_2$ and $\gamma_3 = \langle \theta_{23} \rangle_{23} \eta_3$, where η_i is the canonical homomorphism $e_i R_R \to \overline{e_i R_R}$.

Noting that

$$\gamma_1 \left(\begin{pmatrix} 0 & A_{12} & 0 \\ 0 & 0 & 0 \\ 0 & 0 & 0 \end{pmatrix} \right) = \begin{pmatrix} 0 & 0 & 0 \\ 0 & 0 & 0 \\ 0 & A_{32} & 0 \end{pmatrix},$$

$$\gamma_2 \left(\begin{pmatrix} 0 & 0 & 0 \\ 0 & 0 & A_{23} \\ 0 & 0 & 0 \end{pmatrix} \right) = \begin{pmatrix} 0 & 0 & A_{13} \\ 0 & 0 & 0 \\ 0 & 0 & 0 \end{pmatrix},$$

$$\gamma_3 \left(\begin{pmatrix} 0 & 0 & 0 \\ 0 & 0 & 0 \\ A_{31} & 0 & 0 \end{pmatrix} \right) = \begin{pmatrix} 0 & 0 & 0 \\ A_{21} & 0 & 0 \\ 0 & 0 & 0 \end{pmatrix}$$

and using Lemma B, we can show the following:

Lemma C.

(1) $\{ x \in A_{31} \mid x A_{12} = 0 \} = \{ x \in A_{31} \mid x A_{13} = 0 \}$
$$= \{ x \in A_{31} \mid A_{13} x = 0 \}$$
$$= \{ x \in A_{31} \mid A_{23} x = 0 \}.$$

(2) $\{ x \in A_{23} \mid x A_{31} = 0 \} = \{ x \in A_{23} \mid x A_{32} = 0 \}$
$$= \{ x \in A_{23} \mid A_{32} x = 0 \}$$
$$= \{ x \in A_{23} \mid A_{12} x = 0 \}.$$

(3) $\{ x \in A_{12} \mid x A_{21} = 0 \} = \{ x \in A_{12} \mid x A_{23} = 0 \}$
$$= \{ x \in A_{12} \mid A_{31} x = 0 \}$$
$$= \{ x \in A_{12} \mid A_{21} x = 0 \}.$$

Proof of Lemma C. (1) Put $K_1 = \{\, x \in A_{31} \mid x A_{12} = 0 \,\}$, $K_2 = \{\, x \in A_{31} \mid x A_{13} = 0 \,\}$, $K_3 = \{\, x \in A_{31} \mid A_{13} x = 0 \,\}$ and $K_4 = \{\, x \in A_{31} \mid A_{23} x = 0 \,\}$. By Lemma A, we see $K_2 = K_3$. And using η_2, we see $K_3 = K_4$. To show $K_1 = K_2$, let $x_{31} \in K_1$. If $x_{31} A_{13} \neq 0$, then $x_{31} A_{13} A_{32} \neq 0$ since

$$S(e_3 R) = \begin{pmatrix} 0 & 0 & 0 \\ 0 & 0 & 0 \\ 0 & S(A_{32}) & 0 \end{pmatrix}.$$

But $x_{31} A_{13} A_{32} \subseteq x_{31} A_{12} = 0$, a contradiction. So $x_{31} A_{13} = 0$ and hence $x_{31} \in K_2$. Conversely let $x_{31} \in K_2$; $x_{31} A_{13} = 0$. If $0 \neq x_{31} A_{12} (\subseteq A_{32})$, then $\langle \theta_{31} \rangle_{31}^{-1} (x_{31} A_{12}) \subseteq A_{12}$. Hence $0 \neq \langle \theta_{31} \rangle_{31}^{-1} (x_{31} A_{12}) A_{23}$. But $\langle \theta_{31} \rangle_{31}^{-1} (x_{31} A_{12}) A_{23} = \langle \theta_{31} \rangle_{31}^{-1} (x_{31}) A_{12} A_{23} \subseteq \langle \theta_{31} \rangle_{31}^{-1} (x_{31}) A_{13} = 0$, a contradiction. Hence $x_{31} A_{12} = 0$ and $x_{31} \in K_1$ as desired. (2) and (3) can be proved by the same argument.

We denote the sets in (1), (2) and (3) of Lemma C by A_{31}^*, A_{23}^* and A_{12}^*, respectively, and put

$$X_{31} = \begin{pmatrix} 0 & 0 & 0 \\ 0 & 0 & 0 \\ A_{31}^* & 0 & 0 \end{pmatrix}, \quad X_{23} = \begin{pmatrix} 0 & 0 & 0 \\ 0 & 0 & A_{23}^* \\ 0 & 0 & 0 \end{pmatrix}, \quad X_{12} = \begin{pmatrix} 0 & A_{12}^* & 0 \\ 0 & 0 & 0 \\ 0 & 0 & 0 \end{pmatrix}.$$

Then these X_{ij} are ideals of $\overline{R}$ and

$$\gamma_3(X_{31}) = X_{21}, \quad \gamma_2(X_{23}) = X_{13}, \quad \gamma_1(X_{12}) = X_{32}.$$

Lemma D. *There exist* $\alpha_{12} \in A_{12}$, $\alpha_{21} \in A_{21}$, $c \in J(Q)$ *and* $\sigma \in \mathrm{Aut}(Q)$ *such that*

(1)   (i) $c = \alpha_{12} \alpha_{21} = \alpha_{21} \alpha_{12}$,
     (ii) $\alpha_{12} q = q \alpha_{12}$ *for any* $q \in Q$,
     (iii) $\sigma(q) \alpha_{21} = \alpha_{21} q$ *for any* $q \in Q$,

(2) $\begin{pmatrix} Q & A_{12} \\ A_{21} & Q \end{pmatrix} \simeq \begin{pmatrix} Q & Q \\ Q & Q \end{pmatrix}_{\sigma, c}$ *under* $\begin{pmatrix} q_{11} & q_{12} \alpha_{12} \\ q_{21} \alpha_{21} & q_{22} \end{pmatrix} \longmapsto \begin{pmatrix} q_{11} & q_{12} \\ q_{21} & q_{22} \end{pmatrix},$

(3)   (i) $\mathrm{Im}\, \theta_{23} = \begin{pmatrix} 0 & 0 & 0 \\ A_{21} & cQ & A_{23} \\ 0 & 0 & 0 \end{pmatrix},$

     (ii) $\mathrm{Im}\, \theta_{12} = \begin{pmatrix} cQ & A_{12} & A_{13} \\ 0 & 0 & 0 \\ 0 & 0 & 0 \end{pmatrix},$

$$(iii) \quad \text{Im}\,\theta_{31} = \begin{pmatrix} 0 & 0 & 0 \\ 0 & 0 & 0 \\ A_{31} & A_{32} & cQ \end{pmatrix},$$

(4) $\text{Im}\,\theta_{31}$, $\text{Im}\,\theta_{12}$, $\text{Im}\,\theta_{23}$, $\text{Im}\,\eta_3$, $\text{Im}\,\eta_2$ *and* $\text{Im}\,\eta_1$ *are quasi-injective (or, equivalently, fully invariant) submodules of* $\overline{e_3 R}_R$, $\overline{e_1 R}_R$, $\overline{e_2 R}_R$, $e_1 R_R$, $e_3 R_R$ *and* $e_2 R_R$, *respectively.*

Proof of Lemma D. Considering

$$\begin{pmatrix} Q & A_{12} \\ A_{21} & Q \end{pmatrix},$$

we get $\alpha_{12} \in A_{12}$, $\alpha_{21} \in A_{21}$, $c \in J(Q)$ and $\sigma \in \text{Aut}(Q)$ which satisfy (1) and (2). Furthermore considering

$$\begin{pmatrix} Q & A_{23} \\ A_{32} & Q \end{pmatrix}, \quad \begin{pmatrix} Q & A_{13} \\ A_{31} & Q \end{pmatrix},$$

we get $c_2, c_3 \in J(Q)$ and $\sigma_2, \sigma_3 \in \text{Aut}(Q)$ for which

$$\begin{pmatrix} Q & A_{23} \\ A_{32} & Q \end{pmatrix} \cong \begin{pmatrix} Q & Q \\ Q & Q \end{pmatrix}_{\sigma_2, c_2}, \quad \begin{pmatrix} Q & A_{13} \\ A_{31} & Q \end{pmatrix} \cong \begin{pmatrix} Q & Q \\ Q & Q \end{pmatrix}_{\sigma_3, c_3}.$$

By Lemma A, we have

$$_Q\overline{A_{12}}_Q \cong {_Q}cQ_Q, \quad _Q\overline{A_{21}}_Q \cong {_Q}cQ_Q, \quad _Q\overline{A_{13}}_Q \cong {_Q}c_3 Q_Q,$$

$$_Q\overline{A_{31}}_Q \cong {_Q}c_3 Q_Q, \quad _Q\overline{A_{32}}_Q \cong {_Q}c_2 Q_Q, \quad _Q\overline{A_{23}}_Q \cong {_Q}c_2 Q_Q,$$

where $\overline{A_{ij}} = A_{ij}/A_{ij}^*$. Now set $Y = X + X_{13} + X_{21} + X_{32}$. Then Y is an ideal of R and we can represent the factor ring $\overline{R} = R/Y$ as follows:

$$\overline{R} = \begin{pmatrix} Q_1 & \overline{A_{12}} & \overline{A_{13}} \\ \overline{A_{21}} & Q_2 & \overline{A_{23}} \\ \overline{A_{31}} & \overline{A_{32}} & Q_3 \end{pmatrix}.$$

Since

$$e_1 R/Y = \begin{pmatrix} Q & \overline{A_{12}} & \overline{A_{13}} \\ 0 & 0 & 0 \\ 0 & 0 & 0 \end{pmatrix} \cong \begin{pmatrix} 0 & 0 & 0 \\ 0 & 0 & 0 \\ A_{31} & \overline{A_{32}} & c_3 Q \end{pmatrix} \subseteq e_3 R/X_{32},$$

$$e_2 R/Y = \begin{pmatrix} 0 & 0 & 0 \\ \overline{A_{21}} & Q & \overline{A_{23}} \\ 0 & 0 & 0 \end{pmatrix} \cong \begin{pmatrix} cQ & A_{12} & \overline{A_{13}} \\ 0 & 0 & 0 \\ 0 & 0 & 0 \end{pmatrix} \subseteq e_1 R/X_{13}$$

and

$$e_3R/Y = \begin{pmatrix} 0 & 0 & 0 \\ 0 & 0 & 0 \\ \overline{A_{31}} & \overline{A_{32}} & Q \end{pmatrix} \cong \begin{pmatrix} 0 & 0 & 0 \\ \overline{A_{21}} & c_2Q & A_{23} \\ 0 & 0 & 0 \end{pmatrix} \subseteq e_2R/X_{21},$$

we see that $\overline{A_{ij}}_Q \cong \overline{A_{kj}}_Q$ for $i \neq k$ and $cQ_Q \cong c_2Q_Q \cong c_3Q_Q$. Since cQ_Q, c_2Q_Q and c_3Q_Q are fully invariant submodules of Q, we see that $cQ = c_2Q = c_3Q$. Hence (3) is proved. (4) is clear.

Lemma E.

(1)   (*i*)   *For any $\psi \in (e_3, e_2)$, $\operatorname{Im}\psi \subseteq \operatorname{Im}\theta_{23}$.*
      (*ii*)   *For any $\psi \in (e_2, e_1)$, $\operatorname{Im}\psi \subseteq \operatorname{Im}\theta_{12}$.*
    (*iii*)   *For any $\psi \in (e_1, e_3)$, $\operatorname{Im}\psi \subseteq \operatorname{Im}\theta_{31}$.*
(2)   (*i*)   *For any $\psi \in (e_3, e_1)$, $\operatorname{Im}\psi \subseteq \operatorname{Im}\theta_{12}\theta_{23}$.*
      (*ii*)   *For any $\psi \in (e_2, e_3)$, $\operatorname{Im}\psi \subseteq \operatorname{Im}\theta_{31}\theta_{12}$.*
    (*iii*)   *For any $\psi \in (e_1, e_2)$, $\operatorname{Im}\psi \subseteq \operatorname{Im}\theta_{23}\theta_{31}$.*

Proof of Lemma E. (1). (i). Let $\psi \in (e_3, e_2)$. If $x \in A_{32}^*$ and $\psi(\langle x \rangle_{32}) \neq 0$, then

$$\psi(\langle x \rangle_{32}) = \begin{pmatrix} 0 & 0 & 0 \\ 0 & 0 & A_{23} \\ 0 & 0 & 0 \end{pmatrix} \neq 0 \quad \text{but} \quad \langle x \rangle_{32} \begin{pmatrix} 0 & 0 & 0 \\ 0 & 0 & A_{23} \\ 0 & 0 & 0 \end{pmatrix} = 0,$$

which is impossible. Hence $\psi(\{ \langle x \rangle_{32} \mid x \in A_{32}^* \}) = 0$ and there exists an epimorphism from

$$\operatorname{Im}\theta_3 = \begin{pmatrix} 0 & 0 & 0 \\ A_{21} & cQ & A_{23} \\ 0 & 0 & 0 \end{pmatrix} \quad \text{to} \quad \begin{pmatrix} 0 & 0 & 0 \\ 0 & 0 & 0 \\ A_{31} & A_{32} & Q \end{pmatrix} / \operatorname{Ker}\psi \cong \operatorname{Im}\psi.$$

Since $\operatorname{Im}\theta_3$ is a fully invariant submodule of e_2R, we see $\operatorname{Im}\psi \subseteq \operatorname{Im}\theta_{23}$.
(ii), (iii). These can be shown in the same way as (i).
(2). (i). Let $\psi \in (e_3, e_1)$. Then

$$\psi\left(\begin{pmatrix} 0 & 0 & 0 \\ 0 & 0 & 0 \\ A_{31}^* & A_{32}^* & 0 \end{pmatrix}\right) = 0.$$

This implies $\operatorname{Im}\psi \subseteq \operatorname{Im}\theta_{12}\theta_{23}$.
(ii), (iii). These can be shown in the same way as (i).
This completes the proof of Lemma E.

Now consider the factor ring $\overline{R} = R/X_{32}$ and denote $r + X_{32}$ by $\overline{r}$ for each $r \in R$. We can represent $\overline{R}$ as

$$\overline{R} = \overline{e_1 R} \oplus \overline{e_2 R} \oplus \overline{e_3 R} = \begin{pmatrix} (e_1, e_1) & (e_2, e_1) & (\overline{e_3}, e_1) \\ (e_1, e_2) & (e_2, e_2) & (\overline{e_3}, e_2) \\ (e_1, \overline{e_3}) & (e_2, \overline{e_3}) & (\overline{e_3}, \overline{e_3}) \end{pmatrix} = \begin{pmatrix} Q & A_{12} & A_{13} \\ A_{21} & Q & A_{23} \\ A_{31} & \overline{A_{32}} & Q \end{pmatrix},$$

where $\overline{A_{32}} = A_{32}/A_{32}^*$.

Lemma F. *The map*

$$\tau = \begin{pmatrix} \tau_{11} & \tau_{12} & \tau_{13} \\ \tau_{21} & \tau_{22} & \tau_{23} \\ \tau_{31} & \tau_{32} & \tau_{33} \end{pmatrix} : \begin{pmatrix} Q & A_{12} & A_{12} \\ A_{21} & Q & Q \\ A_{21} & I & Q \end{pmatrix} \longrightarrow \overline{R} = \begin{pmatrix} Q & A_{12} & A_{13} \\ A_{21} & Q & A_{23} \\ A_{31} & \overline{A_{32}} & Q \end{pmatrix},$$

where $I = \theta_3 \overline{A_{32}}$, given by

$$\begin{pmatrix} q_{11} & q_{12} & p_{12} \\ q_{21} & q_{22} & p_{22} \\ t_{21} & t_{22} & y_{22} \end{pmatrix} \longmapsto \begin{pmatrix} q_{11} & q_{12} & p_{12}\theta_3 \\ q_{21} & q_{22} & p_{22}\theta_3 \\ \theta_3^{-1}t_{21} & \theta_3^{-1}t_{22} & \theta_3^{-1}y_{22}\theta_3 \end{pmatrix}$$

is a ring isomorphism.

Proof of Lemma F. By Lemma E, τ is well defined and, furthermore, a ring monomorphism. Noting that $e_1 R_R$ is injective, we see that τ_{13} is an onto map. Noting $e_2 R_R$ is injective, we see that τ_{23} and τ_{33} are also onto maps. It is easy to see that τ_{31} is an onto map and τ_{32} is clearly onto. Hence τ is a ring isomorphism. Hence Lemma F is shown.

By Lemma F, $\overline{A_{32}}Q \cong I_Q$. And hence $I = cQ$. Under the isomorphism

$$\tau : \begin{pmatrix} Q & A_{12} & A_{12} \\ A_{21} & Q & Q \\ A_{21} & I & Q \end{pmatrix} \longrightarrow \begin{pmatrix} Q & A_{12} & A_{13} \\ A_{21} & Q & A_{23} \\ A_{31} & \overline{A_{32}} & Q \end{pmatrix},$$

we put $\langle \alpha_{31} \rangle_{31} = \tau(\langle \alpha_{21} \rangle_{31})$, $\langle \alpha_{13} \rangle_{13} = \tau(\langle \alpha_{12} \rangle_{13})$, $\alpha_{32} = \alpha_{31}\alpha_{12}$ and $\langle \alpha_{23} \rangle_{23} = \tau(\langle 1 \rangle_{23})$. Since A_{32}^* is a small submodule of A_{32}, we see that $\alpha_{32}Q = A_{32}$. Hence R is represented as

$$R \cong \begin{pmatrix} Q & \alpha_{12}Q & \alpha_{12}Q \\ \alpha_{21}Q & Q & \alpha_{23}Q \\ \alpha_{31}Q & \alpha_{32}Q & Q \end{pmatrix}$$

with relations:

(i) $c = \alpha_{21}\alpha_{12} = \alpha_{12}\alpha_{21},$

(*ii*) $\sigma(c) = c$,

(*iii*) $\alpha_{12}q = q\alpha_{12}$ and $\sigma(q)\alpha_{21} = \alpha_{21}q$ for any $q \in Q$.

Putting $\alpha_{ii} = 1$ for each $i = 1, 2, 3$, we obtain the following relations (∗) for any $i, j \in \{1, 2, 3\}$:

$$(*)\cdots\begin{cases} \text{When} \quad i > j, \begin{cases} \sigma(q)\alpha_{ij} = \alpha_{ij}q \ \text{ for any } \ q \in Q, \\ \alpha_{ij}\alpha_{jk} = \begin{cases} \alpha_{ik} & \text{for } i > k \geq j, \\ \alpha_{ik}c & \text{for } k \geq i \text{ or } j > k, \end{cases} \end{cases} \\[2em] \text{When} \quad i = j, \begin{cases} q\alpha_{ij} = \alpha_{ij}q \ \text{ for any } \ q \in Q, \\ \alpha_{ij}\alpha_{jk} = \alpha_{ik}, \end{cases} \\[2em] \text{When} \quad i < j, \begin{cases} q\alpha_{ij} = \alpha_{ij}q \ \text{ for any } \ q \in Q, \\ \alpha_{ij}\alpha_{jk} = \begin{cases} \alpha_{ik}c & \text{for } i \leq k < j, \\ \alpha_{ik} & \text{for } k < i \text{ or } j \leq k. \end{cases} \end{cases} \end{cases}$$

By these relations, we see that

$$R \cong \begin{pmatrix} Q & Q & Q \\ Q & Q & Q \\ Q & Q & Q \end{pmatrix}_{\sigma, c}$$

under the map

$$\begin{pmatrix} q_{11} & q_{12} & q_{13} \\ q_{21} & q_{22} & q_{23} \\ q_{31} & q_{32} & q_{33} \end{pmatrix} \longmapsto \begin{pmatrix} q_{11}\alpha_{11} & q_{12}\alpha_{12} & q_{13}\alpha_{13} \\ q_{21}\alpha_{21} & q_{22}\alpha_{22} & q_{23}\alpha_{23} \\ q_{31}\alpha_{31} & q_{32}\alpha_{32} & q_{33}\alpha_{33} \end{pmatrix}.$$

Proceeding by induction on the cardinality of $Pi(R)$, we may now assume that the statement of our theorem is true for R with $\#Pi(R) = n - 1$ and consider the case of R with $\#Pi(R) = n$, say $Pi(R) = \{e_1, e_2, \ldots, e_n\}$. We may assume that

$$\begin{pmatrix} e_1 & e_2 & \cdots & e_n \\ e_n & e_1 & \cdots & e_{n-1} \end{pmatrix}$$

is the Nakayama permutation. We represent R as

$$R = \begin{pmatrix} Q_1 & A_{12} & A_{13} & \cdots & A_{1n} \\ A_{21} & Q_2 & A_{23} & \cdots & A_{2n} \\ \cdots & \cdots & \cdots & \cdots & \cdots \\ A_{n1} & A_{n2} & A_{n3} & \cdots & Q_n \end{pmatrix},$$

where $Q_i = (e_i, e_i)$ and $A_{ij} = (e_j, e_i)$. By our arguments above, we may assume that $Q_1 = Q_2 = \cdots = Q_n$. Put $Q = Q_i$. Then we note that $_Q A_{ij Q} \cong {}_Q Q_Q$ for each $i, j \in \{1, \ldots, n\}$.

Now we look especially at the first minor matrix

$$R_0 = \begin{pmatrix} Q & A_{12} & \cdots & A_{1,n-1} \\ A_{21} & Q & \cdots & A_{2,n-1} \\ \cdots & \cdots & \cdots & \cdots \\ A_{n-1,1} & A_{n-1,2} & \cdots & Q \end{pmatrix}.$$

By the induction hypothesis, R_0 is isomorphic to a skew matrix ring over a local ring Q with respect to a triple $(\sigma, c, n-1)$, where $\sigma \in \mathrm{Aut}(Q)$ and $c \in J(Q)$. Thus there exist $\alpha_{ij} \in A_{ij}$ and $\alpha_{ii} \in Q$ for each $i, j \in \{1, \ldots, n-1\}$ for which the relations $(*)$ hold. Now we consider an extension ring

$$R_1 = \left(\begin{array}{ccc|c} & & & A_{1,n-1} \\ & R_0 & & \vdots \\ & & & A_{n-2,n-1} \\ & & & Q \\ \hline A_{n-1,1} & \cdots & A_{n-1,n-2}\, cQ & Q \end{array} \right)$$

of R_0. By arguments similar to those used for the case $n = 3$, we infer that there exists a matrix ring isomorphism $\tau = (\tau_{ij})$ from R_1 to

$$R_2 = \left(\begin{array}{ccc|c} & & & A_{1,n} \\ & R_0 & & \vdots \\ & & & A_{n-2,n} \\ & & & A_{n-1,n} \\ \hline A_{n,1}\ A_{n,n-2}\ \overline{A_{n,n-1}} & & & Q \end{array} \right),$$

where $\overline{A_{nn-1}} = A_{nn-1}/A^*_{nn-1}$ and

$$A^*_{nn-1} = \{\, x \in A_{nn-1} \mid x A_{n-1j} = 0 \text{ for any } j \in \{1, \ldots, n\} - \{n-1\} \,\}$$
$$= \{\, x \in A_{nn-1} \mid A_{in} x = 0 \text{ for any } i \in \{1, \ldots, n-1\} \,\}.$$

We put $\langle \alpha_{in} \rangle_{in} = \tau(\langle \alpha_{i,n-1} \rangle_{in})$ and $\langle \alpha_{nj} \rangle_{nj} = \tau(\langle \alpha_{n-1,j} \rangle_{nj})$ for each $i, j \in \{1, \ldots, n-2\}$ and further put $\alpha_{n,n-1} = \alpha_{n,n-2} \alpha_{n-2,n-1} \in A_{n,n-1}$ and $\langle \alpha_{n-1,n} \rangle_{n-1,n} = \tau(\langle 1 \rangle_{n-1,n})$. Since $A^*_{n,n-1}$ is a small submodule of $A_{n,n-1}$, we see that $\alpha_{n,n-1} Q = A_{n,n-1}$.

Since the relations $(*)$ hold for $\{\, \alpha_{ij} \mid 1 \le i, j \le n-1 \,\}$ with respect to σ and c, we can also see that the relations $(*)$ hold for $\{\, \alpha_{ij} \mid 1 \le i, j \le n \,\}$

with respect to σ and c. Accordingly R is isomorphic to the skew matrix ring

$$\begin{pmatrix} Q \cdots Q \\ \cdots \\ Q \cdots Q \end{pmatrix}_{\sigma,c,n}$$

by the map

$$(q_{ij}) \longleftrightarrow (q_{ij}\alpha_{ij}).$$

This has at last completed the proof of Theorem 6.3.1. $\square$

6.4   Strongly QF-Rings

A ring R is said to be a *strongly QF-ring* if eRe is a QF-ring for any idempotent $e \in R$. Using skew matrix rings, we shall give characterizations of strongly QF-rings.

Theorem 6.4.1. *For a basic indecomposable QF-ring R with $Pi(R) = \{e_1, \ldots, e_n\}$, the following are equivalent:*

(1) R *is a* strongly QF-ring.

(2) R *is a weakly symmetric QF-ring or R can be represented as a skew matrix ring over $Q = e_1 R e_1$:*

$$R = \begin{pmatrix} Q \cdots Q \\ \cdots \\ Q \cdots Q \end{pmatrix}_{\sigma,c,n},$$

where $\sigma \in \mathrm{Aut}(Q)$ and $c \in J(Q)$.

(3) R *is either weakly symmetric or, for any idempotent e, eRe is a QF-ring with a cyclic Nakayama permutation.*

(4) *The endomorphism ring of every finitely generated projective right (left) R-module is a QF-ring.*

For a proof (1) $\Rightarrow$ (3), we use the following two lemmas.

Lemma A. (Lemma A in the proof of Theorem 6.3.1.) *Let R be a basic indecomposable QF-ring with $Pi(R) = \{e, f\}$. Assume that hRh is* *a QF-ring for any idempotent h of R and the Nakayama permutation of $\{\,e,\,f\,\}$ is*

$$\begin{pmatrix} e & f \\ f & e \end{pmatrix}.$$

We represent R as

$$R = \begin{pmatrix} Q & A \\ B & T \end{pmatrix},$$

where $Q = eRe$, $T = fRf$, $A = eRf$ and $B = fRe$. Put $X = \{\, a \in A \mid aB = 0 \,\}$ and $Y = \{\, b \in B \mid Ab = 0 \,\}$. Then

(1) $X = \{\, a \in A \mid Ba = 0 \,\}$, $Y = \{\, b \in B \mid bA = 0 \,\}$ *and*

(2) $\langle\, X \,\rangle_{21} = \begin{pmatrix} 0 & 0 \\ X & 0 \end{pmatrix}$ *and* $\langle\, Y \,\rangle_{12} = \begin{pmatrix} 0 & Y \\ 0 & 0 \end{pmatrix}$ *are ideals of R.*

Lemma B. *Let R be a basic indecomposable QF-ring with $Pi(R) = \{\, e_1,\ e_2,\ e_3 \,\}$. If the Nakayama permutation of $Pi(R)$ is not identity and eRe is a QF-ring for any idempotent of R, then the Nakayama permutation of R is cyclic.*

Proof. Since the Nakayama permutation of R is non-identity, we may assume that

$$\begin{pmatrix} e_1 & e_2 & e_3 \\ e_2 & e_1 & e_3 \end{pmatrix}$$

is the Nakayama permutation. We represent R as

$$R = \begin{pmatrix} Q_1 & A_{12} & A_{13} \\ A_{21} & Q_2 & A_{23} \\ A_{31} & A_{32} & Q_3 \end{pmatrix}.$$

Consider the three rings:

$$R(12) = \begin{pmatrix} Q_1 & A_{12} \\ A_{21} & Q_2 \end{pmatrix}, \quad R(23) = \begin{pmatrix} Q_2 & A_{23} \\ A_{32} & Q_3 \end{pmatrix}, \quad R(13) = \begin{pmatrix} Q_1 & A_{13} \\ A_{31} & Q_3 \end{pmatrix}.$$

The socles of these rings are

$$\begin{pmatrix} 0 & S(A_{12}) \\ S(A_{21}) & 0 \end{pmatrix}, \quad \begin{pmatrix} S(Q_2) & 0 \\ 0 & S(Q_3) \end{pmatrix}, \quad \begin{pmatrix} S(Q_1) & 0 \\ 0 & S(Q_3) \end{pmatrix},$$

respectively. Put $X = \{\, x \in A_{21} \mid xA_{12} = 0 \,\}$ and $Y = \{\, y \in A_{12} \mid A_{21}y = 0 \,\}$. Then, by Lemma A, $\langle X \rangle_{21}$ and $\langle Y \rangle_{12}$ in $R(12)$ are ideals of $R(12)$. Noting this fact, together with

$$S\left(\begin{pmatrix} Q_2 & A_{13} \\ 0 & 0 \end{pmatrix}\right) = \begin{pmatrix} S(Q_2) & 0 \\ 0 & 0 \end{pmatrix} \quad \text{and} \quad S\left(\begin{pmatrix} 0 & 0 \\ A_{32} & Q_3 \end{pmatrix}\right) = \begin{pmatrix} 0 & 0 \\ 0 & S(Q_3) \end{pmatrix},$$

we see that

$$\langle X_{21} \rangle_{21} = \begin{pmatrix} 0 & 0 & 0 \\ X_{21} & 0 & 0 \\ 0 & 0 & 0 \end{pmatrix} \quad \text{and} \quad \langle X_{12} \rangle_{12} = \begin{pmatrix} 0 & X_{12} & 0 \\ 0 & 0 & 0 \\ 0 & 0 & 0 \end{pmatrix}$$

are left and right ideals of R. We represent $\overline{e_2R} = e_2R/\langle X_{21} \rangle_{21}$ as

$$\overline{e_2R} = \begin{pmatrix} 0 & 0 & 0 \\ A_{21}/X_{21} & Q_2 & 0 \\ 0 & 0 & 0 \end{pmatrix}.$$

Then

$$S(\overline{e_2R}) = \begin{pmatrix} 0 & 0 & 0 \\ 0 & S(Q) & 0 \\ 0 & 0 & 0 \end{pmatrix}.$$

Since e_1R_R is injective and $S(e_1R_R) \cong S(\overline{e_2R}_R)$, there exists a monomorphism $\varphi_{21} : \overline{e_2R} \to e_1J$. Let $\eta_2 : e_2R \to \overline{e_2R}$ be the canonical epimorphism and put $\pi_{21} = \varphi_{21}\eta_2$. Then $\pi_{21} : e_2R \to e_1J$ is an R-homomorphism and the restriction map $\pi_{21}|_{\langle A_{23} \rangle_{23}}$ is monomorphic. Similarly, considering the factor module $\overline{e_1R} = e_1R/\langle X_{12} \rangle_{12}$, we obtain an R-homomorphism $\pi_{12} : e_1R \to e_2J$ such that $\pi_{12}|_{\langle A_{13} \rangle_{13}}$ is monomorphic. Hence we see that $\pi_{21}(\langle A_{23} \rangle_{23}) = \langle A_{13} \rangle_{13}$ and $\pi_{12}(\langle A_{12} \rangle_{12}) = \langle A_{23} \rangle_{23}$. Since $\pi_{21}(\langle A_{12} \rangle_{12}) \subseteq \langle J(Q_2) \rangle_{22}$ and $\pi_{21}(\langle A_{21} \rangle_{21}) = \langle J(Q_1) \rangle_{11}$, we can obtain m such that

$$(\pi_{21}\pi_{12})^m(e_2R) = \begin{pmatrix} 0 & 0 & A_{13} \\ 0 & 0 & 0 \\ 0 & 0 & 0 \end{pmatrix},$$

which is impossible because $S(eR_R) = \langle S(A_{12}) \rangle_{12}$. Thus the Nakayama permutation of $Pi(R)$ must be cyclic. □

Proof of Theorem 6.4.1. (1) $\Rightarrow$ (3). Suppose that the Nakayama permutation ρ of R is non-identity. We show that ρ is cyclic. Suppose that ρ is not cyclic and write $\rho = \rho_1\rho_2 \cdots \rho_t$, where the ρ_i are cyclic permutations.

We can assume that the size k $(< n)$ of ρ_1 is bigger than 1 and we may express ρ_1 as

$$\rho_1 = \begin{pmatrix} e_1 \ e_2 \ e_3 \ \cdots \ e_n \\ e_k \ e_1 \ e_2 \ \cdots \ e_{k-1} \end{pmatrix}.$$

Put $f_1 = e_1$ and $f_2 = e_2$ and take $f_3 \in Pi(R) - \{\, e_1, \ldots, \ e_k \,\}$. Consider the subring of R:

$$T = \begin{pmatrix} f_1 R f_1 \ f_1 R f_2 \ f_1 R f_3 \\ f_2 R f_1 \ f_2 R f_2 \ f_2 R f_3 \\ f_3 R f_1 \ f_3 R f_2 \ f_3 R f_3 \end{pmatrix}.$$

By (1), this ring is a basic QF-subring of R and we can see that

$$S(T) = \begin{pmatrix} 0 & S(f_1 R f_2) & 0 \\ S(f_2 R f_1) & 0 & 0 \\ 0 & 0 & S(f_3 R f_3) \end{pmatrix}.$$

Furthermore, since R is indecomposable, we can take f_3 such that T is indecomposable. Thus T is a basic indecomposable QF-ring and its Nakayama permutation is neither identity nor cyclic. This contradicts Lemma B.

(3) $\Rightarrow$ (2). This follows from Theorem 6.3.1.

(2) $\Rightarrow$ (1). If R is a weakly symmetric QF-ring, then clearly so is eRe for any idempotent e of R. If R is represented as a skew matrix ring over $Q = e_1 R e_1$ with respect to $\sigma \in \text{Aut}(Q)$ and $c \in J(Q)$, then so is eRe for any idempotent e of R.

(4) $\Rightarrow$ (1). Obvious.

(1) $\Rightarrow$ (4).   Let P be a projective right R-module.   Then P can be expressed as a direct sum of indecomposable projective modules $\{P_i\}_{i\in I}$. We take a subfamily $\{P_i\}_{i\in J}$ of irredundant representatives of $\{P_i\}_{i\in I}$.   Put $T = \oplus_{i\in J} P_i$.   Then $\text{End}(P_R)$ is Morita equivalent to $\text{End}(T_R)$.   Taking an idempotent $e \in R$ such that $eR_R \cong T_R$, $\text{End}(P_R)$ is Morita equivalent to eRe.   Hence $\text{End}(P_R)$ is QF.
□

Remark 6.4.2. Of course, in general QF-rings need not be strongly QF-rings as the following simple example shows.

Let K be a field and consider the skew matrix ring

$$\begin{pmatrix} K \ K \ K \\ K \ K \ K \\ K \ K \ K \end{pmatrix}_{id_K,0}$$

and put

$$R = \begin{pmatrix} K & K & 0 \\ 0 & K & K \\ K & 0 & K \end{pmatrix} = \begin{pmatrix} K & K & K \\ K & K & K \\ K & K & K \end{pmatrix} / \begin{pmatrix} 0 & 0 & K \\ K & 0 & 0 \\ 0 & K & 0 \end{pmatrix}.$$

This factor ring R is QF but not strongly QF.

6.5 Block Extensions of Skew Matrix Rings

Let Q be a local ring and let $c \in J(Q)$ and $\sigma \in \mathrm{Aut}(Q)$ satisfying $\sigma(c) = c$ and $\sigma(q)c = cq$ for any $q \in Q$. We consider the skew matrix ring $R = (Q)_{\sigma,c,n}$. Put $e_i = \langle 1 \rangle_{ii}$ for each $i = 1, \ldots, n$ and represent R as

$$R = \begin{pmatrix} Q_1 & A_{12} & \cdots & & A_{1n} \\ A_{21} & \ddots & \ddots & & \vdots \\ \vdots & \ddots & \ddots & & A_{n-1,n} \\ A_{n1} & \cdots & A_{n,n-1} & & Q_n \end{pmatrix},$$

where $Q_i = e_i R e_i$ for each $i = 1, \ldots, n$ and $A_{ij} = e_i R e_j$ for each distinct $i, j \in \{1, \ldots, n\}$.

For $k(1), \ldots, k(n) \in \mathbb{N}$, we recall that the block extension $R(k(1), \ldots, k(n))$ of R is given by

$$R(k(1), \ldots, k(n)) = \begin{pmatrix} Q(1) & A(1,2) & \cdots & & A(1,n) \\ A(2,1) & \ddots & \ddots & & \vdots \\ \vdots & \ddots & \ddots & & A(n-1,n) \\ A(n,1) & \cdots & A(n,n-1) & & Q(n) \end{pmatrix},$$

where

$$Q(i) = \begin{pmatrix} Q_i & \cdots & \cdots & Q_i \\ J(Q_i) & \ddots & & \vdots \\ \vdots & \ddots & \ddots & \vdots \\ J(Q_i) & \cdots & J(Q_i) & Q_i \end{pmatrix} \quad (k(i) \times k(i)\text{-matrix}),$$

$$A(i,j) = \begin{pmatrix} A_{ij} & \cdots & A_{ij} \\ & \cdots & \\ A_{ij} & \cdots & A_{ij} \end{pmatrix} \quad (k(i) \times k(j)\text{-matrix}).$$

Now we put

$$Q(i;c) = \begin{pmatrix} Q_i & \cdots & \cdots & Q_i \\ Q_ic & \ddots & & \vdots \\ \vdots & \ddots & \ddots & \vdots \\ Q_ic & \cdots & Q_ic & Q_i \end{pmatrix} \quad (\,k(i) \times k(i)\text{-matrix}\,)$$

and

$$R(k(1), \ldots, k(n); c) = \begin{pmatrix} Q(1;c) & A(1,2) & \cdots & & A(1,n) \\ A(2,1) & \ddots & & \ddots & \vdots \\ \vdots & & \ddots & & \ddots & A(n-1,n) \\ A(n,1) & \cdots & & A(n,n-1) & Q(n;c) \end{pmatrix}.$$

Then, as is easily seen, $R(k(1), \ldots, k(n); c)$ is a subring of $R(k(1), \ldots, k(n))$. We say that $R(k(1), \ldots, k(n); c)$ is a *c-block extension* of $R = (Q)_{\sigma,c,n}$.

Under this situation, we show the following theorem which plays an important role in the next chapter for giving a classification of Nakayama rings.

Theorem 6.5.1. $R(k(1), \ldots, k(n); c)$ *is isomorphic to a factor ring of the skew matrix ring* $T = (Q)_{\sigma,c,k(1)+\cdots+k(n)}$.

Proof. We represent T as

$$T = \begin{pmatrix} T(1) & T(1,2) & \cdots & & T(1,n) \\ T(2,1) & \ddots & & \ddots & \vdots \\ \vdots & & \ddots & & \ddots & T(n-1,n) \\ T(n,1) & \cdots & & T(n,n-1) & T(n) \end{pmatrix},$$

where

$$T(i) = \begin{pmatrix} Q & \cdots & Q \\ & \cdots & \\ Q & \cdots & Q \end{pmatrix} \quad (\,k(i) \times k(i)\text{-matrix}\,),$$

$$T(i,j) = \begin{pmatrix} Q & \cdots & Q \\ & \cdots & \\ Q & \cdots & Q \end{pmatrix} \quad (\,k(i) \times k(j)\text{-matrix}\,).$$

Put

$$T = \begin{pmatrix} T(1) & T(1,2) & \cdots & & T(1,n) \\ T(2,1) & \ddots & & \ddots & \vdots \\ \vdots & & \ddots & & \ddots & T(n-1,n) \\ T(n,1) & \cdots & & T(n,n-1) & T(n) \end{pmatrix}.$$

For each $i, j \in \{1, \ldots, n\}$, let $\varphi(i) : T(i) \longrightarrow Q(i; c)$ be the canonical map given by

$$
\begin{pmatrix}
q_{11} & q_{12} & \cdots & & q_{1k} \\
q_{21} & \ddots & \ddots & & \vdots \\
\vdots & \ddots & \ddots & & q_{n-1,n} \\
q_{n1} & \cdots & q_{n,n-1} & & q_{nn}
\end{pmatrix}
\longmapsto
\begin{pmatrix}
q_{11} & q_{12} & \cdots & & q_{1k} \\
q_{21}c & \ddots & \ddots & & \vdots \\
\vdots & \ddots & \ddots & & q_{n-1,n} \\
q_{n1}c & \cdots & q_{n,n-1}c & & q_{nn}
\end{pmatrix}
$$

and let $\varphi(i,j) : T(i,j) \longrightarrow A(i,j)$ be the canonical map given by

$$
\begin{pmatrix}
q_{i1} & \cdots & q_{ik_j} \\
& \cdots & \\
q_{k_i 1} & \cdots & q_{k_i k_j}
\end{pmatrix}
\longmapsto
\begin{pmatrix}
q_{i1}\alpha_{ij} & \cdots & q_{ik_j}\alpha_{ij} \\
& \cdots & \\
q_{k_i 1}\alpha_{ij} & \cdots & q_{k_i k_j}\alpha_{ij}
\end{pmatrix}.
$$

Then it is easy to see that

$$
\Phi =
\begin{pmatrix}
\varphi(1) & \varphi(1,2) & \cdots & & \varphi(1,n) \\
\varphi(2,1) & \ddots & & \ddots & \vdots \\
\vdots & \ddots & & \ddots & \varphi(n-1,n) \\
\varphi(n,1) & \cdots & \varphi(n,n-1) & & \varphi(n)
\end{pmatrix}
$$

is a surjective ring homomorphism from $T = (Q)_{\sigma,c,k(1)+\cdots+k(n)}$ to $R(k(1), \ldots, k(n); c)$. □

Example 6.5.2. (1). Let D be a division ring. Then

$$
\begin{pmatrix}
D & \cdots & \cdots & D \\
D & \ddots & & \vdots \\
\vdots & \ddots & \ddots & \vdots \\
D & \cdots & D & D
\end{pmatrix}_{id,0,k}
\Big/
\begin{pmatrix}
0 & \cdots & \cdots & 0 \\
D & \ddots & & \vdots \\
\vdots & \ddots & \ddots & \vdots \\
D & \cdots & D & 0
\end{pmatrix}
\cong D(k;0) =
\begin{pmatrix}
D & \cdots & \cdots & D \\
0 & \ddots & & \vdots \\
\vdots & \ddots & \ddots & \vdots \\
0 & \cdots & 0 & D
\end{pmatrix}.
$$

(2). For

$$
R = \begin{pmatrix} Q & Q\alpha \\ Q\beta & Q \end{pmatrix}_{\sigma,c,2},
$$

where Q is a local ring and $c \in J(Q)$ and $\sigma \in \mathrm{Aut}(Q)$, we have

$$
R(2,2;c) \cong
\begin{pmatrix}
Q & Q & Q & Q \\
Q & Q & Q & Q \\
Q & Q & Q & Q \\
Q & Q & Q & Q
\end{pmatrix}_{\sigma,c,4}
\Big/
\begin{pmatrix}
0 & 0 & 0 & 0 \\
l_Q(c) & 0 & 0 & 0 \\
0 & 0 & 0 & 0 \\
0 & 0 & l_Q(c) & 0
\end{pmatrix}.
$$

COMMENTS

Most of contents of this chapter are taken from Oshiro [149], Oshiro-Rim [155] and Nagadomi-Oshiro-Uhara-Yamaura [131]. Theorem 6.2.1 is shown in Abe-Hoshino [1], Fujita [53] and Koike [100], independently. However the proof presented here differs from their. Skew matrix rings R with $\#Pi(R) = 2$ are considered in Hannula [59] and Theorem 6.4.1 is shown in this paper for $n = 2$. The naming of a strongly QF-ring is due to Yukimoto [190]. The equivalence of (2) and (3) in Theorem 6.4.1 is due to Oshiro-Rim [155]. The equivalence of (1) and (3) is shown in Yukimoto [190] and Hoshino [78] and the equivalence of (1) and (4) is shown in Hoshino [78].

Chapter 7

The Structure of Nakayama Rings

In this chapter, we discuss the structure of Nakayama rings. It is shown that Nakayama rings are H-rings and hence we can apply the results obtained in previous chapters to Nakayama rings. We give a complete classification of these artinian rings by showing the following results:

Result 1. *Nakayama rings are H-rings. Thus, by the structure of H-rings, if R is a basic indecomposable Nakayama ring, then R can be constructed as an upper staircase factor ring of a block extension of a frame QF-subring $F(R)$ of R.*

Result 2. *A basic indecomposable Nakayama QF-ring with a non-identity Nakayama permutation can be represented as a factor ring of a skew matrix ring over a local Nakayama ring.*

Result 3. *If R is a basic indecomposable Nakayama ring whose frame Nakayama QF-ring $F(R)$ has a non-identity Nakayama permutation, then R can be directly represented as a factor ring of a skew matrix ring over a local Nakayama ring.*

Result 4. *If R is a basic indecomposable Nakayama QF-ring with the identity Nakayama permutation, then $R/S(R)$ can be represented as a skew matrix ring over a local Nakayama ring.*

Result 5. *Every Nakayama ring is self-dual, more precisely, every Nakayama ring has a Nakayama isomorphism.*

In view of these results we see that there are deep connections between QF-rings, left H-rings and Nakayama rings and the essence of Nakayama rings is the skew matrix rings over local Nakayama rings.

7.1 Kupisch Series and Kupisch Well-Indexed Set via Left *H*-Rings

In this section we first recall known results (and record some new results) related to Nakayama rings providing (new) proofs of these from our point of view. The main purpose is to capture the so-called Kupisch series for Nakayama rings through the structure of left *H*-rings.

It is well-known that every projective module over a semiperfect ring has an indecomposable decomposition. And a semiperfect ring R is called a *right (left) QF-2 ring* if every indecomposable projective right (left) R-module has a simple socle, and a left and right QF-2 ring is called a *QF-2 ring*. A ring R is called a *right (left) Kasch ring* if every simple right (left) R-module can be embedded in R, and R is called a *Kasch ring* if it is left and right Kasch.

Lemma 7.1.1. *Let R be a left H-ring and $f \in Pi(R)$. If fR_R is uniserial, then $_RRf$ is also uniserial.*

Proof.   Let $J^i f / J^{i+1} f = \oplus_{j=1}^n \overline{Ru_j}$, where $u_j \in h_j R f$ for some $h_j \in Pi(R)$ with $_R\overline{Ru_j}$ simple. Suppose that $n \geq 2$. We consider the left multiplication map $(u_j)_L : fR \to u_j R$ by u_j. Since fR_R is uniserial, $\mathrm{Ker}(u_j)_L = fJ^{p_j}$ for some $p_j \in \mathbb{N}$. We may assume $p_1 \geq p_2$. Then we have an epimorphism $\varphi : u_1 R \to u_2 R$ defined by $\varphi(u_1) = u_2$.

Now consider the projective module $P = h_1 R \oplus h_2 R$ which contains $u_1 R \oplus u_2 R$. Since R is a left H-ring, P is an extending module with the exchange property by Theorems 1.1.14 and 3.2.10. Hence, by Remark 1.1.17 and Theorem 1.1.18, if φ is not isomorphic, then φ can be extended to $h_1 R \to h_2 R$, whence there exists $r \in h_2 R h_1$ such that $u_2 = r u_1$. Hence $\overline{0} \neq \overline{u_2} = \overline{r u_1}$. This contradicts the fact that $\overline{Ru_1} \cap \overline{Ru_2} = \overline{0}$. On the other hand, if φ is isomorphic (i.e., $p_1 = p_2$), again by Remark 1.1.17, either φ may be extended to $h_1 R \to h_2 R$ or φ^{-1} may be extended to $h_2 R \to h_1 R$. But, by reasoning as above, neither case occurs. Accordingly, $\overline{J^i f}$ is a simple left R-module, and hence $_RRf$ is uniserial. $\square$

Theorem 7.1.2. *If R is a right QF-3 right Nakayama ring, then R is a Nakayama ring.*

Proof.   By Lemma 7.1.1 it suffices to show that R is a left H-ring.

Since R is right QF-2, we can arrange $Pi(R)$ as $Pi(R) = \{e_{ij}\}_{i=1,j=1}^{m \quad n(i)}$ such that, for any $i, j \in \{1, \ldots, m\}$,

(i) $S(e_{i1}R_R) \not\cong S(e_{j1}R_R)$ if $i \neq j$,

(ii) $S(e_{i1}R_R) \cong S(e_{i2}R_R) \cong \cdots \cong S(e_{in(i)}R_R)$ and

(iii) $L(e_{i1}R_R) \geq L(e_{i2}R_R) \geq \cdots \geq L(e_{in(i)}R_R)$.

Then, since R is right QF-3, each $e_{i1}R_R$ is injective by Proposition 3.1.2. Further, because R is right Nakayama, we claim that either $e_{ij}R_R \cong e_{i,j+1}R_R$ or $e_{ij}J_R \cong e_{i,j+1}R_R$ holds. Suppose that there exists $1 \leq s < t$ with $e_{i1}J^s$ is not projective but $e_{i1}J^t$ is projective. Let $\varphi : fR \to e_{i1}J^s$ be a projective cover, where $f \in Pi(R)$. Then $\mathrm{Ker}\, \varphi \neq 0$ and an epimorphism $\varphi|_{fJ^{t-s}} : fJ^{t-s} \to e_{i1}J^t$ splits, i.e., it is an isomorphism, a contradiction. Hence R is a left H-ring. $\square$

Theorem 7.1.3. *If R is a semiprimary right QF-3 ring with $J^2 = 0$, then R is a Nakayama ring.*

Proof. We let $R = e_1R \oplus \cdots e_sR \oplus f_1R \oplus \cdots \oplus f_tR$ be a direct decomposition of indecomposable right ideals with $L(e_iR_R) = 2$ and $L(f_jR_R) = 1$. Then we claim that each e_iR_R is injective. Let $e_iJ = \oplus_{k \in K}S_k$, where S_k is simple. Since R is right QF-3, we have a minimal faithful right ideal gR. Let $gR = \oplus_{k=1}^m g_kR$ be a decomposition into indecomposable submodules. We note that each g_kR is injective. Then there exists a map $\tau : K \to \{1, \ldots m\}$ with $E(S_k) \cong g_{\tau(k)}R_R$ for any $k \in K$ since each S_k is a simple right ideal of R. Hence we have a monomorphism $\xi : E(e_iR_R) = E(\oplus_{k \in K}S_k) \to \Pi_{k \in K}f_{\tau(k)}R$. ξ induces another monomorphism $\xi' : T(e_iR_R) \to \Pi_{k \in K}T(g_{\tau(k)}R_R)$. Therefore there exists $k' \in K$ with $e_iR_R \cong g_{\tau(k')}R_R$, i.e., e_iR_R is injective. Hence each e_iR_R is uniserial. Since each f_jR_R is also uniserial (simple), R is a right Nakayama ring. Hence R is a Nakayama ring by Theorem 7.1.2. $\square$

Lemma 7.1.4. *If R is a left perfect ring such that, for each $e \in Pi(R)$, eJ_R is a hollow module (i.e., $T(eJ_R)$ is simple), then R is a right Nakayama ring.*

Proof. Let $e \in Pi(R)$. Suppose that $eJ \neq 0$. Then, since eJ_R is a cyclic hollow module, there exists $f \in Pi(R)$ such that fR_R is a projective cover of eJ_R. Thus we have an epimorphism $fJ \to eJ^2$ if $eJ^2 \neq 0$. Since fJ_R is cyclic hollow, there exists $g \in Pi(R)$ such that gR_R is a projective cover of fJ_R. Then gR_R is a projective cover of eJ_R^2, and eJ_R^2 is also a cyclic hollow module. By repeating this argument, we eventually find $k \in \mathbb{N}$ for which eJ_R^i is cyclic hollow for $i = 1, \ldots, k$ and $eJ^{k+1} = 0$ since R is left perfect. Hence eR_R is uniserial. $\qquad\square$

Theorem 7.1.5.

    (1)   *If R is a perfect ring such that R/J^2 is right QF-3, then R is a Nakayama ring.*

    (2)   *If R is a semiprimary ring whose factor rings are right QF-3, then R is a Nakayama ring.*

Proof. (1) Since R/J^2 is right *QF-3*, it is a Nakayama ring by Theorem 7.1.3. Hence R is a Nakayama ring by Lemma 7.1.4.

    (2) follows from (1) and Theorem 7.1.3. $\qquad\square$

Theorem 7.1.6. *If R is a semiprimary QF-2 ring with ACC or DCC for right annihilator ideals, then R is QF-3.*

Proof. It suffices to show that R is right *QF-3*. We let $Pi(R) = \{e_{ij}\}_{i=1,j=1}^{m,\ n(i)}$ be as in the proof of Theorem 7.1.2. And we only have to show that each $e_{i1}R_R$ is injective by Proposition 3.1.2. Let $e \in \{e_{i1}\}_{i=1}^m$. We have $f \in Pi(R)$ with fR_R as a projective cover of $S(eR_R)$ since R is *QF-2*. Then $0 \neq S(eR_R)f \subseteq Rf$. Further we claim that $J\,S(eR_R)f = 0$. Suppose that $J\,S(eR_R)f \neq 0$. There exist $g \in Pi(R)$ and $x \in gJ$ with $x\,S(eR_R)f \neq 0$. We consider the left multiplication map $(x)_L : eR \to gR$. Then it is a monomorphism from the assumption $J\,S(eR_R)f \neq 0$ because $S(eR_R)$ is the essential simplle socle. On the other hand, $\operatorname{Im}(x)_L \subseteq gJ$ because $x \in J$. Therefore $L(eR_R) < L(gR_R)$. This contradicts with the maximality of $L(eR_R)$. Hence $0 \neq S(eR_R)f \subseteq S(_RRf)$, i.e., $e\,S(_RRf) \neq 0$. Therefore we have a projective cover $_RRe \to S(_RRf)$ since R is *QF-2*. Thus $(eR; Rf)$ is an *i*-pair. Since R satisfies either *ACC* or *DCC* for right annihilators, we see from Corollary 2.4.6 that eR_R is injective. $\qquad\square$

Corollary 7.1.7. *Let R be a QF-2 ring. If R is left or right artinian, then it is QF-3.*

Corollary 7.1.8. *Nakayama rings are H-rings.*

Proof.   Let R be a Nakayama ring. It suffices to show that R is a left H-ring. Let $Pi(R) = \{e_{ij}\}_{i=1,j=1}^{m,\ n(i)}$ be as in the proof of Theorem 7.1.2. By Corollary 7.1.7, R is a QF-3 Nakayama ring and hence each $e_{i1}R_R$ is a uniserial injective module. Moreover, as is easily seen, either $e_{ij}R_R \cong e_{i,j+1}R_R$ or $e_{ij}J_R \cong e_{i,j+1}R_R$ holds. Hence R is a left H-ring. $\square$

Proposition 7.1.9. *Let R be a basic indecomposable left H-ring and let $e, f \in Pi(R)$ such that fR_R is a projective cover of $S(eR_R)$. If $\mathrm{Hom}_R(fR_R, S_2(eR_R)/S_1(eR_R)) \neq 0$ and fJ_R is a hollow module, then $e = f = 1$. Consequently R is a local Nakayama ring.*

Proof.   We may assume that eR_R is injective since R is a left H-ring and there exists an epimorphism $fR_R \to S(eR_R)$. Because $\mathrm{Hom}_R(fR, S_2(eR_R)/S_1(eR_R)) \neq 0$, fR is a projective cover of $S(eR_R)$ and fJ/fJ_R^2 is simple, there exists an epimorphism $fJ_R \to S(eR_R)$. This, together with an epimorphism $fR_R \to S(eR_R)$, produces an epimorphism $fR_R \to fJ_R$. Therefore we have an epimorphism $fR_R \to fJ^i$ for any $i = 1, 2, \ldots$. Hence fR_R is uniserial such that any of its composition factors is isomorphic to $T(fR_R)$. Hence we remark that $fRg = 0$ for any $g \in Pi(R)$ with $gR_R \not\cong fR_R$.

Next we show that $fR_R \cong eR_R$. Since fR is a projective cover of $S(eR_R)$, $S(fR_R) \cong S(eR_R)$. Therefore there exists a monomorphism $\eta : fR \to eR$ because eR_R is injective. Now $\mathrm{Im}\,\eta = eJ^k$ for some $k \in \mathbb{N}_0$ since R is left H-ring. Then we claim that $k = 0$, i.e., $fR_R \cong eR_R$. Suppose that $k \geq 1$. Since fJ/fJ^2 is simple, $fJ \neq 0$. Hence we have an epimorphism $\varphi : fR \to fJ$. Then, since eR_R is injective, there exists $\tilde{\varphi} : eJ^{k-1} \to eR$ with $\tilde{\varphi}\eta = \eta\varphi$. $\tilde{\varphi}$ is not monic. But, since $\mathrm{Im}\,\tilde{\varphi} = eJ^k \ (\cong fR_R)$, $\tilde{\varphi} : eJ^{k-1} \to eJ^k$ splits and it must be an isomorphism, a contradiction.

Further we claim that $gRf = 0$ for any $g \in Pi(R)$ with $gR_R \not\cong fR_R$. Suppose that $gRf \neq 0$ for some $g \in Pi(R)$ with $gR_R \not\cong fR_R$. Because R is a left H-ring, we may assume that gR_R is injective. Let $0 \neq x \in gRf$. Then we have an epimorphism $(x)_L : fR \to xR$. Since any composition factor of fR_R is isomorphic to $T(fR_R)$, $gR_R \cong E(T(fR_R)) \cong fR \ (\cong eR_R)$, a contradiction.

In consequence, $e = f = 1$ since R is a basic indecomposable ring. And because R is a left H-, right Nakayama ring, R is then a Nakayama ring by Lemma 7.1.1. $\square$

Notation.   Let R be a semiperfect ring and $e, f \in Pi(R)$. We use $eR \to fR$ or $fR \leftarrow eR$ to mean that there exists an epimorphism from eR_R to fJ_R, i.e., fJ_R is (cyclic) hollow and eR_R is a projective cover of fJ_R. Similarly we use $Re \to Rf$ to mean that there exists an epimorphism from $_RRe$ to $_RJf$. Indeed, instead of $eR \to fR$ or $fR \leftarrow eR$ (or $Re \to Rf$ or $Rf \leftarrow Re$), we will simply write $e \to f$ or $f \leftarrow e$ if no confusion arises. For $f_1, \ldots, f_k \in Pi(R)$,

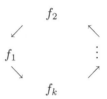

denotes $f_1 \leftarrow \cdots \leftarrow f_k \leftarrow f_1$ and we refer to this as a *cyclic quiver*.

Proposition 7.1.10.   *Let R be a basic semiperfect ring and $f_1, \ldots, f_k \in Pi(R)$ with a cyclic quiver:*

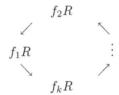

Let $f = f_1 + \cdots + f_k$. Then the following hold:

(1) $f_i R_R$ *is uniserial for each $i = 1, \ldots, k$.*
(2)   *For each $i = 1, \ldots, n$ and $t \in \mathbb{N}$ with $0 \neq f_i J^t \subsetneq f_i R$, there exists $f_j \in \{f_1, \ldots, f_k\}$ such that $f_j R_R$ is a projective cover of $f_i J_R^t$.*
(3) $\operatorname{Hom}_R(gR_R, fR_R) = 0$ *for any $g \in Pi(R) - \{f_i\}_{i=1}^k$.*
(4)   *If R is a left H-ring, then $_RRf_i$ is also uniserial with a cyclic quiver:*

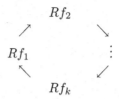

Then, from these cyclic quivers, we see that

(i) $\mathrm{Hom}_R(fR_R, gR_R) = 0$ and $\mathrm{Hom}_R(_RRf, _RRg) = 0$ for any $g \in Pi(R) - \{f_i\}_{i=1}^k$ and

(ii) fRf is a Nakayama ring which is a direct summand of R as a ring.

Proof. (1), (2), (3). From the cyclic quiver

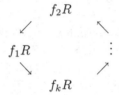

we see that there are epimorphisms $f_2R \to f_kJ^2$, $f_3R \to f_kJ^3$, ..., $f_kR \to f_kJ^k$, $f_1R \to f_kJ^{k+1}$, ... Hence we see that f_kR (and hence each f_iR) is uniserial and there is an epimorphism $f_sR \to f_kJ^j$ for any pair $(s, j) \in \{1, \ldots, k\} \times \{0, \ldots, k-1\}$ with $s \equiv j \pmod{k}$. Since R is basic, the uniqueness of projective covers up to isomorphism implies (3).

(4). Since f_iR_R is uniserial, $_RRf_i$ is also uniserial for each $i = 1, \ldots, m$ by Lemma 7.1.1. Furthermore, because $f_{i+1}R \leftarrow f_iR$, we can take $f_ir f_{i+1} \in J - J^2$ such that $f_ir f_{i+1}R = f_iJ \nsubseteq J^2$. Thus $f_ir f_{i+1}$ induces an epimorphism $Rf_i \to Jf_{i+1}$. Similarly we have an epimorphism $Rf_k \to Jf_1$ and hence we obtain a cyclic quiver:

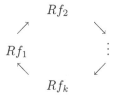

Therefore each $_RRf_i$ is uniserial and $\operatorname{Hom}_R(_RRf_i, {}_RRg) = 0$ for any $g \in Pi(R) - \{f_j\}_{j=1}^k$. □

Proposition 7.1.11. *Let R be a basic left H-ring with a left well-indexed set $Pi(R) = \{e_{ij}\}_{i=1,j=1}^{m \ \ n(i)}$. For each $i = 1, \ldots, m$, a projective cover P of $e_{in(i)}J_R$ is injective.*

Proof.   We write $P = P_1 \oplus \cdots \oplus P_k$, where each P_i is indecomposable. And let $\varphi : P \to e_{in(i)}J$ be an epimorphism. Then φ extends to $\varphi^* : E(P) \to E(e_{in(i)}J_R)$. Note that $E(e_{in(i)}J_R) \cong e_{j1}R_R$ for some j. Suppose that $\varphi^*(E(P)) \nsubseteq e_{in(i)}J$. Then $\varphi^*(E(P)) \supseteq e_{in(i)}R$ Hence $\varphi^*(E(P))$ is projective, and thus $\varphi^* : E(P) \to \varphi^*(E(P))$ splits. This implies that some $E(P_i)$ must be isomorphic to $\varphi^*(E(P))$. Hence $e_{in(i)}J$ contains a projective submodule which is isomorphic to P_i, a contradiction. Therefore $\varphi^*(E(P)) \subseteq e_{in(i)}J$ and hence $\varphi^*(E(P)) = e_{in(i)}J$. Since $E(P)$ is projective, a projective cover P of $e_{in(i)}J$ is isomorphic to a direct summand of $E(P)$ and hence P is injective. □

Theorem 7.1.12. *Let R be a basic indecomposable left H-ring with a left well-indexed set $Pi(R) = \{e_{ij}\}_{i=1,j=1}^{m \ \ n(i)}$. For $k \leq m$, we consider the situation that $e_{11}R \leftarrow \cdots \leftarrow e_{1n(1)}R \leftarrow \cdots \leftarrow e_{k1}R \leftarrow \cdots \leftarrow e_{kn(k)}R$, where $e_{kn(k)}J_R$ is cyclic hollow. Let $\varphi : fR_R \to e_{km(k)}J_R$ be a projective cover, where $f \in Pi(R)$. Then the following hold:*

(1) *fR_R is injective, i.e., $f = e_{s1}$ for some s.*
(2) *If $s \in \{1, \ldots, k\}$, then $s = 1$, $k = m$ and R is a Nakayama ring.*

Proof.   (1). This follows from Proposition 7.1.11.
(2). This follows from Proposition 7.1.10. □

We now come to the main purpose of this section, namely, the Kupisch series on a Nakayama ring as seen from our vantage point.

Let R be a basic Nakayama ring. Since R is a basic H-ring, $Pi(R)$ can be arranged as a left well-indexed set $\{e_{ij}\}_{i=1,j=1}^{m,\ n(i)}$. In particular, by Theorem 7.1.12, we may arrange $Pi(R)$ as

$$e_{11} \leftarrow e_{12} \leftarrow \cdots \leftarrow e_{1n(1)} \leftarrow e_{21} \leftarrow e_{22} \leftarrow \cdots \leftarrow e_{m-1,n(m-1)} \leftarrow e_{m1} \cdots \leftarrow e_{mn(m)}$$

and, if $e_{mn(m)}R$ is not simple, then

$$e_{mn(m)} \leftarrow e_{11}.$$

Thus we have either a quiver

$$e_{11} \leftarrow e_{12} \leftarrow \cdots \leftarrow e_{m1} \leftarrow e_{mn(m)}$$

or a cyclic quiver

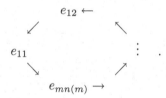

In this case, the sequence $e_{mn(m)}R, \ldots, e_{m1}R, \ldots, e_{1m(1)}R, \ldots, e_{11}R$ is called the *Kupisch series* of R.

Notation. For a basic indecomposable Nakayama ring R with a left well-indexed set $Pi(R) = \{e_{ij}\}_{i=1,j=1}^{m,\ n(i)}$, we say that $Pi(R)$ is a *Kupisch left well-indexed set* if the sequence $e_{mn(m)}R, \ldots, e_{m1}R, \ldots, e_{1n(1)}R, \ldots, e_{11}R$ is the Kupisch series of R.

Theorem 7.1.13. *Let R be a basic indecomposable left H-ring. If R has a simple projective right R-module, then R can be represented as a factor ring of an upper triangular matrix ring over a division ring as follows:*

$$R \cong \begin{pmatrix} D & & & & \\ & \ddots & & & \\ & & D & & 0 \\ & & & \ddots & \\ & 0 & & & \\ & & & & D \end{pmatrix}$$

Proof. We arrange $Pi(R)$ as a left well-indexed set $\{e_{ij}\}_{i=1,j=1}^{m \ \ n(i)}$. We may assume that $e_{mn(m)}R_R$ is simple. We recall that the maps $\sigma, \rho :$ $\{1, \ldots, m\} \to \mathbb{N}$ in Chapter 3. These maps mean that $(e_{i1}R; Re_{\sigma(i)\rho(i)})$ is an i-pair for each $i = 1, \ldots, m$. Since $e_{mn(m)}R_R$ is simple, $\sigma(m) = m$ and $\rho(m) = n(m)$. Let $\{e_{i_j1}\}_{j=1}^{y}$ be a subset of $\{e_{i1}\}_{i=1}^{m}$ such that $\{e_{i_j1}\}_{j=1}^{y} = \{e_{\sigma(i_j)1}\}_{j=1}^{y}$ and the frame QF-subring of R is

$$F(R) = \begin{pmatrix} e_{i_11}Re_{i_11} & \cdots & e_{i_11}Re_{i_y1} \\ \cdots & \cdots & \cdots \\ e_{i_y1}Re_{i_11} & \cdots & e_{i_y1}Re_{i_y1} \end{pmatrix}.$$

For the making of $F(R)$, we may take $e_{i_y} = e_{m1}$. For $F(R)$, there exist $k(1), k(2), \ldots, k(y) \in \mathbb{N}$ such that R can be represented as an upper staircase factor ring of the block extension $P = F(R)(k(1), \ldots, k(y))$ (see §4.2), say

$$\overline{P} = \begin{pmatrix} P(i_1,i_1) \cdots & & \overline{P(i_1,\sigma(i_1))} & \cdots & P(i_1,i_y) \\ & & \cdots & & \\ P(i_j,i_1) & \cdots & \overline{P(i_j,\sigma(i_j))} & \cdots & \\ & & \cdots & & \cdots \\ P(i_y,i_1) & & \cdots & & \overline{P(i_y,i_y)} \end{pmatrix}.$$

Here we note that $\sigma(m) = m$ and $e_{mn(m)}R$ appears as the last row in the last row block of this representation of P. Since $e_{mn(m)}R_R$ is simple, we see that

$$\left(P(i_y,i_1) \cdots P(i_{y-1},i_y)\right) = 0,$$

$$\overline{P(i_y,i_y)} = \begin{pmatrix} D & & & \\ & \ddots & D & 0 \\ & & & \ddots \\ & 0 & & \ddots \\ & & & D \end{pmatrix},$$

where $D = \text{End}(e_{mn(m)}R_R) = \text{End}(e_{mn(m)}R_R)$. Then we claim that $P(i_j,i_y) = 0$ for any $i_j \in \{i_1, \ldots, i_{y-1}\}$. Since $i_j \neq i_y$, $\sigma(i_j) \neq i_y$. if $P(i_j,i_y) \neq 0$, then $P(i_j.j_y)P(i_y,\sigma(i_j)) \neq 0$, which is impossible. Since R

is an indecomposable ring, we see that $i_y = i_1$, $F(R) = D$ and hence

$$R = \overline{P(i_1, i_1)} \cong \begin{pmatrix} D & & & \rceil & & \\ & \ddots & D & 0 & & \\ & & & \lfloor & & \\ & 0 & & & & \\ & & & & \ddots & \\ & & & & & D \end{pmatrix}.$$ □

Corollary 7.1.14. *Let R be a basic indecomposable Nakayama ring. If R has a simple projective right R-module, then R can be represented as a factor ring of an upper triangular matrix ring over a division ring.*

7.2 Nakayama QF-Rings

Theorem 7.2.1. *Let R be a basic indecomposable Nakayama ring with a Kupisch left well-indexed set $Pi(R) = \{f_i\}_{i=1}^{k} = \{e_{ij}\}_{i=1,j=1}^{m,\ n(i)}$. Then the following are equivalent:*

(1) *R is QF.*
(2) *$|f_1 R_R| = \cdots = |f_k R_R|$.*

Proof. (1) $\Rightarrow$ (2). Since R is QF, $\{f_i\}_{i=1}^{k} = \{e_{i1}\}_{i=1}^{m}$. If $e_{m1}R_R$ is simple, then $m = 1$ and R is a division ring. If $e_{m1}R_R$ is not simple, then R has a cyclic quiver and hence (2) follows.

(2) $\Rightarrow$ (1). This implication is clear from $\{f_i\}_{i=1}^{k} = \{e_{i1}\}_{i=1}^{m}$. □

Corollary 7.2.2. *If R is a Nakayama QF-ring, then so is $R/S_i(R_R)$ for any $i = 0, 1, 2, \ldots$.*

Lemma 7.2.3. *Let R be a left H-ring. If R is right Kasch and $R/S(R_R)$ is right QF-2, then R is a Nakayama QF-ring.*

Proof. Let $e \in Pi(R)$. Since $\overline{S_2(eR_R)} = S_2(eR_R)/S(eR_R)$ is a simple right R-module, we can take $f \in E$ such that fR is injective and

$$\overline{S_2(eR_R)} \overset{\varphi}{\cong} S(fR_R)$$
$$\cap \qquad\qquad \cap$$
$$eR/S(eR_R) \qquad fR$$

Because fR_R is injective, φ extends to $\varphi^* : \overline{eR} = eR/S(eR_R) \to fR$. Hence $\varphi^* : S_2(\overline{eR}_R) \cong S_2(fR_R)$. Since $R/S(R_R)$ is right QF-2, $S_2(fR_R)/S(fR_R)$ is simple. Moreover, since $S_2(\overline{eR}_R)/S(\overline{eR}_R) \cong S_3(eR_R)/S_2(eR_R)$, we see that $S_3(eR_R)/S_2(eR_R)$ is simple. Proceeding inductively, we obtain that $S_i(eR_R)/S_{i-1}(eR_R)$ is simple for any $i = 1, 2, \ldots$. Since there exists j such that $S_j(eR_R) = eR$, this implies that eR_R is uniserial.

Next we show that R is QF. Again let $e \in Pi(R)$. Then $T(eR_R)$ is simple and can be embedded in R. Because R is a left H-ring, a projective cover of $T(eR_R)$ is injective, whence eR_R is injective as desired. $\square$

Since QF-rings are right Kasch and right QF-2, we obtain the following theorem from Theorem 7.2.1, Corollary 7.2.2 and Lemma 7.2.3.

Theorem 7.2.4. *For a left H-ring R, the following are equivalent:*

(1) R *is a Nakayama QF-ring.*
(2) $R/S_i(R_R)$ *is a Nakayama QF-ring for $i = 0, 1$.*
(3) $R/S_i(R_R)$ *is a Nakayama QF-ring for each $i = 0, 1, 2, \ldots$.*
(4) R *is right Kasch and $R/S(R_R)$ is right QF-2.*
(5) $R/S_i(R_R)$ *is a QF-ring for $i = 0, 1$.*
(6) $R/S_i(R_R)$ *is a QF-ring for each $i = 0, 1, 2, \ldots$.*

Proposition 7.2.5. *Let R be a basic indecomposable Nakayama QF-ring with the Kupisch series $e_n R$, $e_{n-1} R$, $\ldots$, $e_1 R$. Then the Nakayama permutation of R is cyclic or identity, that is, there exists i for which the Nakayama permutation is*

$$\begin{pmatrix} e_1 & \cdots & e_{n-i} & e_{n-i+1} & \cdots & e_n \\ e_{i+1} & \cdots & e_n & e_1 & \cdots & e_i \end{pmatrix}.$$

Proof. By Theorem 7.2.1, $|e_1 R_R| = \cdots = |e_n R_R|$. Our proof is by induction on the composition length $k(R) = |e_1 R_R|$. If either $n = 1$ or $k(R) = 1$, then the statement trivially holds. Therefore we consider the case that $n \geq 2$ and $k(R) \geq 2$. Assuming the Nakayama permutation of R is not the identity, we show the statement by induction on $k(R)$. Let $k(R) = 2$. Then, since each $e_i J_R$ is simple, we can easily see that a projective cover of $e_i J_R$ is $e_{i+1} R$ for $i = 1, \ldots n - 1$ and a projective cover of $e_n J_R$ is $e_1 R$. Hence

$$\begin{pmatrix} e_1 & \cdots & e_{n-1} & e_n \\ e_2 & \cdots & e_n & e_1 \end{pmatrix}$$

is the Nakayama permutation. Now suppose that $t \geq 3$ and the statement holds for all rings R with $k(R) < t$. And we consider R with $k(R) = t$. By Corollary 7.2.2, $\overline{R} = R/S(R)$ is also a Nakayama QF-ring. Put $\overline{e_j} = e_j + S(R)$ for each $j = 1, \ldots, n$. Then $|e_1 R_R| = |\overline{e_1 R}_R| + 1$. Now, by induction, we may show that, if

$$\begin{pmatrix} \overline{e_1} & \cdots & \overline{e_{n-i+1}} & \overline{e_{n-i+2}} & \cdots & \overline{e_n} \\ \overline{e_i} & \cdots & \overline{e_n} & \overline{e_1} & \cdots & \overline{e_{i-1}} \end{pmatrix}$$

is a Nakayama permutation of $\overline{R}$, then

$$\begin{pmatrix} e_1 & \cdots & e_{n-i} & e_{n-i+1} & \cdots & e_n \\ e_{i+1} & \cdots & e_n & e_1 & \cdots & e_i \end{pmatrix}$$

is a Nakayama permutation of R.

Since $S(e_1 R_R) \cong S(\overline{e_2 R}_{\overline{R}}) = T(\overline{e_{i+1} R}_{\overline{R}}) \cong T(e_{i+1} R_R)$, we see that $S(e_1 R_R) \cong T(e_{i+1} R_R)$. Similarly we have $S(e_2 R_R) \cong T(e_{i+2} R_R), \ldots, S(e_{n-i} R_R) \cong T(e_n R_R)$, $S(e_{n-i+1} R_R) \cong T(e_1 R_R), \ldots, S(e_n R_R) \cong T(e_i R_R)$ as desired. □

7.3 A Classification of Nakayama Rings

In this section, we give a classification of Nakayama rings by proving Results 1–4 as detailed in the introduction of this chapter.

We first note that, by the structure theorem of a left H-ring, every basic indecomposable Nakayama ring R has a frame Nakayama QF-subring $F(R)$ and R can be represented as an upper staircase factor ring of a block extension of $F(R)$. Therefore, to understand the structure of Nakayama rings, we should first focus on basic indecomposable Nakayama QF-rings.

Notation. Let R be a basic indecomposable Nakayama QF-ring with the Kupisch series $e_n R, \ldots, e_1 R$. Then, by Proposition 7.2.5, there exists i

for which

$$\begin{pmatrix} e_1 & e_2 & \cdots & e_{n-i+1} & e_{n-i+2} & \cdots & e_n \\ e_i & e_{i+1} & \cdots & e_n & e_1 & \cdots & e_{i-1} \end{pmatrix}$$

is a Nakayama permutation. In this case, we say that R is a ring of $KNP(1 \to i)$-type for $Pi(R) = \{e_1, \ldots, e_n\}$. We will see later that the rings of $KNP(1 \to n)$-type　$KNP(1 \to 2)$-type and $KNP(1 \to n)$-type are fundamental types in the class of basic indecomposable Nakayama QF-rings.

Proposition 7.3.1. *Let R be a basic indecomposable Nakayama QF-ring with the Kupisch series $e_n R, \ldots, e_1 R$ and put $\overline{R} = R/S(R_R)$. Then the following hold:*

(1)　*For each $i \in \{2, \ldots, n\}$, if R is of $KNP(1 \to i)$-type for $Pi(R) = \{e_1, \ldots, e_n\}$, then $\overline{R}$ is of $KNP(1 \to i - 1)$-type for $Pi(\overline{R}) = \{\overline{e_1}, \ldots, \overline{e_n}\}$.*

(2)　*If R is of $KNP(1 \to 1)$-type for $Pi(R) = \{e_1, \ldots, e_n\}$, then $\overline{R}$ is of $KNP(1 \to n)$-type for $Pi(\overline{R}) = \{\overline{e_1}, \ldots, \overline{e_n}\}$.*

Proof. This follows from the proof of Proposition 7.2.5.　　　　□

Consider a local QF-ring Q. Let $c \in J(Q)$ and $\sigma \in \mathrm{Aut}(Q)$ satisfying $\sigma(c) = c$ and $\sigma(q)c = cq$ for any $q \in Q$. Set

$$R = \begin{pmatrix} Q & \cdots & Q \\ & \cdots & \\ Q & \cdots & Q \end{pmatrix}_{\sigma, c, n}$$

and $e_i = \langle 1 \rangle_{ii}$ for each $i = 1, \ldots, n$. Then, by Theorem 6.1.4, R is a basic indecomposable QF-ring with the Nakayama permutation

$$\begin{pmatrix} e_1 & e_2 & \cdots & e_n \\ e_n & e_1 & \cdots & e_{n-1} \end{pmatrix}.$$

For this ring R, we show the following theorem.

Theorem 7.3.2. *The following are equivalent:*

(1)　*R is a Nakayama ring.*

(2)　*$cQ = J(Q)$.*

In this case, for each $i = 1, \ldots, n$, $R/S_i(R_R)$ is a basic indecomposable Nakayama QF-ring of $KNP(1 \to n - i + 1)$-type for $Pi(R) = \{e_1, \ldots, e_n\}$.

Proof. (1) $\Rightarrow$ (2). Assume that R is a Nakayama ring and let $a \in Q - cQ$. Then

$$
I = \begin{pmatrix} cQ & Q & \cdots & Q \\ 0 & \cdots & \cdots & 0 \\ \vdots & & & \vdots \\ 0 & \cdots & \cdots & 0 \end{pmatrix}, \quad L = \begin{pmatrix} aQ & \cdots & \cdots & aQ \\ 0 & \cdots & \cdots & 0 \\ \vdots & & & \vdots \\ 0 & \cdots & \cdots & 0 \end{pmatrix}
$$

are right ideals of R. Then $L \not\subseteq I$ implies $I \subsetneq L$ and it follows that $Q = aQ$ and hence we see that $cQ = J$.

(2) $\Rightarrow$ (1). Suppose that $cQ = J(Q)$. Then, since $J(Q) \cong Q/S(Q)$ as a right and left Q-module, we see that $Q/J(Q)^2$ is a Nakayama ring, whence Q is a Nakayama ring by Lemma 7.1.4. As noted above, R is a basic indecomposable QF-ring and

$$
\begin{pmatrix} e_1 & e_2 & \cdots & e_n \\ e_n & e_1 & \cdots & e_{n-1} \end{pmatrix}
$$

is the Nakayama permutation of R. Further, since $Qc = J(Q)$, by Proposition 6.1.3 R has a cyclic quiver:

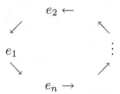

Hence R is a Nakayama QF-ring of $KNP(1 \to n)$-type for $Pi(R) = \{e_1, \ldots, e_n\}$. Moreover, in this case, we can inductively reduce that

$$
S(R) = \begin{pmatrix} 0 & \cdots & \cdots & \cdots & 0 & S(Q) \\ S(Q) & \ddots & & & & 0 \\ 0 & \ddots & \ddots & & & \vdots \\ \vdots & \ddots & \ddots & \ddots & & \vdots \\ \vdots & & \ddots & \ddots & \ddots & \vdots \\ 0 & \cdots & \cdots & 0 & S(Q) & 0 \end{pmatrix},
$$

$$
S_2(R) = \begin{pmatrix}
0 & \cdots \cdots \cdots & 0 & S(Q) & S(Q) \\
S(Q) & \ddots & & \ddots & S(Q) \\
S(Q) & \ddots & \ddots & & 0 \\
0 & \ddots & \ddots & \ddots & \vdots \\
\vdots & \ddots & \ddots & \ddots & \ddots & \vdots \\
\vdots & & \ddots & \ddots & \ddots & \ddots & \vdots \\
0 & \cdots \cdots & 0 & S(Q) & S(Q) & 0
\end{pmatrix},
$$

$$
S_i(R) = \begin{pmatrix}
0 & \cdots \cdots & 0 & \overset{\overset{n-i+1}{\vee}}{S(Q)} & \cdots & S(Q) \\
S(Q) & \ddots & & & \ddots & \ddots & \vdots \\
\vdots & \ddots & \ddots & & & \ddots & S(Q) \\
S(Q) & & \ddots & \ddots & & & 0 \\
0 & \ddots & & \ddots & \ddots & & \vdots \\
\vdots & \ddots & \ddots & & \ddots & \ddots & \vdots \\
0 & \cdots & 0 & S(Q) & \cdots & S(Q) & 0
\end{pmatrix},
$$

$$
S_{n-1}(R) = \begin{pmatrix}
0 & S(Q) & \cdots \cdots \cdots & \cdots & S(Q) \\
S(Q) & \ddots & \ddots & & \vdots \\
\vdots & \ddots & \ddots & \ddots & \vdots \\
\vdots & & \ddots & \ddots & \ddots & \vdots \\
\vdots & & & \ddots & \ddots & S(Q) \\
S(Q) & \cdots & \cdots \cdots \cdots & S(Q) & 0
\end{pmatrix}.
$$

We look at the form of $S_i(R)$. By applying Fuller's Theorem (Corollary 2.4.6), we see that $R/S_i(R)$ is also a QF-ring, and hence $R/S_i(R)$ is a basic indecomposable Nakayama QF-ring by Theorem 7.2.4. Furthermore the form of $S_i(R)$ shows that R is of $KNP(1 \to n - i + 1)$-type for $Pi(R) = \{e_1, \ldots, e_n\}$.   □

Theorem 7.3.3. *Let R be a basic indecomposable Nakayama QF-ring of $KNP(1 \rightarrow n)$-type for $Pi(R) = \{e_1, \ldots, e_n\}$. Then R can be represented as a skew matrix ring over a local Nakayama QF-ring $Q = e_1 R e_1$, namely*

$$R = \begin{pmatrix} Q & \cdots & Q \\ & \cdots & \\ Q & \cdots & Q \end{pmatrix}_{\sigma, c, n},$$

where $\sigma \in \mathrm{Aut}(Q)$ and $c \in J(Q)$ with $cQ = J(Q)$.

Proof. Let $\{i_1, i_2, \ldots, i_k\} \subseteq \{1, 2, \ldots, n\}$ with $i_1 < i_2 < \cdots < i_k$. Put $e = e_{i_1} + \cdots + e_{i_k}$. Then, since R is a Nakayama ring of $KNP(1 \rightarrow n)$-type for $Pi(R) = \{e_1, \ldots, e_n\}$, we see that eRe is a basic indecomposable Nakayama QF-ring of $KNP(1 \rightarrow k)$-type for $\{e_{i_1}, \ldots, e_{i_k}\}$, and hence eRe has a cyclic Nakayama permutation. Thus, putting $Q = e_1 R e_1$, by Theorem 6.4.1, there exist $\sigma \in \mathrm{Aut}(Q)$ and $c \in J(Q)$ for which R can be represented as a skew matrix ring over Q. Further, by Theorem 7.3.2, we see that $cQ = J(Q)$. $\square$

By Proposition 7.3.1 and Theorem 7.3.3, we obtain the following theorem.

Theorem 7.3.4. *Let R be a basic indecomposable Nakayama QF-ring of $KNP(1 \rightarrow 1)$-type for $Pi(R) = \{e_1, \ldots, e_n\}$ and put $Q = e_1 R e_1$. Then $\overline{R} = R/S(R)$ is a ring of $KNP(1 \rightarrow n)$-type for $Pi(\overline{R}) = \{\overline{e_1}, \ldots, \overline{e_n}\}$ and it can be represented as a skew matrix ring over the local Nakayama QF-ring $\overline{Q} = Q/S(R)$ with respect to some $\sigma \in \mathrm{Aut}(\overline{Q})$ and $\overline{c} \in J(\overline{Q})$.*

For later use, we illustrate an example of a Nakayama QF-ring of $KNP(1 \rightarrow 2)$-type.

Example 7.3.5. Let Q be a local Nakayama ring and let $c \in Q$ such that $J(Q) = Qc$. Let $\sigma \in \mathrm{Aut}(Q)$ satisfying $\sigma(c) = c$ and $\sigma(q)c = cq$ for all $q \in Q$. We consider the skew matrix ring

$$R = \begin{pmatrix} Q & Q \\ Q & Q \end{pmatrix}_{\sigma, c}$$

and represent R by the skew matrix units $\{\alpha_{11}, \alpha_{12}, \beta_{21}, \alpha_{22}\}$:

$$R = \begin{pmatrix} Q\alpha_{11} & Q\alpha_{12} \\ Q\beta_{21} & Q\alpha_{22} \end{pmatrix}_{\sigma, c} = \begin{pmatrix} Q & Q\alpha \\ Q\beta & Q \end{pmatrix}_{\sigma, c},$$

where $\alpha = \alpha_{12}$ and $\beta = \beta_{21}$. Now we make the following block extension of R with the matrix size n:

$$T = \begin{pmatrix} Q & Q\alpha & \cdots & \cdots & Q\alpha \\ Q\beta & Q & \cdots & \cdots & Q \\ \vdots & Qc & \ddots & & \vdots \\ \vdots & \vdots & \ddots & \ddots & \vdots \\ Q\beta & Qc & \cdots & Qc & Q \end{pmatrix}.$$

Then, by Theorem 6.5.1, there exists a canonical surjective matrix ring homomorphism

$$\gamma : W = \begin{pmatrix} Q & \cdots & Q \\ & \cdots & \\ Q & \cdots & Q \end{pmatrix}_{\sigma,c,n} \rightarrow T = \begin{pmatrix} Q & Q\alpha & \cdots & \cdots & Q\alpha \\ Q\beta & Q & \cdots & \cdots & Q \\ \vdots & Qc & \ddots & & \vdots \\ \vdots & \vdots & & \ddots & \vdots \\ Q\beta & Qc & \cdots & Qc & Q \end{pmatrix}$$

with

$$\mathrm{Ker}\,\gamma = \begin{pmatrix} 0 & \cdots & \cdots & \cdots & 0 \\ \vdots & 0 & & & \vdots \\ \vdots & S(Q) & \ddots & & \vdots \\ \vdots & \vdots & \ddots & \ddots & \vdots \\ 0 & S(Q) & \cdots & S(Q) & 0 \end{pmatrix}.$$

We represent $W/\mathrm{Ker}\,\gamma$ as

$$W/\mathrm{Ker}\,\gamma = \begin{pmatrix} Q & \cdots & \cdots & \cdots & Q \\ \vdots & Q & & & \vdots \\ \vdots & \overline{Q} & \ddots & & \vdots \\ \vdots & \vdots & \ddots & \ddots & \vdots \\ Q & \overline{Q} & \cdots & \overline{Q} & Q \end{pmatrix}.$$

Since W is a Nakayama ring, T is also a Nakayama ring and

$$T = \begin{pmatrix} Q & Q\alpha & \cdots & \cdots & Q\alpha \\ Q\beta & Q & \cdots & \cdots & Q \\ \vdots & Qc & \ddots & & \vdots \\ \vdots & \vdots & \ddots & \ddots & \vdots \\ Q\beta & Qc & \cdots & Qc & Q \end{pmatrix} \cong \begin{pmatrix} Q & \cdots & \cdots & \cdots & Q \\ \vdots & Q & \cdots & \cdots & Q \\ \vdots & \overline{Q} & \ddots & & \vdots \\ \vdots & \vdots & \ddots & \ddots & \vdots \\ Q & \overline{Q} & \cdots & \overline{Q} & Q \end{pmatrix} = W/\mathrm{Ker}\,\gamma.$$

Now consider the following ideals V of W and U of T:

$$V = \begin{pmatrix} 0 & 0 & S(Q) & \cdots & \cdots & S(Q) \\ S(Q) & 0 & \ddots & S(Q) & \cdots & S(Q) \\ \vdots & S(Q) & \ddots & \ddots & \ddots & \vdots \\ \vdots & \vdots & \ddots & \ddots & \ddots & S(Q) \\ S(Q) & \vdots & & \ddots & \ddots & 0 \\ 0 & S(Q) & \cdots & \cdots & S(Q) & 0 \end{pmatrix},$$

$$U = \begin{pmatrix} 0 & 0 & S(Q\alpha) & \cdots & \cdots & S(Q\alpha) \\ S(Q\beta) & 0 & \ddots & S(Q) & \cdots & S(Q) \\ S(Q\beta) & S(Q) & \ddots & \ddots & \ddots & \vdots \\ \vdots & \vdots & \ddots & \ddots & \ddots & S(Q) \\ S(Q\beta) & \vdots & & \ddots & \ddots & 0 \\ 0 & S(Q) & \cdots & \cdots & S(Q) & 0 \end{pmatrix}$$

and represent $\overline{T} = T/U$ and $\overline{W} = W/V$ as follows:

$$\overline{T} = \begin{pmatrix} Q & Q\alpha & \overline{Q\alpha} & \cdots & \cdots & \overline{Q\alpha} \\ \overline{Q\beta} & Q & Q & \overline{Q} & \cdots & \overline{Q} \\ \vdots & \overline{J(Q)} & \ddots & \ddots & \ddots & \vdots \\ \vdots & \vdots & \ddots & \ddots & \ddots & \overline{Q} \\ \overline{Q\beta} & \vdots & & \ddots & \ddots & Q \\ Q\beta & \overline{J(Q)} & \cdots & \cdots & \overline{J(Q)} & Q \end{pmatrix},$$

$$\overline{W} = \begin{pmatrix} Q & Q & \overline{Q} & \cdots & \cdots & \overline{Q} \\ \overline{Q} & Q & \ddots & \ddots & & \vdots \\ \vdots & \overline{Q} & \ddots & \ddots & \ddots & \vdots \\ \vdots & \vdots & \ddots & \ddots & \ddots & \overline{Q} \\ \overline{Q} & \vdots & & \ddots & \ddots & Q \\ Q & \overline{Q} & \cdots & \cdots & \overline{Q} & Q \end{pmatrix}.$$

Then $\overline{W} \cong \overline{T}$ and these rings are basic indecomposable Nakayama QF-rings of $KNP(1 \to 2)$-type.

In summary, starting from a skew matrix ring

$$R = \begin{pmatrix} Q & Q \\ Q & Q \end{pmatrix}_{\sigma,c} = \begin{pmatrix} Q & Q\alpha \\ Q\beta & Q \end{pmatrix}_{\sigma,c}$$

with $cQ = J(Q)$, we make a block extension T of R with matrix size n. Then T is a homomorphic image of a skew matrix ring W over Q with the same matrix size n. Furthermore we can construct isomorphic canonical factor rings $\overline{W}$ and $\overline{T}$ which are basic indecomposable Nakayama QF-rings of $KNP(1 \to 2)$-type. As we see later, this observation plays an important role.

We need further details on a basic indecomposable Nakayama QF-ring of $KNP(1 \to 2)$-type. Let R be such a ring for $Pi(R) = \{e_1, \ldots, e_n\}$. In order to investigate R, we prepare some notations. For each $i, j \in \{1, \ldots, n\}$ with $1 \le i \le j \le n$, we define a homomorphism $\theta_{ij} : e_j R \to e_i R$ as follows:

1. θ_{ii} is the identity map of $e_i R$.
2. $\theta_{i,i+1}: e_{i+1}R \to e_i J$ is an epimorphism for $1 \le i < n$ (of course such $\theta_{i,i+1}$ exists).
3. $\theta_{n1} : e_1 R \to e_n J$ is an epimorphism.
4. $\theta_{ij} = \theta_{i,i+1}\theta_{i+1,i+2} \cdots \theta_{j-1,j}$ for $1 \le i < j \le n$.

Then the following hold:

(a) $\operatorname{Ker} \theta_{ij} = S_{j-i}(e_j R_R)$.

(b) $\operatorname{Im} \theta_{ij} = e_i J^{j-i}$, so θ_{ij} induces the isomorphism:

$$e_j R/S_{j-i}(e_j R_R) \overset{\overline{\theta_{ij}}}{\cong} e_i J^{j-i}, \text{ where } J^0 = R.$$

(c)   For any homomorphism $\alpha : e_t R \to e_j R/S_k(e_j R_R)$, there exists $\beta : e_t R \to e_j R$ satisfying $\alpha = \eta\beta$, where $\eta : e_j R \to e_j R/S_k(e_j R_R)$ is the canonical epimorphism. Here, if $\operatorname{Hom}_R(e_t R, S_k(e_j R_R)) = 0$, we see that this β is uniquely determined, and we denote it by $[\alpha]_{j(k)t}$ or, simply, by $[\alpha]$ if no confusion arises:

$$
\begin{array}{ccc}
 & & e_t R \\
 & [\alpha]\swarrow & \downarrow \alpha \\
e_j R & \overset{\eta}{\longrightarrow} & e_j R/S_k(e_j R_R) \longrightarrow 0
\end{array}
$$

(d)   For a homomorphism $e_t R \overset{\gamma}{\to} e_i J^{j-i} \overset{\overline{\theta_{ij}}^{-1}}{\cong} e_j R/S_{j-i}(e_j R_R)$, where

$i \le j \le t$, if $\operatorname{Hom}_R(e_t R, S_{j-i}(e_j R)) = 0$, then $[\,\overline{\theta}_{ij}^{\,-1}\ \gamma\,] \in \operatorname{Hom}_R(e_t R, e_j R)$ can be defined, and we see that

$$\theta_{ij}[\,\overline{\theta}_{ij}^{\,-1}\ \gamma\,] = \gamma,$$

(e)   For $\theta_{ij}\alpha = \theta_{ij}\beta$, where $\alpha, \beta \in \operatorname{Hom}_R(e_t R, e_j R)$ and $i \le j \le t$, if $\operatorname{Hom}(e_t R, S_{j-i}(e_j R_R)) = 0$, then $\alpha = \beta$ because $\theta_{ij}(\alpha - \beta) = 0$ implies $\operatorname{Im}(\alpha - \beta) \subseteq S_{j-i}(e_j R_R)$, whence $\alpha - \beta = 0$ by assumption.

We put $A_{ij} = e_i R e_j = (e_i, e_j) = \operatorname{Hom}_R(e_j R, e_i R)$ for each $i, j \in \{1, \dots, n\}$ and use the following expressions for R:

$$\begin{pmatrix} e_1 R e_1 & \cdots & e_1 R e_n \\ & \cdots & \\ e_n R e_1 & \cdots & e_n R e_n \end{pmatrix} = \begin{pmatrix} (e_1, e_1) & \cdots & (e_1, e_n) \\ & \cdots & \\ (e_n, e_1) & \cdots & (e_n, e_n) \end{pmatrix} = \begin{pmatrix} A_{11} & \cdots & A_{1n} \\ & \cdots & \\ A_{n1} & \cdots & A_{nn} \end{pmatrix}.$$

Under the conditions above, we show the following lemma.

Lemma 7.3.6.   *Let R be a basic indecomposable Nakayama QF-ring with $Pi(R) = \{e_1, e_2, e_3\}$. Suppose that R is of KNP $(1 \to 2)$-type for $Pi(R) = \{e_1, e_2, e_3\}$. Put $Q = e_1 R e_1$. Then the following hold:*

(1)   *There exist $c \in J(Q)$ and $\sigma \in Q$ satisfying $Qc = J(Q)$, $\sigma(c) = c$, and $\sigma(q)c = cq$ for all $q \in Q$.*

(2)   *R is isomorphic to a factor ring of a skew matrix ring $(Q)_{\sigma,c,n}$. More precisely, R can be represented as*

$$R \cong \begin{pmatrix} Q & Q & \overline{Q} \\ \overline{Q} & Q & Q \\ Q & \overline{Q} & Q \end{pmatrix}_{\sigma,c,3} = \begin{pmatrix} Q & Q & Q \\ Q & Q & Q \\ Q & Q & Q \end{pmatrix}_{\sigma,c,3} \Big/ \begin{pmatrix} 0 & 0 & S(Q) \\ S(Q) & 0 & 0 \\ 0 & S(Q) & 0 \end{pmatrix},$$

where $\overline{Q} = Q/S(Q)$.

Proof.   We represent R as

$$R = \begin{pmatrix} (e_1, e_1) & (e_2, e_1) & (e_3, e_1) \\ (e_1, e_2) & (e_2, e_2) & (e_3, e_2) \\ (e_1, e_3) & (e_2, e_3) & (e_3, e_3) \end{pmatrix} = \begin{pmatrix} Q & A_{12} & A_{13} \\ B_{21} & Q_2 & A_{23} \\ B_{31} & B_{32} & Q_3 \end{pmatrix},$$

where $Q_i = (e_i, e_i)$ for $i = 2, 3$, and $A_{ij} = (e_j, e_i)$, $B_{ji} = (e_i, e_j)$ for each $i, j \in \{1, 2, 3\}$ with $i < j$. Put

$$Z = \begin{pmatrix} (e_1, e_1) & (e_2, e_1) \\ (e_1, e_3) & (e_2, e_2) \end{pmatrix}.$$

For the sake of convenience, in Z, we use $\langle x \rangle_{31}$ instead of $\langle x \rangle_{21}$, that is, $\langle x \rangle_{31}$ is the matrix in Z whose $(2,1)$ entry is x and the others are 0.

We take an addition on Z as the usual matrix sum and define a multiplication on Z as follows. For

$$\alpha = \begin{pmatrix} x_{11} & x_{12} \\ x_{31} & x_{22} \end{pmatrix}, \quad \beta = \begin{pmatrix} y_{11} & y_{12} \\ y_{31} & y_{22} \end{pmatrix},$$

we set

$$\alpha\beta = \begin{pmatrix} x_{11}y_{11} + x_{12}\theta_{23}y_{31} & x_{11}y_{12} + x_{12}y_{22} \\ x_{31}y_{11} + [\overline{\theta_{23}}^{-1}x_{22}\theta_{23}]y_{31} & \theta_{23}x_{31}y_{12} + x_{22}y_{22} \end{pmatrix}.$$

Of course, $[\overline{\theta_{23}}^{-1}x_{22}\theta_{23}] = [\overline{\theta_{23}}^{-1}x_{22}\theta_{23}]_{3(1)3}$ is well defined since $\operatorname{Hom}_R(e_3R, S(e_3R_R)) = 0$.

For Z, we show the following Claim 1:

Claim 1. *Z is an artinian ring.*

Proof of Claim 1. To check the distributivity of elements of Z is easy. For the associativity of the multiplication, we show that

$$(\langle x \rangle_{ij}\langle y \rangle_{jk})\langle z \rangle_{kl} = \langle x \rangle_{ij}(\langle y \rangle_{jk}\langle z \rangle_{kl})$$

for any ij, jk, kl. There is no need to check the case that $31 \notin \{ij, jk, kl\}$.

Let $ij = 31$. Then we should observe the following cases of $\{ij, jk, kl\}$: $\{31, 11, 11\}$, $\{31, 11, 12\}$, $\{31, 12, 31\}$, $\{31, 12, 21\}$, $\{31, 12, 22\}$. We can easily check the associative law except $\{ij, jk, kl\} = \{31, 12, 31\}$. Let $\{ij, jk, kl\} = \{31, 12, 31\}$. Then $(\langle x \rangle_{31}\langle y \rangle_{12})\langle z \rangle_{31} = \langle \theta_{23}xy \rangle_{22}\langle z \rangle_{31} = \langle [\overline{\theta_{23}}^{-1}\theta_{23}xy\theta_{23}]z \rangle_{31}$. $\langle x \rangle_{31}(\langle y \rangle_{12}\langle z \rangle_{31}) = \langle x \rangle_{31}\langle y\theta_{23}z \rangle_{31} = \langle xy\theta_{23}z \rangle_{31}$.

Since $\theta_{23}[\overline{\theta_{23}}^{-1}\theta_{23}xy\theta_{23}] = \theta_{23}xy\theta_{23}$, we obtain $[\overline{\theta_{23}}^{-1}\theta_{23}xy\theta_{23}] = xy\theta_{23}$. Hence $(\langle x \rangle_{31}\langle y \rangle_{12})\langle z \rangle_{31} = \langle x \rangle_{31}(\langle y \rangle_{12}\langle z \rangle_{31})$.

Let $jk = 31$. Then we need to check for the cases of $\{ij, jk, kl\}$: $\{12, 31, 11\}$, $\{22, 31, 11\}$, $\{12, 31, 12\}$, $\{22, 31, 12\}$. It is easy to check the associative law except $\{ij, jk, kl\} = \{22, 31, 12\}$. For $\{22, 31, 12\}$, $(\langle x \rangle_{22}\langle y \rangle_{31})\langle z \rangle_{12} = \langle [\overline{\theta_{23}}^{-1}]x\theta_{23}y \rangle_{31}\langle z \rangle_{12} = \langle \theta_{23}[\overline{\theta_{23}}^{-1}x\theta_{23}]yz \rangle_{22} = \langle x\theta_{23}yz \rangle_{22}$ and $\langle x \rangle_{22}(\langle y \rangle_{31}\langle z \rangle_{12}) = \langle x \rangle_{22}\langle \theta_{23}yz \rangle_{22} = \langle x\theta_{23}yz \rangle_{22}$. Hence $(\langle x \rangle_{22}\langle y \rangle_{31})\langle z \rangle_{12} = \langle x \rangle_{22}(\langle y \rangle_{31}\langle z \rangle_{12})$.

Let $kl = 31$. Then we need to consider the following cases of $\{ij, jk, kl\}$: $\{11, 12, 31\}$, $\{21, 12, 31\}$, $\{12, 22, 31\}$, $\{22, 22, 31\}$, $\{31, 12, 31\}$. For these except $\{31, 12, 31\}$, it is easy to check. For $\{31, 12, 31\}$, we have $(\langle x \rangle_{31}\langle y \rangle_{12})\langle z \rangle_{31} = \langle \theta_{23}xy \rangle_{22}\langle z \rangle_{31} =$

$\langle\,[\,\overline{\theta_{23}}^{-1}(\theta_{23}xy)\theta_{23}\,]z\,\rangle_{31} = \langle\,xy\theta_{23}z\,\rangle_{31}$ since $\theta_{23}([\,\overline{\theta_{23}}^{-1}\theta_{23}xy\theta_{23}\,]z) = \theta_{23}(xy\theta_{23}z)$. $\langle\,x\,\rangle_{31}(\langle\,y\,\rangle_{12}\langle\,z\,\rangle_{31}) = \langle\,x\,\rangle_{31}\langle\,y\theta_{23}z\,\rangle_{11} = \langle\,xy\theta_{23}z\,\rangle_{31}$. Hence $(\langle\,x\,\rangle_{31}\langle\,y\,\rangle_{12})\langle\,z\,\rangle_{31} = \langle\,x\,\rangle_{31}(\langle\,y\,\rangle_{12}\langle\,z\,\rangle_{31})$.

From these observations, we see that Z becomes a ring and it is an artinian ring. Thus Claim 1 is shown.

In Z, let $f_i = \langle\,1\,\rangle_{ii}$ for $i = 1, 2$. Then $\{f_1, f_2\}$ is a complete set of orthogonal primitive idempotents of Z.

Next we claim the following:

Claim 2.

(I)   *The map*

$$\tau = \begin{pmatrix} \tau_{11} & \tau_{12} \\ \tau_{21} & \tau_{22} \end{pmatrix} : Z = \begin{pmatrix} (e_1, e_1) & (e_2, e_1) \\ (e_1, e_3) & (e_2, e_2) \end{pmatrix} \longrightarrow \begin{pmatrix} (e_1, e_1) & (e_2, e_1) \\ (e_1, e_2) & (e_2, e_2) \end{pmatrix}$$

given by

$$\begin{pmatrix} \alpha_{11} & \alpha_{12} \\ \alpha_{21} & \alpha_{22} \end{pmatrix} \longmapsto \begin{pmatrix} \alpha_{11} & \alpha_{12} \\ \theta_{23}\alpha_{31} & \alpha_{22} \end{pmatrix}$$

is a surjective ring homomorphism with

$$\mathrm{Ker}\,\tau = \begin{pmatrix} 0 & 0 \\ S((e_1, e_3)) & 0 \end{pmatrix}.$$

(II) (f_1Z, Zf_2) *and* (f_2Z, Zf_1) *are i-pairs. Thus* Z *is a basic indecomposable Nakayama QF-ring with a cyclic Nakayama permutation.*

Proof of Claim 2. (I). To show that τ_{21} is an onto map, let $x_{21} \in (e_1, e_2)$. Then $[\theta_{23}^{-1}x_{21}] \in (e_1, e_3)$ and $\theta_{23}[\theta_{23}^{-1}x_{21}] = x_{21}$, whence τ_{21} is an onto map. Since $\mathrm{Ker}\,\tau_{21} = \{\,\alpha_{31}\mid \theta_{23}\alpha_{31} = 0\,\} = \{\,\alpha_{31}\mid \mathrm{Im}\,\alpha_{31} \subseteq S(e_3R_R)\,\} = S((e_1, e_3))$, we have

$$\mathrm{Ker}\,\tau = \begin{pmatrix} 0 & 0 \\ S((e_1, e_3)) & 0 \end{pmatrix}.$$

(II). Since R is a ring of $KNP(1 \to 2)$-type, we see that

$$\begin{cases} S(f_1Z_Z) = \begin{pmatrix} 0 & S((e_1, e_2)) \\ 0 & 0 \end{pmatrix} = S(_ZZf_2), \\[2mm] \begin{pmatrix} 0 & 0 \\ S((e_1, e_3)) & 0 \end{pmatrix} = S(_ZZf_1). \end{cases}$$

In order to show

$$S(f_2Z_Z) = \begin{pmatrix} 0 & 0 \\ S((e_1, e_3)) & 0 \end{pmatrix} = S(_ZZf_1),$$

let $0 \neq \alpha_{22} \in (e_2, e_2)$ and assume $[\overline{\theta_{23}}^{-1}\alpha_{22}\theta_{23}](e_1, e_3) = 0$. Then, noting $[\overline{\theta_{23}}^{-1}\alpha_{22}\theta_{23}] \in (e_3, e_3)$ and

$$S(e_3 R_R) = \begin{pmatrix} 0 & 0 & 0 \\ 0 & 0 & 0 \\ S((e_1, e_3)) & 0 & 0 \end{pmatrix},$$

we see that $[\overline{\theta_{23}}^{-1}\alpha_{22}\theta_{23}] = 0$, whence $\alpha_{22}e_2J = 0$. Since $\mathrm{Hom}_R(e_2 R, S(e_2 R_R)) = 0$, this implies $\alpha_{22} = 0$, a contradiction. Therefore $[\overline{\theta_{23}}^{-1}\alpha\theta_{23}](e_1, e_3) \neq 0$ for any $0 \neq \alpha_{22} \in (e_2, e_2)$. This fact, together with $\theta_{23}S((e_1, e_3)) = \theta_{23}\{\beta \in (e_1, e_3) \mid \mathrm{Im}\,\beta \subseteq S(e_3 R_R)\} = 0$, shows

$$S(f_2 Z_Z) = \begin{pmatrix} 0 & 0 \\ S((e_1, e_3)) & 0 \end{pmatrix}.$$

Accordingly $(f_1 Z; Z f_2)$ and $(f_2 Z; Z f_1)$ are i-pairs and hence Z is a basic indecomposable QF-ring. Moreover it follows from (I) that Z is a Nakayama ring with the Nakayama permutation

$$\begin{pmatrix} f_1 & f_2 \\ f_2 & f_1 \end{pmatrix}.$$

Thus Claim 2 is shown.

Now we shall show (1) and (2) in our statement. By Theorem 7.3.3, there exist $c \in J(f_1 Z f_1) = J(Q)$ and $\sigma \in Q$ satisfying $Qc = J(Q)$, $\sigma(c) = c$ and $\sigma(q)c = cq$ for any $q \in Q$, and there exists a matrix ring isomorphism

$$Z \cong P = \begin{pmatrix} Q & Q \\ Q & Q \end{pmatrix}_{\sigma, c}.$$

We return to

$$R = \begin{pmatrix} (e_1, e_1) & (e_2, e_1) & (e_3, e_1) \\ (e_1, e_2) & (e_2, e_2) & (e_3, e_2) \\ (e_1, e_3) & (e_2, e_3) & (e_3, e_3) \end{pmatrix} = \begin{pmatrix} Q & A_{12} & A_{13} \\ B_{21} & Q_2 & A_{23} \\ B_{31} & B_{32} & Q_3 \end{pmatrix}.$$

By the form of the Nakayama permutation of R, we see that

$$S(R) = \begin{pmatrix} 0 & S(A_{12}) & 0 \\ 0 & 0 & S(A_{23}) \\ S(B_{31}) & 0 & 0 \end{pmatrix}.$$

We put

$$\overline{R} = R/S(e_3 R_R) = \begin{pmatrix} Q & A_{12} & A_{13} \\ B_{21} & Q_2 & A_{23} \\ \overline{B_{31}} & B_{32} & Q_3 \end{pmatrix}.$$

Then $\overline{e_1 R_{\overline{R}}}$ and $\overline{e_2 R_{\overline{R}}}$ are injective and $\overline{e_2 J_{\overline{R}}} \cong \overline{e_3 R_{\overline{R}}}$. Since $\overline{R}$ is a left H-ring and $S(\overline{e_2 R_{\overline{R}}}) = \langle S(A_{23}) \rangle_{23}$, using the matrix representative theorem of left H-rings discussed in Theorem 4.3.1, we see that $\overline{R}$ is represented as a factor ring:

$$\overline{R} = \begin{pmatrix} Q & A_{12} & A_{13} \\ B_{21} & Q_2 & A_{23} \\ \overline{B}_{31} & B_{32} & Q_3 \end{pmatrix} \cong \begin{pmatrix} Q & A_{12} & \overline{A}_{12} \\ B_{21} & Q_2 & Q_2 \\ B_{21} & J(Q_2) & Q_2 \end{pmatrix}.$$

Put

$$P = \begin{pmatrix} Q & A_{12} & A_{12} \\ B_{21} & Q_2 & Q_2 \\ B_{21} & J(Q_2) & Q_2 \end{pmatrix}.$$

Then

$$S(P) = \begin{pmatrix} 0 & S(A_{12}) & 0 \\ 0 & 0 & S(Q_2) \\ 0 & 0 & S(Q_2) \end{pmatrix}.$$

Hence, again from the representative theorem of left H-rings, P can be expressed as

$$\overline{R} \cong \begin{pmatrix} Q & Q & \overline{Q} \\ J(Q) & Q & Q \\ J(Q) & J(Q) & Q \end{pmatrix} = \begin{pmatrix} Q & Q & Q \\ J(Q) & Q & Q \\ J(Q) & J(Q) & Q \end{pmatrix} / \begin{pmatrix} 0 & 0 & S(Q) \\ 0 & 0 & 0 \\ 0 & 0 & 0 \end{pmatrix}.$$

Consequently there exists a matrix ring isomorphism $\varphi = (\varphi_{ij})$:

$$\overline{R} = \begin{pmatrix} Q & A_{12} & A_{13} \\ B_{21} & Q_2 & A_{23} \\ \overline{B}_{31} & B_{32} & Q_3 \end{pmatrix} \overset{\varphi}{\cong} \begin{pmatrix} Q & Q & \overline{Q} \\ J(Q) & Q & Q \\ J(Q) & J(Q) & Q \end{pmatrix} = \begin{pmatrix} Q & Q & Q \\ J(Q) & Q & Q \\ J(Q) & J(Q) & Q \end{pmatrix} / \begin{pmatrix} 0 & 0 & S(Q) \\ 0 & 0 & 0 \\ 0 & 0 & 0 \end{pmatrix}$$

As noted above, there exist $c \in J(Q)$ and $\sigma \in J(Q)$ satisfying $Qc = J(Q), \sigma(c) = c$ and $\sigma(q) = cq$ for all $q \in Q$. Thus we can make $(Q)_{\sigma,c,3}$. Using c, we have a matrix isomorphism $\varphi = (\varphi_{ij})$:

$$\overline{R} = \begin{pmatrix} Q & A_{12} & A_{13} \\ B_{21} & Q & A_{23} \\ \overline{B}_{31} & B_{32} & Q \end{pmatrix} \overset{\varphi}{\cong} \begin{pmatrix} Q & Q & \overline{Q} \\ Qc & Q & Q \\ Qc & Qc & Q \end{pmatrix}.$$

We put

$$\begin{cases} a_{12} = \varphi_{12}^{-1}(1), \\ a_{23} = \varphi_{23}^{-1}(1), \\ b_{21} = \varphi_{21}^{-1}(c), \\ b_{32} = \varphi_{32}^{-1}(c), \\ a_{13} = a_{12} a_{23}. \end{cases}$$

We can take $b_{31} \in B_{31}$ such that $cb_{31} = b_{32}b_{21}$ in R and $\overline{cb_{31}} = \overline{b_{32}b_{21}} = \varphi_{31}^{-1}(c^2)$. Then $c\varphi_{31}^{-1}(c) = \varphi_{11}^{-1}(c)\varphi_{31}^{-1}(c) = \varphi_{31}^{-1}(c^2) = \varphi_{32}^{-1}(c)\varphi_{21}^{-1}(c) = b_{32}b_{21} = cb_{31}$. Hence $c(\varphi_{31}^{-1}(c) - b_{31}) = 0$ and this implies that $\varphi_{31}^{-1}(c) = \overline{b_{31}}$ and $Qa_{ij} = A_{ij}$ and $Qb_{ij} = B_{ij}$ for any $i, j \in \{1, 2, 3\}$ with $i < j$. Thus

$$\overline{R} = \begin{pmatrix} Q & Qa_{12} & Qa_{13} \\ Qb_{21} & Q & Qa_{23} \\ \overline{Qb_{31}} & Qb_{32} & Q \end{pmatrix} \overset{\varphi}{\cong} \begin{pmatrix} Q & Q & \overline{Q} \\ Qc & Q & Q \\ Qc & Qc & Q \end{pmatrix}.$$

Now consider the skew matrix ring

$$T = (Q)_{\sigma,c,3} = \begin{pmatrix} Q & Q\alpha_{12} & Q\alpha_{13} \\ Q\beta_{21} & Q & Q\alpha_{23} \\ Q\beta_{31} & Q\beta_{32} & Q \end{pmatrix},$$

where $\{\alpha_{ij}\} \cup \{\beta_{ji}\}$ are the skew matrix units. By Theorem 6.5.1, the map

$$\tau : T \longrightarrow \begin{pmatrix} Q & Q & Q \\ Qc & Q & Q \\ Qc & Qc & Q \end{pmatrix}$$

given by

$$\begin{pmatrix} q_{11} & q_{12}\alpha_{12} & q_{13}\alpha_{13} \\ q_{21}\beta_{21} & q_{22} & q_{23}\alpha_{23} \\ q_{31}\beta_{31} & q_{32}\beta_{32} & q_{33} \end{pmatrix} \longmapsto \begin{pmatrix} q_{11} & q_{12}a_{12} & q_{13} \\ q_{21}c & q_{22} & q_{23} \\ q_{31}c & q_{32}c & q_{33} \end{pmatrix}$$

is a surjective ring homomorphism with

$$\mathrm{Ker}\,\tau = \begin{pmatrix} 0 & 0 & 0 \\ S(Q) & 0 & 0 \\ S(Q) & S(Q) & 0 \end{pmatrix}.$$

Thus the composite map $\varphi^{-1}\tau$ is given by

$$\begin{pmatrix} q_{11} & q_{12}\alpha_{12} & q_{13}\alpha_{13} \\ q_{21}\beta_{21} & q_{22} & q_{23}\alpha_{23} \\ q_{31}\beta_{31} & q_{32}\beta_{32} & q_{33} \end{pmatrix} \longmapsto \begin{pmatrix} q_{11} & q_{12}a_{12} & q_{13}a_{13} \\ q_{21}b_{21} & q_{22} & q_{23}a_{23} \\ q_{31}b_{31} & q_{32}b_{32} & q_{33} \end{pmatrix}$$

and

$$\mathrm{Ker}\,\varphi^{-1}\tau = \begin{pmatrix} 0 & 0 & S(Q) \\ S(Q) & 0 & 0 \\ 0 & S(Q) & 0 \end{pmatrix}.$$

By this epimorphism together with $cb_{31} = b_{32}b_{21}$, we can easily check that the map $\Phi : T \to R$ given by

$$\begin{pmatrix} q_{11} & q_{12}\alpha_{12} & q_{13}\alpha_{13} \\ q_{21}\beta_{21} & q_{22} & q_{23}\alpha_{23} \\ q_{31}\beta_{31} & q_{32}\beta_{32} & q_{33} \end{pmatrix} \longmapsto \begin{pmatrix} q_{11} & q_{12}a_{12} & q_{13}a_{13} \\ q_{21}b_{21} & q_{22} & q_{23}a_{23} \\ q_{31}b_{31} & q_{32}b_{32} & q_{33} \end{pmatrix}$$

is a surjective ring homomorphism with

$$
\mathrm{Ker}\,\Phi = \begin{pmatrix} 0 & 0 & S(Q) \\ S(Q) & 0 & 0 \\ 0 & S(Q) & 0 \end{pmatrix}.
$$

Consequently

$$
R \cong \begin{pmatrix} Q & Q & \overline{Q} \\ \overline{Q} & Q & Q \\ Q & \overline{Q} & Q \end{pmatrix}_{\sigma,c,n} \qquad \square
$$

Theorem 7.3.7. *Let R be a basic indecomposable Nakayama QF-ring of $KNP(1 \to 2)$-type for $Pi(R) = \{e_1, \ldots, e_n\}$. Then R can be represented as a factor ring of a skew matrix ring $(Q)_{\sigma,c,n}$, where $Q = e_1 R e_1$, $\sigma \in \mathrm{Aut}(Q)$ and $cQ = J(Q)$. More presicely, R can be represented as*

$$
R \cong \begin{pmatrix} Q & Q & \overline{Q} & \cdots & \cdots & \overline{Q} \\ \overline{Q} & Q & \ddots & \ddots & & \vdots \\ \vdots & \ddots & \ddots & \ddots & \ddots & \vdots \\ \vdots & & \ddots & \ddots & \ddots & \overline{Q} \\ \overline{Q} & & & \ddots & \ddots & Q \\ Q & \overline{Q} & \cdots & \cdots & \overline{Q} & Q \end{pmatrix}_{\sigma,c,n}
$$

$$
= \begin{pmatrix} Q & Q & Q & \cdots & \cdots & Q \\ Q & \ddots & \ddots & \ddots & & \vdots \\ \vdots & \ddots & \ddots & \ddots & & \vdots \\ \vdots & & \ddots & \ddots & \ddots & Q \\ Q & & & \ddots & \ddots & Q \\ Q & Q & \cdots & \cdots & Q & Q \end{pmatrix}_{\sigma,c,n} \Big/ \begin{pmatrix} 0 & 0 & S(Q) & \cdots & \cdots & S(Q) \\ S(Q) & \ddots & \ddots & \ddots & \ddots & \vdots \\ \vdots & \ddots & \ddots & \ddots & \ddots & \vdots \\ \vdots & & \ddots & \ddots & \ddots & S(Q) \\ S(Q) & & & \ddots & \ddots & 0 \\ 0 & S(Q) & \cdots & \cdots & S(Q) & 0 \end{pmatrix},
$$

where $\overline{Q} = Q/S(Q)$.

Proof. When $n = 1$, $R = Q$. For $n = 2, 3$, the statement follows from Theorem 6.4.1 and Lemma 7.3.6.

Now, by arguing as in the proof of Lemma 7.3.6, we will establish the case $n = 4$. Here, since R is $KNP(1 \to 2)$-type for $Pi(R) = \{e_1, e_2, e_3, e_4\}$,

$$
\begin{pmatrix} e_1 & e_2 & e_3 & e_4 \\ e_2 & e_1 & e_2 & e_3 \end{pmatrix}
$$

is the Nakayama permutation. We represent R as

$$R = \begin{pmatrix} (e_1, e_1) & (e_2, e_1) & (e_3, e_1) & (e_4, e_1) \\ (e_1, e_2) & (e_2, e_2) & (e_3, e_2) & (e_4, e_2) \\ (e_1, e_3) & (e_2, e_3) & (e_3, e_3) & (e_4, e_3) \\ (e_1, e_4) & (e_2, e_4) & (e_3, e_4) & (e_4, e_4) \end{pmatrix} = \begin{pmatrix} Q & A_{12} & A_{13} & A_{14} \\ B_{21} & Q_2 & A_{23} & A_{24} \\ B_{31} & B_{32} & Q_3 & A_{34} \\ B_{41} & B_{42} & B_{43} & Q_4 \end{pmatrix},$$

where $Q_i = (e_i, e_i)$ for each $i = 1, 2, 3, 4$ and $A_{ij} = (e_j, e_i)$ and $B_{ji} = (e_i, e_j)$ for each $i, j \in \{1, 2, 3, 4\}$ with $i < j$. By the form of the Nakayama permutation above, we see that

$$S(R) = \begin{pmatrix} 0 & S(A_{12}) & 0 & 0 \\ 0 & 0 & S(A_{23}) & 0 \\ 0 & 0 & 0 & S(A_{34}) \\ S(B_{41}) & 0 & 0 & 0 \end{pmatrix}.$$

Put $Q = Q_1$. First we show that there exist $c \in J(Q)$ and $\sigma \in \mathrm{Aut}(Q)$ satisfying $Qc = J(Q)$, $\sigma(c) = c$ and $\sigma(q)c = cq$ for all $q \in Q$. We put

$$W = \begin{pmatrix} (e_1, e_1) & (e_2, e_1) \\ (e_1, e_4) & (e_2, e_2) \end{pmatrix}.$$

By using θ_{24}, the similar way above for Z, we can show that W becomes a basic indecomposable Nakayama QF-ring of KNP(1-2)-type and is expressed as

$$W = \begin{pmatrix} Q & Q \\ Q & Q \end{pmatrix}_{\sigma, c}$$

for some $c \in J(Q)$ and $\sigma \in \mathrm{Aut}(Q)$ satisfying $Qc = J(Q), \sigma(c) = c$ and $\sigma(q)c = cq$ for all $q \in Q$. We put

$$\overline{R} = R/S(e_4 R_R) = \begin{pmatrix} Q & A_{12} & A_{13} & A_{14} \\ B_{21} & Q_2 & A_{23} & A_{24} \\ B_{31} & B_{32} & Q_3 & A_{34} \\ \overline{B_{41}} & B_{42} & B_{43} & Q_4 \end{pmatrix},$$

where we let $\overline{B_{41}} = B_{41}/S(B_{41})$. Then $\overline{e}_1 \overline{R_{\overline{R}}}$, $\overline{e}_2 \overline{R_{\overline{R}}}$ and $\overline{e}_3 \overline{R_{\overline{R}}}$ are injective, $\overline{e}_3 \overline{J_{\overline{R}}} \cong \overline{e}_4 \overline{R_{\overline{R}}}$ and the socle of $\overline{e}_2 \overline{R} + \overline{e}_3 \overline{R} + \overline{e}_4 \overline{R}$ is

$$\begin{pmatrix} 0 & 0 & 0 & 0 \\ 0 & 0 & S(A_{23}) & 0 \\ 0 & 0 & 0 & S(A_{34}) \\ 0 & 0 & 0 & S(Q_4) \end{pmatrix}.$$

By this form of $\overline{R}$ as a left H- ring, $\overline{R}$ can be represented as a factor ring

$$\overline{R} \cong \begin{pmatrix} Q & A_{12} & A_{13} & A_{13} \\ B_{21} & Q_2 & A_{23} & \overline{A}_{23} \\ B_{31} & B_{32} & Q_3 & Q_3 \\ B_{31} & B_{32} & J(Q_3) & Q_3 \end{pmatrix}$$

of

$$P = \begin{pmatrix} Q & A_{12} & A_{13} & A_{13} \\ B_{21} & Q_2 & A_{23} & A_{23} \\ B_{31} & B_{32} & Q_3 & Q_3 \\ B_{31} & B_{32} & J(Q_3) & Q_3 \end{pmatrix}.$$

We also look at the form of P. Noting its socle is

$$\begin{pmatrix} 0 & S(A_{12}) & 0 & 0 \\ 0 & 0 & 0 & S(A_{23}) \\ 0 & 0 & 0 & S(Q_3)) \\ 0 & 0 & 0 & S(Q_3) \end{pmatrix},$$

we see that P can be expressed as the factor ring

$$P \cong \begin{pmatrix} Q & A_{12} & \overline{A}_{12} & \overline{A}_{12} \\ B_{21} & Q_2 & Q_2 & Q_2 \\ B_{21} & J(Q_2) & Q_2 & Q_2 \\ B_{21} & J(Q_2) & J(Q_2) & Q_2 \end{pmatrix},$$

of the left H-ring

$$T = \begin{pmatrix} Q & A_{12} & A_{12} & A_{12} \\ B_{21} & Q_2 & Q_2 & Q_2 \\ B_{21} & J(Q_2) & Q_2 & Q_2 \\ B_{21} & J(Q_2) & J(Q_2) & Q_2 \end{pmatrix},$$

where we let $\overline{A}_{12} = A_{12}/S(A_{12})$. Furthermore, since the socle of T is

$$\begin{pmatrix} 0 & S(A_{12}) & 0 & 0 \\ 0 & 0 & 0 & S(Q_2) \\ 0 & 0 & 0 & S(Q_2) \\ 0 & 0 & 0 & S(Q_2) \end{pmatrix},$$

T can be expressed as

$$T \cong \begin{pmatrix} Q & Q & \overline{Q} & \overline{Q} \\ J(Q) & Q & Q & Q \\ J(Q) & J(Q) & Q & Q \\ J(Q) & J(Q) & J(Q) & Q \end{pmatrix}.$$

Accordingly $\overline{R}$ can be expressed as follows:

$$\overline{R} = \begin{pmatrix} Q & A_{12} & A_{13} & A_{14} \\ B_{21} & Q_2 & A_{23} & A_{24} \\ B_{31} & B_{32} & Q_3 & A_{34} \\ \overline{B_{41}} & B_{42} & B_{43} & Q_4 \end{pmatrix}$$

$$\cong \begin{pmatrix} Q & Q & \overline{Q} & \overline{Q} \\ J(Q) & Q & Q & \overline{Q} \\ J(Q) & J(Q) & Q & Q \\ J(Q) & J(Q) & J(Q) & Q \end{pmatrix}$$

$$= \begin{pmatrix} Q & Q & Q & Q \\ J(Q) & Q & Q & Q \\ J(Q) & J(Q) & Q & Q \\ J(Q) & J(Q) & J(Q) & Q \end{pmatrix} \Big/ \begin{pmatrix} 0 & 0 & S(Q) & S(Q) \\ 0 & 0 & 0 & S(Q) \\ 0 & 0 & 0 & 0 \\ 0 & 0 & 0 & 0 \end{pmatrix}.$$

Now, as noted above, there exist $c \in J(Q)$ and $\sigma \in J(Q)$ satisfying $Qc = J(Q)$, $\sigma(c) = c$ and $\sigma(q) = cq$ for all $q \in Q$. Using c, we represent $\overline{R}$ by a matrix isomorphism $\varphi = (\varphi_{ij})$ as follows:

$$\overline{R} = \begin{pmatrix} Q & A_{12} & A_{13} & A_{14} \\ B_{21} & Q & A_{23} & A_{24} \\ B_{31} & B_{32} & Q & A_{34} \\ \overline{B_{41}} & B_{42} & B_{43} & Q \end{pmatrix} \overset{\varphi}{\cong} \begin{pmatrix} Q & Q & \overline{Q} & \overline{Q} \\ Qc & Q & Q & \overline{Q} \\ Qc & Qc & Q & Q \\ Qc & Qc & Qc & Q \end{pmatrix}.$$

Put

$a_{12} = \varphi_{12}^{-1}(1)$, $a_{23} = \varphi_{23}^{-1}(1)$, $a_{34} = \varphi_{34}^{-1}(1)$,
$b_{21} = \varphi_{21}^{-1}(c)$, $b_{31} = \varphi_{31}^{-1}(c)$, $b_{32} = \varphi_{32}^{-1}(c)$, $\quad b_{42} = \varphi_{42}^{-1}(c)$, $b_{43} = \varphi_{43}^{-1}(c)$,
$a_{13} = a_{12}a_{23}$, $a_{24} = a_{23}a_{34}$, $a_{14} = a_{12}a_{23}a_{34}$.

We take $b_{41} \in B_{41}$ such that $cb_{41} = b_{42}b_{21}$. Then $\overline{b_{41}} = \varphi_{41}^{-1}(c)$, $cb_{41} = b_{43}b_{31}$, $Qa_{ij} = A_{ij}$ and $Qb_{ji} = B_{ji}$ for any $i, j \in \{1, 2, 3, 4\}$ with $i < j$. Therefore

$$\overline{R} = \begin{pmatrix} Q & Qa_{12} & Qa_{13} & Qa_{14} \\ Qb_{21} & Q & Qa_{23} & Qa_{24} \\ Qb_{31} & Qb_{32} & Q & Qb_{34} \\ \overline{Qb_{41}} & Qb_{42} & Qb_{43} & Q \end{pmatrix} \overset{\varphi}{\cong} \begin{pmatrix} Q & Q & \overline{Q} & \overline{Q} \\ Qc & Q & Q & \overline{Q} \\ Qc & Qc & Q & Q \\ Qc & Qc & Qc & Q \end{pmatrix}.$$

Now consider the skew matrix ring

$$U = (Q)_{\sigma,c,4} = \begin{pmatrix} Q & Q\alpha_{12} & Q\alpha_{13} & Q\alpha_{14} \\ Q\beta_{21} & Q & Q\alpha_{23} & Q\alpha_{24} \\ Q\beta_{31} & Q\beta_{32} & Q & Q\beta_{34} \\ Q\beta_{41} & Q\beta_{42} & Q\beta_{43} & Q \end{pmatrix}_{\sigma,c,4},$$

where $\{\alpha_{ij}\} \cup \{\beta_{ji}\}$ are skew matrix units. By Theorem 6.5.1, the map

$$\tau : T \longrightarrow \begin{pmatrix} Q & Q & Q & Q \\ Qc & Q & Q & Q \\ Qc & Qc & Q & Q \\ Qc & Qc & Qc & Q \end{pmatrix}$$

given by

$$\begin{pmatrix} q_{11} & q_{12}\alpha_{12} & q_{13}\alpha_{13} & q_{14}\alpha_{14} \\ q_{21}\beta_{21} & q_{22} & q_{23}\alpha_{23} & q_{24}\alpha_{24} \\ q_{31}\beta_{31} & q_{32}\beta_{32} & q_{33} & q_{34}\alpha_{34} \\ q_{41}\beta_{41} & q_{42}\beta_{42} & q_{43}\beta_{43} & q_{44} \end{pmatrix} \longmapsto \begin{pmatrix} q_{11} & q_{12}a_{12} & q_{13} & q_{14} \\ q_{21}c & q_{22} & q_{23} & q_{24} \\ q_{31}c & q_{32}c & q_{33} & q_{34} \\ q_{41}c & q_{42}c & q_{43}c & q_{44} \end{pmatrix}$$

is a surjective ring homomorphism with

$$\operatorname{Ker}\tau = \begin{pmatrix} 0 & 0 & 0 & 0 \\ S(Q) & 0 & 0 & 0 \\ S(Q) & S(Q) & 0 & 0 \\ S(Q) & S(Q) & S(Q) & 0 \end{pmatrix}.$$

Thus the composite map $\varphi^{-1}\tau$ is a surjective ring homomorphism given by

$$\begin{pmatrix} q_{11} & q_{12}\alpha_{12} & q_{13}\alpha_{13} & q_{14}\alpha_{14} \\ q_{21}\beta_{21} & q_{22} & q_{23}\alpha_{23} & q_{24}\alpha_{24} \\ q_{31}\beta_{31} & q_{32}\beta_{32} & q_{33} & q_{34}\alpha_{34} \\ q_{41}\beta_{41} & q_{42}\beta_{42} & q_{43}\beta_{43} & q_{44} \end{pmatrix} \longmapsto \begin{pmatrix} q_{11} & q_{12}a_{12} & q_{13}a_{13} & q_{14}a_{14} \\ q_{21}b_{21} & q_{22} & q_{23}a_{23} & q_{24}a_{24} \\ q_{31}b_{31} & q_{32}b_{32} & q_{33} & q_{34}a_{34} \\ \overline{q_{41}b_{41}} & q_{42}b_{42} & q_{43}b_{43} & q_{44} \end{pmatrix}$$

and

$$\operatorname{Ker}\varphi^{-1}\tau = \begin{pmatrix} 0 & 0 & S(Q) & S(Q) \\ S(Q) & 0 & 0 & S(Q) \\ S(Q) & S(Q) & 0 & 0 \\ S(Q) & S(Q) & S(Q) & 0 \end{pmatrix}.$$

By this homomorphism together with $cb_{41} = b_{42}b_{21} = b_{43}b_{31}$, we can easily check that the map $\Phi : U \to R$ given by

$$\begin{pmatrix} q_{11} & q_{12}\alpha_{12} & q_{13}\alpha_{13} & q_{14}\alpha_{14} \\ q_{21}\beta_{21} & q_{22} & q_{23}\alpha_{23} & q_{24}\alpha_{24} \\ q_{31}\beta_{31} & q_{32}\beta_{32} & q_{33} & q_{34}\alpha_{34} \\ q_{41}\beta_{41} & q_{42}\beta_{42} & q_{43}\beta_{43} & q_{44} \end{pmatrix} \longmapsto \begin{pmatrix} q_{11} & q_{12}a_{12} & q_{13}a_{13} & q_{14}a_{14} \\ q_{21}b_{21} & q_{22} & q_{23}a_{23} & q_{24}a_{24} \\ q_{31}b_{31} & q_{32}b_{32} & q_{33} & q_{34}a_{34} \\ q_{41}b_{41} & q_{42}b_{42} & q_{43}b_{43} & q_{44} \end{pmatrix}$$

is a surjective ring homomorphism with

$$\operatorname{Ker}\Phi = \begin{pmatrix} 0 & 0 & S(Q) & S(Q) \\ S(Q) & 0 & 0 & S(Q) \\ S(Q) & S(Q) & 0 & 0 \\ 0 & S(Q) & S(Q) & 0 \end{pmatrix}.$$

In view of the proof above, we can prove our statement for any n by quite the same argument. □

　We extend Theorem 7.3.7 to the following form.

Theorem 7.3.8. *Let R be a basic indecomposable Nakayama QF-ring of KNP$(1 \to k)$-type for $Pi(R) = \{e_1, \ldots, e_n\}$ with $k > 1$. Put $Q = e_1 Re_1$. Then there exist $c \in J(Q)$ and $\sigma \in \mathrm{Aut}(Q)$ satisfying $Qc = J(Q)$, $\sigma(c) = c$ and $\sigma(q)c = cq$ for all $q \in Q$ and R can be represented as a factor ring of a skew matrix ring $T = (Q)_{\sigma,c,n}$ by the $(n-k)$-th socle $S_{n-k}(T)$ of T, that is, R can be represented as*

$$
R \cong
\begin{pmatrix}
Q & \cdots & \cdots & Q & \overset{\overset{k}{\vee}}{\overline{Q}} & \cdots & \overline{Q} \\
\overline{Q} & Q & & & & \ddots & \vdots \\
\vdots & \ddots & \ddots & & & & \overline{Q} \\
\overline{Q} & & \ddots & \ddots & & & Q \\
Q & \ddots & & \ddots & \ddots & & \vdots \\
\vdots & \ddots & \ddots & & \ddots & \ddots & \vdots \\
Q & \cdots & Q & \overline{Q} & \cdots & \overline{Q} & Q
\end{pmatrix}
$$

$$
=
\begin{pmatrix}
Q & \cdots & \cdots & Q & \overset{\overset{k}{\vee}}{Q} & \cdots & Q \\
Q & \ddots & & & & \ddots & \vdots \\
\vdots & \ddots & \ddots & & & & Q \\
Q & & \ddots & \ddots & & & Q \\
Q & \ddots & & \ddots & \ddots & & \vdots \\
\vdots & \ddots & \ddots & & \ddots & \ddots & \vdots \\
Q & \cdots & Q & Q & \cdots & Q & Q
\end{pmatrix}_{\sigma,c,n}
\Bigg/
\begin{pmatrix}
0 & \cdots & \cdots & 0 & \overset{\overset{k}{\vee}}{S(Q)} & \cdots & S(Q) \\
S(Q) & \ddots & & & & \ddots & \vdots \\
\vdots & \ddots & \ddots & & & & S(Q) \\
S(Q) & & \ddots & \ddots & & & 0 \\
0 & \ddots & & \ddots & \ddots & & \vdots \\
\vdots & \ddots & \ddots & & \ddots & \ddots & \vdots \\
0 & \cdots & 0 & S(Q) & \cdots & S(Q) & 0
\end{pmatrix},
$$

where we put $\overline{Q} = Q/S(Q)$.

　Proof. We use induction on k. When $k = 2$, this statement is Theorem 7.3.7. Assume that the statement holds for $k = i - 1$ and let R be of

$KNP(1 \to i)$-type. We represent R as follows:

$$R = \begin{pmatrix} A_{11} & \cdots & \cdots & A_{1,i-1} & A_{1i} & \cdots & \cdots & & A_{1n} \\ A_{21} & \ddots & & & & \ddots & \ddots & & \vdots \\ \vdots & & \ddots & & & & \ddots & \ddots & \vdots \\ \vdots & & & \ddots & & & & \ddots & A_{n-i+1,n} \\ A_{n-i+2,1} & & & & \ddots & & & & A_{n-i+2,n} \\ A_{n-i+3} & \ddots & & & & \ddots & & & \vdots \\ \vdots & \ddots & \ddots & & & & \ddots & & \vdots \\ A_{n1} & \cdots & A_{n,i-2} & A_{n,i-1} & \cdots & \cdots & A_{n,n-1} & & A_{nn} \end{pmatrix}.$$

Then

$$S(R) = \begin{pmatrix} 0 & \cdots\cdots\cdots & 0 & S(A_{1i}) & 0 & \cdots & & 0 \\ \vdots & \ddots & & & \ddots & \ddots & \ddots & \vdots \\ \vdots & & \ddots & & & \ddots & \ddots & 0 \\ 0 & & & \ddots & & & \ddots & S(A_{n-i+1,n}) \\ S(A_{n-i+2,n}) & \ddots & & & \ddots & & & 0 \\ 0 & \ddots & \ddots & & & \ddots & & \vdots \\ \vdots & & \ddots\ddots & & & & \ddots & \vdots \\ 0 & & \cdots & 0 & S(A_{n,i-1}) & 0 & \cdots\cdots & 0 \end{pmatrix}.$$

We put

$$\overline{R} = R/S(R) = \begin{pmatrix} A_{11} & \cdots & \cdots & A_{1,i-1} & \overline{A_{1i}} & A_{1,i+1} & \cdots & A_{1n} \\ \vdots & \ddots & & & \ddots & \ddots & & \vdots \\ \vdots & & \ddots & & & \ddots & \ddots & A_{n-i,n} \\ A_{n-i+1,1} & & & \ddots & & & \ddots & \overline{A_{n-i+1,n}} \\ \overline{A_{n-i+2,1}} & \ddots & & & \ddots & & & A_{n-i+2,n} \\ A_{n-i+3} & \ddots & \ddots & \ddots & & \ddots & & \vdots \\ \vdots & \ddots & \ddots & & & & \ddots & \vdots \\ A_{n1} & \cdots & A_{n,i-2} & \overline{A_{n,i-1}} & A_{n,i} & \cdots & \cdots & A_{nn} \end{pmatrix}.$$

Then we see that $\overline{R}$ is a basic indecomposable Nakayama QF-ring of

$KNP(1 \to i-1)$-type with

$$S(\overline{R}) = \begin{pmatrix} 0 & \cdots \cdots & 0 & S(A_{1,i-1}) & 0 & \cdots & & 0 \\ \vdots & \ddots & \ddots & & \ddots & \ddots & & \vdots \\ \vdots & & \ddots & & & \ddots & \ddots & 0 \\ 0 & & & \ddots & & & \ddots & S(A_{n-i+2,n}) \\ S(A_{n-i+3,n}) & \ddots & & & \ddots & & & 0 \\ 0 & \ddots & \ddots & & & \ddots & & \vdots \\ \vdots & \ddots & \ddots & \ddots & & & \ddots & \vdots \\ 0 & \cdots & 0 & S(A_{n,i-2}) & 0 & \cdots \cdots & & 0 \end{pmatrix}.$$

Hence, by induction, there exist $c \in J(Q)$ and $\sigma \in \mathrm{Aut}(Q)$ satisfying $Qc = J(Q)$, $\sigma(c) = c$ and $\sigma(q)c = cq$ for all $q \in Q$ and a surjective matrix ring homomorphism $\varphi = \{\varphi_{ij}\}$ from the skew matrix ring $(Q)_{\sigma,c,n}$ to $\overline{R}$ with

$$\mathrm{Ker}\,\varphi = \begin{pmatrix} 0 & \cdots \cdots & 0 & \overset{\overset{i-1}{\vee}}{S(Q)} & \cdots & S(Q) \\ S(Q) & \ddots & & & \ddots & \vdots \\ \vdots & \ddots & \ddots & & \ddots & S(Q) \\ S(Q) & & \ddots & \ddots & & 0 \\ 0 & \ddots & & \ddots & \ddots & \vdots \\ \vdots & \ddots & \ddots & & \ddots & \ddots & \vdots \\ 0 & \cdots & 0 & S(Q) & \cdots & S(Q) & 0 \end{pmatrix}.$$

Therefore we obtain

$$\overline{R} \cong (\overline{Q})_{\sigma,c,n} = \begin{pmatrix} Q & \cdots \cdots & Q & \overset{\overset{i-1}{\vee}}{\overline{Q}} & \cdots & \overline{Q} \\ \overline{Q} & \ddots & & & \ddots & \ddots & \vdots \\ \vdots & \ddots & \ddots & & & \ddots & \overline{Q} \\ \overline{Q} & & \ddots & \ddots & & & Q \\ Q & \ddots & & \ddots & \ddots & \vdots \\ \vdots & \ddots & \ddots & & \ddots & \ddots & \vdots \\ Q & \cdots & Q & \overline{Q} & \cdots & \overline{Q} & Q \end{pmatrix}.$$

Now let $\{\alpha_{ij}\}_{i,j\in\{1,...,n\}}$ be the skew matrix units of $(Q)_{\sigma,c,n}$. Thus

$$(*) \quad \begin{cases} \text{When} \quad i > j, & \begin{cases} \sigma(q)\alpha_{ij} = \alpha_{ij}q \text{ for all } q \in Q, \\ \alpha_{ij}\alpha_{jk} = \begin{cases} \alpha_{ik} & \text{if } i > k \geq j, \\ \alpha_{ik}c & \text{if } k \geq i \text{ or } j > k, \end{cases} \end{cases} \\ \text{When} \quad i = j, & \begin{cases} q\alpha_{ij} = \alpha_{ij}q \text{ for all } q \in Q, \\ \alpha_{ij}\alpha_{jk} = \alpha_{ik}, \end{cases} \\ \text{When} \quad i < j, & \begin{cases} q\alpha_{ij} = \alpha_{ij}q \text{ for all } q \in Q, \\ \alpha_{ij}\alpha_{jk} = \begin{cases} \alpha_{ik}c & \text{if } i \leq k < j, \\ \alpha_{ik} & \text{if } k < i \text{ or } j \leq k. \end{cases} \end{cases} \end{cases}$$

We represent $(Q)_{\sigma,c,n}$ by $\{\alpha_{ij}\}_{i,j\in\{1,...,n\}}$:

$$(Q)_{\sigma,c,n} = \begin{pmatrix} Q\alpha_{11} & \cdots & \cdots & Q\alpha_{1,i-1} & Q\alpha_{1i} & \cdots & \cdots & Q\alpha_{1n} \\ Q\alpha_{21} & \ddots & & & & \ddots & \ddots & \vdots \\ \vdots & & \ddots & & & & \ddots & \ddots & \vdots \\ \vdots & & & \ddots & & & & \ddots & Q\alpha_{n-i+1,n} \\ Q\alpha_{n-i+2,1} & & & & \ddots & & & & Q\alpha_{n-i+2,n} \\ Q\alpha_{n-i+3,1} & \ddots & & & & \ddots & & & \vdots \\ \vdots & & \ddots & \ddots & & & & \ddots & \vdots \\ Q\alpha_{n1} & \cdots & Q\alpha_{n,i-2} & Q\alpha_{n,i-1} & \cdots & \cdots & \cdots & Q\alpha_{nn} \end{pmatrix}.$$

We put

$$\beta_{pq} = \varphi(\alpha_{pq})$$

for each $(p,q) \notin D = \{(1,i), (2,i+1), \ldots, (n-i+1,n), (n-i+2,1), \ldots, (n,i-1)\}$ and put

$$\begin{cases} x_{1i} = \varphi(\alpha_{1i}), \\ x_{2,i+1} = \varphi(\alpha_{2,i+1}), \\ \quad \cdots \\ x_{n-i+1,n} = \varphi(\alpha_{n-i+1,n}), \\ x_{n-i+2,1} = \varphi(\alpha_{n-i+2,1}), \\ x_{n-i+3,2} = \varphi(\alpha_{n-i+3,2}), \\ \quad \cdots \\ x_{n,i-1} = \varphi(\alpha_{n,i-1}). \end{cases}$$

Then we have

$$\begin{cases}
\beta_{pq} \in A_{pq}, \\
x_{1i} \in \overline{A_{1i}}, \\
x_{2,i+1} \in \overline{A_{2,i+1}}, \\
\qquad \cdots \\
x_{n-i+1,n} \in \overline{A_{n-i+1,n}}, \\
x_{n-i+2,1} \in \overline{A_{n-i+2,1}}, \\
x_{n-i+3,2} \in \overline{A_{n-i+3,2}}, \\
\qquad \cdots \\
x_{n,i-1} \in \overline{A_{n,i-1}}.
\end{cases}$$

Next, for each $(p, q) \in D$, we define the following β_{pq} corresponding to x_{pq}

$$\begin{cases}
\beta_{1i} = \beta_{12}\beta_{2i}, \\
\beta_{2,i+1} = \beta_{23}\beta_{3,i+1}, \\
\qquad \cdots \\
\beta_{n-i+1,n} = \beta_{n-i+1,n-i+2}\beta_{n-i+2,n}, \\
\beta_{n-i+2,1} = \beta_{n-i+2,n-i+3}\beta_{n-i+3,1}, \\
\beta_{n-i+3,2} = \beta_{n-i+3,n-i+4}\beta_{n-i+4,2}, \\
\qquad \cdots \\
\beta_{n,i-1} = \beta_{n1}\beta_{1,i-1}.
\end{cases}$$

Then it is straightforward to verify the following:

(1) $Q\beta_{pq} = A_{pq}$ for all $(p, q) \notin D$.
(2) $\overline{\beta}_{pq} = x_{pq}$ for all $(p, q) \in D$.
(3) $Q\alpha_{pq} \cong Q\beta_{pq}$ by the map $q\alpha_{pq} \mapsto q\beta_{pq}$ for all $(p, q) \in D$.
(4) R can be represented as:

$$R = \begin{pmatrix} Q\beta_{11} & \cdots & Q\beta_{1n} \\ & \cdots & \\ Q\beta_{n1} & \cdots & Q\beta_{nn} \end{pmatrix}$$

(5) $\{\beta_{ij}\}_{i,\,j \in \{1,\ldots,n\}}$ satisfies the same relations (*) as recorded above for $\{\alpha_{ij}\}_{i,\,j \in \{1,\ldots,n\}}$.

Thus, by (3), (4) and (5), the map

$$\tau : (Q)_{\sigma,c,n} = \begin{pmatrix} Q\alpha_{11} & \cdots & Q\alpha_{1n} \\ \vdots & & \vdots \\ Q\alpha_{n1} & \cdots & Q\alpha_{nn} \end{pmatrix} \longrightarrow R = \begin{pmatrix} Q\beta_{11} & \cdots & Q\beta_{1n} \\ & \cdots & \\ Q\beta_{n1} & \cdots & Q\beta_{nn} \end{pmatrix}$$

given by

$$\tau : \begin{pmatrix} q_{11}\alpha_{11} & \cdots & q_{1n}\alpha_{1n} \\ & \cdots & \\ q_{n1}\alpha_{n1} & \cdots & q_{nn}\alpha_{nn} \end{pmatrix} \longmapsto \begin{pmatrix} q_{11}\beta_{11} & \cdots & q_{1n}\beta_{1n} \\ & \cdots & \\ q_{n1}\beta_{n1} & \cdots & q_{nn} \end{pmatrix}$$

is a surjective ring homomorphism and, moreover,

$$\mathrm{Ker}\,\tau = \begin{pmatrix} 0 & & \cdots\cdots & 0 & S(Q\alpha_{1i}) & \cdots & S(Q\alpha_{1n}) \\ S(Q\alpha_{21}) & \ddots & & & & & \vdots \\ \vdots & & \ddots\ddots & & & \ddots & \vdots \\ \vdots & & & \ddots & \ddots & \ddots S(Q\alpha_{n-i+1,n}) \\ S(Q\alpha_{n-i+2,n}) & & & \ddots & \ddots & & 0 \\ 0 & \ddots & & & \ddots & \ddots & \vdots \\ \vdots & & \ddots\ddots & & & & \vdots \\ 0 & & \cdots & 0\,S(Q\alpha_{n,i-1}) & \cdots S(Q\alpha_{n,n-1}) & & 0 \end{pmatrix}.$$

This completes the proof.   □

Remark 7.3.9. Let R be a basic indecomposable Nakayama ring with a Kupisch left well-indexed set $Pi(R) = \{e_{ij}\}_{i=1,j=1}^{m,\ n(i)}$. We recall the construction of the frame QF-subring $F(R)$ of R introduced in Chapter 4. Firstly $F(R)$ is derived from a suitable subset $\{f_j = e_{i_j 1}\}_{j=1}^{y} \subseteq \{e_{i1}\}_{i=1}^{m}$ and $F(R) = (f_1 + \cdots + f_s)R(f_1 + \cdots + f_s)$. Then, without loss of generality, we may take $\{e_{i_j 1}\}_{j=1}^{y}$ so that $1 = i_1 < i_2 < \cdots < i_y \leq m$, and $F(R)$ is a basic indecomposable Nakayama QF-ring with $Pi(F(R)) = \{f_i\}_{i=1}^{y}$ a Kupisch left well-indexed set. Hence, by Proposition 7.2.5, there exists i such that

$$\begin{pmatrix} e_1 & e_2 & \cdots & e_{y-i+1} & e_{y-i+2} & \cdots & e_y \\ e_i & e_{i+1} & \cdots & e_y & e_1 & \cdots & e_{i-1} \end{pmatrix}$$

is a Nakayama permutation of $F(R)$. We put $Q = f_1 R f_1$, $\overline{Q} = Q/S(Q)$ and $\overline{R} = R/S(R)$. Since $\overline{R}$ is also a basic indecomposable Nakayama ring, we can consider its frame Nakayama QF-subring $F(\overline{R})$. Then it is easy to see that $F(\overline{R}) = \overline{F(R)}$.

Under this setting, we now provide a structure theorem for Nakayama rings.

Theorem 7.3.10. (**Classification**)   *Let R be a basic indecomposable Nakayama ring with a Kupisch left well-indexed set $Pi(R) = \{e_{ij}\}_{i=1,j=1}^{m\ \ n(i)}$. And let $F(R)$, $Pi(F(R)) = \{f_i\}_{i=1}^{y}$, Q, $\overline{Q}$ and $\overline{R}$ be as above. Then the following statements hold.*

(1) *R can be represented as an upper staircase factor ring of a block extension of $F(R)$, that is, there exist a suitable block extension $B(F(R)) = F(R)(k(1), \ldots, k(y))$ with $k(1) + \cdots + k(y) = \#Pi(R)$ and a surjective matrix ring homomorphism φ of $B(F(R))$ to R such that $B(F(R))/\operatorname{Ker}\varphi$ forms an upper staircase factor ring.*

(2) *When $F(R)$ is a local Nakayama ring, then $F(R) = Q$ and R can be represented as an upper staircase factor ring of a block extension of Q.*

(3) *When $y = \#Pi(F(R)) \geq 2$ and the Nakayama permutation of $F(R)$ is cyclic, the following hold:*

(a) *There exist $c \in Q$ and $\sigma \in \operatorname{Aut}(Q)$ satisfying $cQ = J(Q)$, $\sigma(c) = c$ and $\sigma(q)c = cq$ for all $q \in Q$. (Thus skew matrix rings over Q with respect to $\{\sigma, c, y\}$ can be constructed.)*

(b) *There exists a surjective matrix ring homomorphism $\tau : V = (Q)_{\sigma,c,y} \to F(R)$.*

(c) *τ can be canonically extended to a surjective matrix ring homomorphism $\tau^* : B(V) = V(k(1), \ldots, k(t); c) \to B(F(R)) = F(R)(k(1), \ldots, k(t); c)$.*

(d) *There exists a canonical surjective matrix ring homomorphism $\pi : Y = (Q)_{\sigma,c,s} \to B(V)$, where $s = \sum_{i=1}^{y} k(i) = \sum_{i=1}^{m} n(i)$.*

(4) *If $F(R)$ has the identity Nakayama permutation, then the basic indecomposable Nakayama QF-ring $F(\overline{R})$ is of $KNP(1 \to y)$-type, and we can make a suitable block extension of $F(\overline{R})$ so that $\overline{R}$ is represented as a factor ring of a skew matrix ring over $\overline{Q}$ with respect to $\{\sigma, c, s\}$. (Consequently, in this case, we can have enough information on the structure of R through $\overline{R}$.)*

Proof.   (1), (2). These follow from Theorem 4.3.5.

(3). (a) follows from Theorem 7.3.8. (b) is clear while (c) and (d) follow from Theorem 6.5.1.

(4). This follows from (3) and Theorem 6.4.1. □

7.4   An Example of a Nakayama QF-Ring of $KNP(1 \to 1)$-Type

In general, for a local Nakayama ring Q, we do not know whether there exist $\sigma \in \mathrm{Aut}(Q)$ and $c \in J(Q)$ satisfying $Qc = J(Q)$, $\sigma(c) = c$ and $\sigma(q)c = cq$ for all $q \in Q$. However, for $\overline{Q} = Q/S(Q)$, such elements do exist as the following shows.

Proposition 7.4.1. Let Q be a local Nakayama ring and put $\overline{Q} = Q/J(Q)$. Then there exist $\tau \in \mathrm{Aut}(\overline{Q})$ and $c \in Q$ satisfying

(1) $Qc = J(Q)$, so $\overline{Q}\overline{c} = J(\overline{Q})$ and
(2) $\tau(\overline{c}) = \overline{c}$ and $\tau(\overline{q})c = c\overline{q} = cq$ for all $q \in Q$.   (Note that $Qc = J(Q)$ is a $\overline{Q}$-bimodule.)

Proof.   Consider the Nakayama ring

$$R = \begin{pmatrix} Q & Q \\ J(Q) & Q \end{pmatrix}.$$

Put $e = \langle 1 \rangle_{11}$ and $f = \langle 1 \rangle_{22}$. Then

$$S(eR_R) = \begin{pmatrix} 0 & S(Q) \\ 0 & 0 \end{pmatrix} = S(_RRf).$$

Next set

$$\overline{R} = \begin{pmatrix} Q & \overline{Q} \\ J(Q) & Q \end{pmatrix} = \begin{pmatrix} Q & Q \\ J(Q) & Q \end{pmatrix} / \begin{pmatrix} 0 & S(Q) \\ 0 & 0 \end{pmatrix}.$$

Then $\overline{R}$ is a basic indecomposable Nakayama QF-ring with the identity Nakayama permutation. It follows that $T = \overline{R}/S(\overline{R})$ is a basic indecomposable Nakayama QF-ring with a cyclic Nakayama permutation by Theorem 7.3.4. Noting that

$$T = \begin{pmatrix} Q & Q \\ J(Q) & Q \end{pmatrix} / \begin{pmatrix} S(Q) & S(Q) \\ 0 & S(Q) \end{pmatrix} = \begin{pmatrix} \overline{Q} & \overline{Q} \\ J(Q) & \overline{Q} \end{pmatrix},$$

there exist $\tau \in \mathrm{Aut}(\overline{Q})$ and $c \in Q$ satisfying $Qc = J(Q)$, $\tau(\overline{c}) = \overline{c}$ and $\tau(\overline{q})c = c\overline{q}$ for all $q \in Q$ by Theorem 7.3.3. □

Now consider a basic indecomposable Nakayama QF-ring R of type $(1 \to 1)$ for $Pi(R) = \{e, f\}$. Setting $eRe = Q$, $eRf = A$, $fRe = B$ and $fRf = T$, we have

$$R = \begin{pmatrix} Q & A \\ B & T \end{pmatrix}.$$

Then, since $S(Q)A = AS(T) = 0$ and $S(T)B = BS(Q) = 0$, A is a $(\overline{Q}, \overline{T})$-bimodule and B is a $(\overline{T}, \overline{Q})$-bimodule, where $\overline{Q} = Q/S(Q)$ and $\overline{T} = T/S(T)$. We put

$$\overline{R} = R/S(R) = \begin{pmatrix} \overline{Q} & A \\ B & \overline{T} \end{pmatrix} = \begin{pmatrix} Q & A \\ B & T \end{pmatrix} \Big/ \begin{pmatrix} S(Q) & 0 \\ 0 & S(T) \end{pmatrix}.$$

By Proposition 7.3.1, $\overline{R}$ is of type $(1 \to 2)$ and, by Theorem 7.3.3, $\overline{R}$ is represented as a skew matrix ring $\overline{R} \cong (\overline{Q})_{\tau, \overline{c}, 2}$, where $\tau \in \mathrm{Aut}(\overline{Q})$ and $c \in Q$ satisfying $Qc = J(Q)$, $\tau(\overline{c}) = \overline{c}$ and $\tau(\overline{q})c = cq$ for all $q \in Q$. More precisely, letting $\{\alpha, \beta\}$ be the skew matrix units of $\overline{R}$, there exists a matrix ring isomorphism:

$$\varphi = \begin{pmatrix} \varphi_{11} & \varphi_{12} \\ \varphi_{21} & \varphi_{22} \end{pmatrix} : \begin{pmatrix} \overline{Q} & A \\ B & \overline{T} \end{pmatrix} \longrightarrow \begin{pmatrix} \overline{Q} & \overline{Q}\alpha \\ \overline{Q}\beta & \overline{Q} \end{pmatrix}_{\tau, \overline{c}},$$

where φ_{11} is the identity map of $\overline{Q}$. Put $a = \varphi_{12}^{-1}(\alpha)$, $b = \varphi_{21}^{-1}(\beta)$, $c = ab$ and $d = ba$. Then, as is easily seen,

(1) $A = Qa = aT$ and $B = Tb = bQ$,
(2) $at = \varphi_{22}(\overline{t})a$ for any $t \in T$,
(3) $bq = \pi(\overline{q})b$ for any $q \in Q$, where $\pi = \varphi_{22}^{-1}\tau$,
(4) $\overline{d} = \varphi_{22}^{-1}(\overline{c})$,
(5) for any two elements

$$X = \begin{pmatrix} x_{11} & x_{12}a \\ x_{21}b & x_{22} \end{pmatrix} \quad \text{and} \quad Y = \begin{pmatrix} y_{11} & y_{12}a \\ y_{21}b & y_{22} \end{pmatrix},$$

we have

$$XY = \begin{pmatrix} x_{11}y_{11} + x_{12}\varphi_{22}^{-1}(\overline{y_{21}})c & x_{11}y_{12}a + x_{12}\varphi_{22}^{-1}(\overline{y_{22}})a \\ x_{21}\pi(\overline{y_{11}})b + x_{22}y_{21}b & x_{21}\pi(\overline{y_{12}})d + x_{22}y_{22} \end{pmatrix}.$$

Accordingly R can be represented as

$$R = \begin{pmatrix} Q & Qa \\ Qb & T \end{pmatrix}$$

with the conditions (1) - (5).

Noting this observation, we shall construct a basic indecomposable QF-ring with the identity Nakayama permutation.

Example 7.4.2. Let Q and T be local Nakayama rings. Put $\overline{Q} = Q/S(Q)$ and $\overline{T} = T/S(T)$. Suppose that there exists a ring isomorphism $\varphi : \overline{T} \to \overline{Q}$. By Proposition 7.4.1, there exist $c \in J(Q)$ and $\tau \in \mathrm{Aut}(\overline{Q})$ satisfying $Qc = J(Q)$, $\tau(\overline{c}) = \overline{c}$ and $\overline{q}c = c\overline{q} = cq$ for all $q \in Q$. (Note that $Qc = J(Q)$ is a $\overline{Q}$-bimodule.) Put $A = \overline{Q}$, $a = \overline{1} \in A$, $B = \overline{T}$ and $b = \overline{1} \in B$. And put

$$R = \begin{pmatrix} Q & A \\ B & T \end{pmatrix} = \begin{pmatrix} Q & Qa \\ Qb & T \end{pmatrix}.$$

We define a multiplication on R as follows: For any two elements

$$X = \begin{pmatrix} x_{11} & x_{12}a \\ x_{21}b & x_{22} \end{pmatrix} \quad \text{and} \quad Y = \begin{pmatrix} y_{11} & y_{12}a \\ y_{21}b & y_{22} \end{pmatrix},$$

we let

$$XY = \begin{pmatrix} x_{11}y_{11} + x_{12}\varphi^{-1}(\overline{y_{21}})c & x_{11}y_{12}a + x_{12}\varphi^{-1}(\overline{y_{22}})a \\ x_{21}\pi(\overline{y_{11}})b + x_{22}y_{21}b & x_{21}\pi(\overline{y_{12}})c + x_{22}y_{22} \end{pmatrix},$$

where $\pi = \varphi^{-1}\tau$. Then, with this multiplication and the usual matrix addition, R becomes a basic indecomposable QF-ring with the identity Nakayama permutation and, moreover,

$$R/S(R) \cong \begin{pmatrix} \overline{Q} & \overline{Q} \\ \overline{Q} & \overline{Q} \end{pmatrix}_{\overline{c},\tau,2}.$$

Remark 7.4.3. (Cf. [162, Example 8.13]) In the Example above, if we take $Q = \mathbb{Z}/4\mathbb{Z}$ and $T = \mathbb{Z}_2[x]/(x^2)$, we see that R is a basic indecomposable Nakayama QF-ring with a cyclic Nakayama permutation but its diagonal rings Q and T are not isomorphic.

Remark 7.4.4. For a basic indecomposable Nakayama QF-ring R of $KPN(1 \to 1)$-type, $R/S(R)$ can be represented as a skew matrix ring over $Q/S(Q)$, where $Q = eRe$ for any $e \in Pi(R)$. Then, as we saw in Proposition 7.4.1, we can obtain enough information on the structure of R. However a more precise description of R is conditional on the following two questions:

    1. Let Q be a local Nakayama ring and let τ be an automorphism of $Q/S(Q)$. Then, does there exist an automorphism σ of Q which induces τ?

    2. Let Q be a local Nakayama ring with $|Q| = n$. Then does there exist another local Nakayama ring T with $|T| = n+1$, for which $R \cong T/S(T)$?

7.5   The Self-Duality of Nakayama Rings

In the introduction of Chapter 5, we described the background of self-duality for Nakayama rings. By Theorem 5.2.7, in order to confirm the self-duality of these rings, it suffices to show that every basic indecomposable Nakayama QF-ring has a Nakayama automorphism.

Let R be a basic indecomposable Nakayama QF-ring of $KNP(1 \to k)$-type with $Pi(R) = \{e_i\}_{i=1}^n$. We may assume that $k \neq 1$, i.e., the Nakayama permutation of R is not the identity. Put $Q = e_1 R e_1$. Then, by Theorem 7.3.8, there exist $c \in J(Q)$ and $\sigma \in \mathrm{Aut}(Q)$ satisfying $Qc = J(Q)$ and $\sigma(q)c = cq$ for all $q \in Q$ and $R \cong \overline{T} = T/S_{n-k}(T_T)$, where

$$
T = \begin{pmatrix} Q & \cdots & Q \\ & \cdots & \\ Q & \cdots & Q \end{pmatrix}_{\sigma,c,n} \quad .
$$

Note that

$$
S_{n-k}(T_T) = \begin{pmatrix}
0 & \cdots\cdots & & 0 & \overset{\overset{k}{\vee}}{S(Q)} & \cdots & S(Q) \\
S(Q) & \ddots & & & & \ddots & \vdots \\
\vdots & \ddots & \ddots & & & & S(Q) \\
S(Q) & & \ddots & \ddots & & & 0 \\
0 & \ddots & & \ddots & \ddots & & \vdots \\
\vdots & \ddots & \ddots & & & \ddots & \vdots \\
0 & \cdots & 0 & S(Q) & \cdots & S(Q) & 0
\end{pmatrix} \quad .
$$

Put $f_i = \langle 1 \rangle_{11}$ in T for each $i = 1, \ldots, n$. By the matrix isomorphism

$$
R \cong \overline{T} = \begin{pmatrix}
Q & \cdots\cdots & Q & \overset{\overset{k}{\vee}}{\overline{Q}} & \cdots & \overline{Q} \\
\overline{Q} & \ddots & & & \ddots & \vdots \\
\vdots & \ddots & \ddots & & & \overline{Q} \\
\overline{Q} & & \ddots & \ddots & & Q \\
Q & \ddots & & \ddots & \ddots & \vdots \\
\vdots & \ddots & \ddots & & \ddots & \vdots \\
Q & \cdots & Q & \overline{Q} & \cdots & \overline{Q} & Q
\end{pmatrix} ,
$$

where $\overline{Q} = Q/S(Q)$, we see that the Nakayama permutation of $\overline{T}$ is

$$\begin{pmatrix} \overline{f_1} & \overline{f_2} & \cdots & \overline{f_{n-k+1}} & \overline{f_{n-k+2}} & \cdots & \overline{f_n} \\ \overline{f_k} & \overline{f_{k+1}} & \cdots & \overline{f_n} & \overline{f_1} & \cdots & \overline{f_{k-1}} \end{pmatrix}.$$

By Theorem 6.1.4, the Nakayama permutation of T is

$$\begin{pmatrix} f_1 & f_2 & f_3 & \cdots & \cdots & f_n \\ f_n & f_1 & f_2 & \cdots & \cdots & f_1 \end{pmatrix}$$

and T has a Nakayama automorphism, say τ. Since $\tau(S_{n-k}(T_T)) = S_{n-k}(T_T)$, τ induces the Nakayama automorphism $\overline{\tau}$ of $\overline{T}$.

Thus our theory provides a conceptual proof of the following well-known fact.

Theorem 7.5.1. *Every Nakayama ring has self-duality.*

Remark 7.5.2. Since Nakayama algebras over a field are self-dual, we see that basic Nakayama QF-algebras have a Nakayama automorphism (as rings) by Theorem 5.2.7. However this is a known fact since Frobenius algebras have Nakayama automorphisms as algebras (cf. Section 1.3).

Remark 7.5.3. Let F be a basic indecomposable QF-ring such that $Pi(F) = \{e, f\}$ and

$$\begin{pmatrix} e & f \\ f & e \end{pmatrix}$$

is the Nakayama permutation. Assume that F has the Nakayama automorphism φ. Put $Q = eRe, T = fRf, A = eRf$ and $B = fRe$ and express F as

$$F = \begin{pmatrix} Q & A \\ B & T \end{pmatrix}.$$

Consider the following two canonical ring extensions of F with the same matrix size $k + 1$:

$$R_1 = \begin{pmatrix} Q & \cdots & Q & A \\ & \cdots & & \vdots \\ Q & \cdots & Q & A \\ B & \cdots & & B & T \end{pmatrix}, \quad R_2 = \begin{pmatrix} Q & A & \cdots & A \\ B & T & \cdots & T \\ \vdots & & \cdots & \\ B & T & \cdots & T \end{pmatrix}.$$

Then φ induces a canonical ring isomorphism between R_1 and R_2 and also between $R_1/S(R_1)$ and $R_2/S(R_2)$:

$$
R_1 = \begin{pmatrix} Q & \cdots & Q & \overline{A} \\ & \cdots & & \vdots \\ Q & \cdots & Q & \overline{A} \\ \overline{B} & \cdots & \overline{B} & T \end{pmatrix} \cong \begin{pmatrix} Q & \overline{A} & \cdots & \overline{A} \\ \overline{B} & T & \cdots & T \\ \vdots & & \cdots & \\ \overline{B} & T & \cdots & T \end{pmatrix} = R_2,
$$

where $\overline{A} = A/S(A)$ and $B = S(B)$.

COMMENTS

Lemma 7.1.1 is due to Nonomura [143]. Theorem 7.1.2 is due to Oshiro [150]. Part (2) of Theorem 7.1.5 is due to Kupisch [105], and is also shown in Vanaja-Purav [161] under the assumption that R is left and right *QF-3*. Corollary 7.1.7 is well-known (cf. Anderson-Fuller [5]). Theorem 7.1.13 is due to Iwase [85], while Corollary 7.1.14 is due to Kupisch [105]. Almost all other results of Section 7.1 are new. Theorem 7.2.1 and Proposition 7.2.5 are folklore while Theorem 7.2.4 is new. The classification (Theorem 7.3.10) of Nakayama rings is based on Oshiro [143] which was published in 1987. The study of Nakayama rings was initiated by Kupisch [104] and Murase [125]–[127]. Their works are important. In particular, Kupisch introduced not only the so-called Kupisch series but also the rudiments of skew matrix rings. He showed the following fact: If R is a Nakayama ring such that $|e| \not\equiv 1 \pmod{n}$ for any $e \in Pi(R)$, then R can be represented as a factor ring of a skew matrix ring over a local Nakayama ring. For local Nakayama rings, one may refer to Mano [111] and for more information on Nakayama rings the reader is referred to Puninski's book "Serial Rings" [162]. Eisenbud-Griffith-Robson ([43], [44]) showed that every proper factor ring of a hereditary noetherian prime ring is a Nakayama ring. This classical theorem turned the spotlight of Nakayama rings in the history of ring theory.

Chapter 8

Modules over Nakayama Rings

In this chapter, we study modules over Nakayama rings and give module-theoretic characterizations of these rings using lifting and extending properties.

8.1 Characterizations of Nakayama Rings by Lifting and Extending Properties

Proposition 8.1.1. *If M is a quasi-injective right R-module, then so is $J_\alpha(M)$ for every ordinal α. Dually, if M is a quasi-projective right R-module, then so is $M/S_\alpha(M)$ for every ordinal α.*

Proof. Obvious. □

The following fundamental result is due to Nakayama [134]. We give a proof as an application of H-rings.

Theorem 8.1.2. *Let R be a Nakayama ring. Then any right R-module can be expressed as a direct sum of uniserial modules.*

Proof. We use induction on the composition length $|R_R|$. Assume that the statement is true for rings R with $|R_R| < k$ and consider R with $|R_R| = k$. Let M be a right R-module. Then, by Theorem 3.2.10, we may write $M = P \oplus Z$, where P is a projective module and Z is a singular module. Then, since P is projective, it can be expressed as a direct sum of uniserial modules. On the other hand, since Z is a singular module, $ZS(R_R) = 0$, and hence Z is a right $R/S(R_R)$-module. By our induction hypothesis, Z can be expressed as a direct sum of uniserial modules as a right $R/S(R_R)$-module and hence as an R-module.   □

We use Harada's work on the final version of Krull-Remak-Schmidt-Azumaya Theorem. Let us recall his deep results.

Let M be a right R-module with two infinite direct sum decompositions $M = \oplus_{\alpha \in I} M_\alpha = \oplus_{\beta \in J} N_\beta$, where each M_α and N_β are indecomposable modules with local endomorphism rings. (Such decompositions are called *completely indecomposable decompositions*.) Then the following are equivalent:

(1) For any subset $K \subseteq I$, there exists $L \subseteq J$ such that $M = (\oplus_{\alpha \in K} M_\alpha) \oplus (\oplus_{\beta \in L} N_\beta)$.

(2) $\{M_\alpha\}_{\alpha \in I}$ satisfies the $lsTn$ condition, i.e., for any subfamily $\{M_{\alpha_i}\}_{i \in \mathbb{N}}$ with distinct $\{\alpha_i\}_{i \in \mathbb{N}}$, any countable family of non-isomorphisms $\{f_i \colon M_{\alpha_i} \to M_{\alpha_{i+1}}\}_{i \in \mathbb{N}}$ and any $x \in M_{\alpha_1}$, there exists n (depending on x) such that $f_n \cdots f_2 f_1(x) = 0$.

(3) M satisfies the LSS condition, i.e., if $A = \oplus_{\alpha \in T} A_\alpha$ is a submodule of M such that $\sum_{\alpha \in F} A_\alpha$ is a direct summand of M for any finite subset F of T, then A is a direct summand of M.

(4) M satisfies the (finite) exchange property.

Although this result is essentially due to Harada (see [64]), for more detailed contributions, the reader is referred to Harada and Kanbara [71], Harada and Ishii [70], Yamagata [188] and Zimmerman-Huisgen and Zimmerman [191].

Proposition 8.1.3. *Let R be a Nakayama ring and $\{M_i\}_{i \in \mathbb{N}}$ a family of indecomposable modules. Then the following hold:*

(1) *Each M_i is uniserial and $max\{\,|\,M_i\,|\,\}_{i \in \mathbb{N}} < \infty$.*

(2) *There does not exist a family of monomorphisms $M_1 \xrightarrow{f_1} M_2 \xrightarrow{f_2} M_3 \xrightarrow{f_2} \cdots$ which are not epimorphisms.*

(3) *If $\{\,f_i \colon M_i \to M_{i+1}\,\}_{i \in \mathbb{N}}$ is a family of non-monomorphisms, then there exists $k \in \mathbb{N}$ for which $f_k f_{k-1} \cdots f_1 = 0$.*

(4) *For $j > 1$, if there exists a family $\{\,f_i \colon M_i \to M_{i+1}\,\}_{i=1}^{j}$ of homomorphisms with $f_j f_{j-1} \cdots f_1$ a monomorphism, then f_i is a monomorphism for each $i = 1, \ldots, j$.*

Proof. (1). This is obvious since each M_i is uniserial and its projective cover is isomorphic to eR_R for some $e \in Pi(R)$.

(2). This follows from (1).

(3). Since M_1 is uniserial, there exist $m \in M_1$ and a finite subset $\{r_i\}_{i=1}^t$ of R such that $0 \subsetneq mr_tR \subsetneq mr_{t-1}R \subsetneq \cdots \subsetneq mr_1R \subsetneq mR = M_1$ is its unique composition series. Since $0 \neq \operatorname{Ker} f_1$, there exists $t(1) \in \{1, \ldots, t-1\}$ such that $\operatorname{Ker} f_1 = mr_{t(1)}R$. If $t(1) = 1$, then $f_1(M_1) = 0$. If $t(1) > 1$, then $\operatorname{Ker} f_2 f_1 = mr_{t(2)}R$ for some $t(2) \in \{1, \ldots, t(1) - 1\}$. If $t(2) = 1$, then $f_2 f_1(M_1) = 0$. This procedure stops after a finite number of steps. Thus (3) holds.

(4). It suffices to show this for $j = 2$. Thus assume that $f_2 f_1$ is monomorphic. Then f_1 is trivially monomorphic. Suppose that $\operatorname{Ker} f_2 \neq 0$. Since $\operatorname{Im} f_1$ is an essential submodule of M_2, $\operatorname{Im} f_1 \cap \operatorname{Ker} f_2 \neq 0$. Take $0 \neq f_1(x) \in \operatorname{Im} f_1 \cap \operatorname{Ker} f_2$. Then $f_2 f_1(x) = 0$, whence $x = 0$, a contradiction. Thus $\operatorname{Ker} f_2 = 0$ as desired. $\square$

As a corollary of Proposition 8.1.3, we obtain the following theorem.

Theorem 8.1.4. *Let R be a Nakayama ring and $M = \oplus_{\alpha \in I} M_\alpha$ an infinite direct sum of uniserial right R-modules. Then $\{M_\alpha\}_{\alpha \in I}$ satisfies the condition lsTn (and hence satisfies LSS).*

Lemma 8.1.5. *Let $M = M_1 \oplus M_2$ be a module, $\pi : M = M_1 \oplus M_2 \to M_2$ the projection and A_1, A_2 submodules of M with $A_1 \cap A_2 = 0$. If $A_1 \subseteq_e M_1$ and $\pi(A_2) \subseteq_e M_2$, then $A_1 \oplus A_2 \subseteq_e M_1 \oplus M_2$.*

Proof. Let $\pi : M = M_1 \oplus M_2 \to M_2$ be the projection. Then $A_1 \oplus A_2 \subseteq_e M_1 \oplus \pi(A_2) \subseteq_e M_1 \oplus M_2$. Hence $A_1 \oplus A_2 \subseteq_e M_1 \oplus M_2$. $\square$

Lemma 8.1.6. *Let R be a Nakayama ring and M a lifting right R-module. Then any uniform submodule of M can be essentially extended to a direct summand of M.*

Proof. M can be expressed as a direct sum of uniserial modules by Theorem 8.1.2 and any uniform submodule of M is contained in a finite direct sum of uniserial modules which appear in the direct sum. Hence, by Theorem 1.1.18, it suffices to show that, for any two uniserial modules M_1

and M_2, if $M_1 \oplus M_2$ is a lifting module, then it is an extending module or, equivalently, M_1 and M_2 are relative generalized injective.

Therefore we consider submodules $A_1 \subseteq M_1$ and $A_2 \subseteq M_2$ and a homomorphism $\varphi : A_1 \to A_2$, and show the following:

(1) If $\operatorname{Ker} \varphi \neq 0$, then φ can be extended to $M_1 \to M_2$.
(2) If $\operatorname{Ker} \varphi = 0$, then either φ can be extended to $M_1 \to M_2$ or φ^{-1} can be extended to $M_2 \to M_1$.

(1). We extend φ to a homomorphism $\varphi^* : M_1 \to E(M_2)$. Suppose that $\varphi^*(M_1) \not\subseteq M_2$. Then $\varphi^*(M_1) \supsetneq M_2$. Put $T = \{\, x \in M_1 \mid \varphi^*(x) \in M_2 \,\}$. Then $A \subseteq T \subsetneq M_1$ and $\varphi^*(T) = M_2$. Put $X = \{\, x + \varphi^*(x) \mid x \in T \,\}$. Since $\operatorname{Ker} \varphi^* \subseteq_e X$, X is a uniform submodule of M. Furthermore we see, from $\varphi^*(T) = M_2$, that X is not a small submodule of M. Since M is a lifting module, X must be a direct summand of M. Hence $M = X \oplus M_1$ or $M = X \oplus M_2$ by the exchange property of X. Since $X \cap M_1 \neq 0$, we get $M = X \oplus M_2$, from which we see that $T = M_1$, a contradiction. Thus $\varphi^*(M_1) \subseteq M_2$, and hence $\varphi^*|_{M_1}$ is a desired extension map of φ.

(2). We may assume that $|M_1| \leq |M_2|$. We extend φ to a homomorphism $\varphi^* : M_1 \to E(M_2)$. Then, since $E(M_1)\,(\cong E(M_2)\,)$ is uniserial and $|M_1| \leq |M_2|$, we see that $\varphi^*(M_1) \subseteq M_2$. $\square$

We are now in a position to give module theoretic characterizations of Nakayama rings.

Theorem 8.1.7. *The following are equivalent for a given ring R:*

(1) *Every extending <u>left</u> R-module is a lifting module.*
(2) *Every quasi-injective <u>left</u> R-module is a lifting module.*
(3) *R is a right perfect <u>ring</u> and every lifting <u>right</u> R-module is an extending module.*
(4) *Every quasi-projective <u>right</u> R-module is an extending module.*
(5) *R is a Nakayama ring.*

Since Nakayama rings are left-right symmetric, the following are also equivalent to (1) – (5):

(1′) *Every extending right R-module is a lifting module.*
(2′) *Every quasi-injective right R-module is a lifting module.*
(3′) *R is a left perfect ring and every lifting left R-module is an extending module.*

($4'$)   *Every quasi-projective left R-module is an extending module.*

Proof.   (1) $\Rightarrow$ (2). This is obvious.

(3) $\Rightarrow$ (4). This follows from Theorem 1.2.17.

(2) $\Rightarrow$ (5). By (2), R is a left H-ring, and hence R is a QF-3 ring (see Theorem 3.2.10). To show (5), by Theorem 7.1.2, it suffices to show that R is a left serial ring. Let $e \in Pi(R)$ such that $_RRe$ is injective. Since $S(_RRe)$ is simple and $_RJe$ is quasi-injective, $_RJe$ is a lifting module. Since $_RJe$ is a uniform module, we see that $_RJe$ is a cyclic hollow module, and hence $_RJ^ke$ is cyclic hollow for any $k \in \mathbb{N}$. Thus $_RRe$ is uniserial. Let $f \in Pi(R)$ such that $_RRf$ is not simple. Since R is QF-3, we can write $E(_RRf)$ as a finite direct sum of indecomposable injective projective uniserial modules, say $E(_RRf) = Rx_1 \oplus \cdots \oplus Rx_n$. Let $\pi_i : Rf \to Rx_i$ be the projection for each $i = 1, \ldots, n$. Then $2 \leq |\pi_k(Rf)|$ for some $k \in \{1, \ldots, n\}$. Therefore $\mathrm{Ker}\,\pi_k \subseteq Rf$, and this implies that $_RJf$ is uniserial. Hence $_RRf$ is also uniserial.

(4) $\Rightarrow$ (5). By (4), R is a right co-H-ring. Hence, since R is a QF-3 ring, we suffices to show that R is a right serial ring. Let $e \in Pi(R)$. Since R is right co-H, $S(eR_R)$ is simple. Suppose that $eR \supsetneq S_2(eR_R)$. Since $eR/S(eR_R)$ is quasi-projective, $eR/S(eR_R)$ is an extending module, from which we can see that $S_2(eR_R)/S_1(eR_R)$ is simple. Similarly we can see that $S_k(eR_R)/S_{k-1}(eR_R)$ are simple for any $k = 2, 3, \ldots$. Hence eR_R is right serial as desired.

(5) $\Rightarrow$ (3). Let M be a lifting left R-module and A a submodule of M. We express A as a direct sum of uniserial modules, say $A = \oplus_{\alpha \in I} A_\alpha$. Consider the family Ω of all pairs $(J, \oplus_{\beta \in J} M_\beta)$ such that J is a subset of I and $\{M_\beta\}_{\beta \in J}$ is an independent family of direct summands of M such that $\oplus_{\beta \in J} M_\beta$ is a locally direct summand of M with $\oplus_{\beta \in J} A_\beta \subseteq_e \oplus_{\beta \in J} M_\beta$. Then $\Omega \neq \emptyset$ by Lemma 8.1.6. Using Zorn's Lemma, we can take a maximal element $(J_0, \oplus_{\beta \in J_0} M_\beta)$ in Ω. Then $\oplus_{\beta \in J_0} M_\beta$ is a direct summand of M by Theorem 8.1.4, say $M = (\oplus_{\beta \in J_0} M_\beta) \oplus M'$. Note that M' is also a lifting module. Now let $\pi : M = (\oplus_{\beta \in J_0} M_\beta) \oplus M' \to M'$ be the projection. Suppose that $J_0 \subsetneq I$ and take $\alpha \in I - J_0$. Then we see from $\pi(A_\alpha) \cong A_\alpha$ that $\pi(A_\alpha)$ is a uniform submodule. By Lemma 8.1.6, we have a direct summand M_α of M' such that $\pi(A_\alpha) \subseteq_e M_\alpha$. Then, by Lemma 8.1.5, $(\oplus_{\beta \in J_0} A_\beta) \oplus A_\alpha \subseteq_e (\oplus_{\beta \in I} M_\beta) \oplus M_\alpha \oplus M$. This contradicts the maximality of $(J_0, \oplus_{\beta \in J_0} M_\beta)$.

(5) $\Rightarrow$ (1). Let M be an extending right R-module and A a submodule of M. We show that A can be co-essentially lifted to a direct summand

of M. It suffices to consider the case that A is not a small submodule of M. As above, we may express A as a direct sum of uniserial modules, say $A = \oplus_{\alpha \in I} A_\alpha$. Since M is an extending module, we can easily see that each A_α is a direct summand or a small submodule of M. Using Zorn's Lemma, we can take a maximal subset J_0 of I such that $\oplus_{\beta \in J_0} A_\beta$ is a locally direct summand of M. (Of course, 'maximal' here means that, if $J_0 \subseteq J \subseteq I$ and $\oplus_{\beta \in J} A_\beta$ is a locally direct summand of M, then $J_0 = J$.) By Harada's result mentioned above, $M = (\oplus_{\beta \in J_0} A_\beta) \oplus M'$ for some submodule M'. This implies that $A = (\oplus_{\beta \in J_0} A_\beta) \oplus (M' \cap A)$.

Thus it suffices to show that $M' \cap A$ is a small submodule of M. If $J_0 = I$, then there is nothing to prove. Thus suppose that $I - J_0 \neq \emptyset$. And let $\pi : A = (\oplus_{\beta \in J_0} A_\beta) \oplus (M' \cap A) \to M' \cap A$ be the projection. Take any $\alpha \in I - J_0$. Then $A_\alpha \cong \pi(A_\alpha)$, and hence $\pi(A_\alpha)$ is a uniform submodule. Since M' is an extending module, A_α extends to a direct summand M_α of M'. If $\pi(A_\alpha) = M_\alpha$, then $(\oplus_{\beta \in J_0} A_\beta) \oplus A_\alpha \triangleleft^\oplus M$. But this contradicts the choice of J_0. Thus $\pi(A_\alpha) \subsetneq M_\alpha$, and hence $\pi(A_\alpha)$ is small in M. Accordingly $A = (\oplus_{\beta \in J_0} A_\beta) \oplus (\oplus_{\alpha \in I - J_0} \pi(A_\alpha))$ and $\oplus_{\alpha \in I - J_0} \pi(A_\alpha)$ is a small submodule of M. $\qquad \square$

Corollary 8.1.8. *For a given ring R, the following are equivalent:*

(1) *Every right R-module is an extending module.*
(2) *Every right R-module is a lifting module.*
(3) *R is a Nakayama ring with Jacobson radical square zero.*

Proof. In view of Theorem 8.1.7, we need only show (1) $\Leftrightarrow$ (3).

(1) $\Rightarrow$ (3). Assume that there exists $e \in Pi(R)$ with $eJ^2 \neq 0$. Let $M = eR \oplus (eJ/eJ^2)$. Since M is an extending module, the canonical map $\varphi : eJ \to eJ/eJ^2$ extends to $\varphi^* : eR \to eJ/eJ^2$ by Remark 1.1.17. But, since eJ/eJ^2 is simple, this implies $\operatorname{Ker} \varphi^* = eJ$, and hence $\varphi = 0$, a contradiction.

(3) $\Rightarrow$ (1). Let M be a right R-module and A a submodule of M. Then A can be expressed as a direct sum of uniserial modules $\{A_\alpha\}_{\alpha \in I}$. We note that, since R is right QF-3 and $J^2 = 0$, indecomposable R-modules are either injective or simple. Since M is an extending module, each A_α extends to a direct summand A_α^* of M. By Zorn's Lemma, there exists a maximal subset J of I such that $\oplus_{\beta \in J} A_\beta^*$ is a locally direct summand of M. Then, by Theorem 8.1.4, $X = \oplus_{\beta \in J} A_\beta^*$ is a direct summand of M, say $M = X \oplus Y$. Let $\pi : M \to Y$ be the projection. Then, for any $\alpha \in I - J$,

$\pi(A_\alpha)$ must be simple and $A_\alpha \subsetneq A_\alpha^*$ by the maximality of J. Thus A_α^* is injective for any $\alpha \in I - J$, and hence $Z = \oplus_{\alpha \in I-J} A_\alpha^*$ is injective and direct summand of Y. Thus $A \subseteq_e X \oplus Z \oplus M$ as desired. $\qquad\square$

Remark 8.1.9. Consider the following conditions:

(a) Every projective right R-module is an injective module.

(a') Every injective right R-module is a projective module.

(b) R is a right perfect ring and every lifting right R-module is an extending module.

(b') Every extending right R-module is a lifting module.

(c) Every projective right R-module is an extending module.

(c') Every injective left R-module is a lifting module.

As is well-known, R is a QF-ring $\Leftrightarrow$ (a) $\Leftrightarrow$ (a'). By Theorem 8.1.7, R is a Nakayama ring $\Leftrightarrow$ (b) $\Leftrightarrow$ (b'). By Theorems 3.1.12, 3.2.10 and 3.3.3, R is a left H-ring $\Leftrightarrow$ R is a right co-H-ring $\Leftrightarrow$ (c) $\Leftrightarrow$ (c').

We summarize the conditions in the following:

Figure

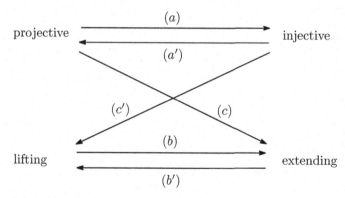

COMMENTS

Most of the material in Chapter 8 is taken from Oshiro [148]. Theorem 8.1.2 is due to Nakayama [133]. This result shows that Nakayama rings are typical examples of rings of finite representation type. For more information on Corollary 8.1.8, the reader is referred to Oshiro-Wisbauer [158], Dung-Smith [40] and Purav-Vanaja [161].

Chapter 9

Nakayama Algebras

This chapter is concerned with Nakayama algebras over algebraically closed fields and Nakayama group algebras. We show the following:

> **Result.** *Let R be a basic indecomposable Nakayama algebra over an algebraically closed field K. Then R can be represented as a factor ring of a skew matrix ring $(Q)_{id,\bar{x},n}$ with $Q = K[x]/(x^{d+1})$, $\sharp Pi(R) = n$ and $|Q| = d + 1$.*

We then apply this result to the study of Nakayama group algebras over algebraically closed fields.

9.1 Nakayama Algebras over Algebraically Closed Fields

By the structure theorems in Section 7.3, all indecomposable basic Nakayama algebras can be constructed as upper staircase factor rings of block extensions of Nakayama QF-algebras. In this section, we give more detailed descriptions of this construction for Nakayama algebras over algebraically closed fields. In this section, we assume that K is an algebraically closed field.

Proposition 9.1.1. *Let Q be an algebra over K and consider any skew matrix ring*

$$R = \begin{pmatrix} Q & \cdots & Q \\ & \cdots & \\ Q & \cdots & Q \end{pmatrix}_{\sigma,c,n} \, ,$$

where σ is a ring endomorphism of Q. We treat K as a subring of R by *the map*

$$k \longmapsto \begin{pmatrix} k & 0 & \cdots & 0 \\ 0 & \ddots & \ddots & \vdots \\ \vdots & \ddots & \ddots & 0 \\ 0 & \cdots & 0 & k \end{pmatrix}.$$

Then the following hold:

(1) *When $n = 1$, R is Q itself.*

(2) *When $n \geq 2$, R becomes a K-algebra if and only if σ is a K-algebra endomorphism, i.e., $\sigma(k) = k$ for any $k \in K$.*

Proof. (1). This is trivial since σ has no impact.

(2). ($\Rightarrow$). Let $\{\alpha_{ij}\}_{i,j \in \{1,\ldots,n\}}$ be the skew matrix units of R. Since $\langle \sigma(k)\alpha_{21} \rangle_{21} = \langle \alpha_{21}k \rangle_{21} = \langle \alpha_{21} \rangle_{21}k = k\langle \alpha_{21} \rangle_{21} = \langle k\alpha_{21} \rangle_{21}$, we see that $\sigma(k)\alpha_{21} = k\alpha_{21}$, whence $\sigma(k) = k$ for any $k \in K$.

($\Leftarrow$). This is obvious. $\qquad\square$

Proposition 9.1.2. *Let Q be a local Nakayama algebra over K. Put $J = J(Q) = cQ$ and let $d + 1 = |Q|$, so $J^d \neq 0$ and $J^{d+1} = 0$. Then $\{1, c, c^2, \ldots, c^d\}$ is an independent set over K and $Q = K \oplus cK \oplus \cdots \oplus c^d K$. Hence Q can be represented as the following factor ring of the polynomial ring $K[x]$:*

$$Q \cong K[x]/(x^{d+1}).$$

Proof. Since any division K-algebra is isomorphic to K and Q is uniserial, we see that $\dim_K(J^i/J^{i+1}) = 1$ for each $i = 1, \ldots, d$, where $\dim_K(X)$ denotes the dimension of the K-vector space X. Hence $\dim_K(Q) = d + 1$. On the other hand, $\{1, c, c^2, \ldots, c^d\}$ is a linearly independent set. Thus we obtain that $Q = \{ k_0 + k_1 c + \cdots + k_d c^d \mid k_i \in K \}$, and hence Q is isomorphic to the factor ring $K[x]/(x^{d+1})$. $\qquad\square$

Proposition 9.1.3. *Let Q be a local Nakayama algebra over K and put $J = J(Q) = cQ$. If σ is an algebra automorphism of Q satisfying $\sigma(c) = c$ and $\sigma(q)c = cq$ for all $q \in Q$, then σ is the identity map id_Q of Q; whence, by Proposition 9.1.2,*

$$(Q)_{id,c,n} = \begin{pmatrix} Q & \cdots & Q \\ & \cdots & \\ Q & \cdots & Q \end{pmatrix}_{id,c,n} \cong \begin{pmatrix} K[x]/(x^{d+1}) & \cdots & K[x]/(x^{d+1}) \\ & \cdots & \\ K[x]/(x^{d+1}) & \cdots & K[x]/(x^{d+1}) \end{pmatrix}_{id,\overline{x},n},$$

where $d + 1 = |Q|$ and $\bar{x} = x + K[x]/(x^{d+1})$.

Proof.   Since $\sigma(c) = c$ and $\sigma(k) = k$ for any $k \in K$, we see from Proposition 9.1.2 that $\sigma(q) = q$ for any $q \in Q$, i.e., σ is the identity map id_Q of Q.   □

Remark 9.1.4. Let Q be a local QF-ring, $\sigma \in \mathrm{Aut}(Q)$ and $c \in J = J(Q)$ satisfying $\sigma(c) = c$ and $\sigma(q)c = cq$ for all $q \in Q$. Set

$$R = \begin{pmatrix} Q & \cdots & Q \\ & \cdots & \\ Q & \cdots & Q \end{pmatrix}_{\sigma, c, n}.$$

Then

$$I = \begin{pmatrix} 0 & S(Q) & \cdots & S(Q) \\ S(Q) & \ddots & \ddots & \vdots \\ \vdots & \ddots & \ddots & S(Q) \\ S(Q) & \cdots & S(Q) & 0 \end{pmatrix}$$

is an ideal of R and R/I is a basic indecomposable QF-ring with the identity Nakayama permutation.

The following theorem shows that the basic indecomposable Nakayama QF-algebras over K of $KNP(1 \to 1)$-type are those algebras mentioned in the remark above.

Theorem 9.1.5. *Let R be a basic indecomposable Nakayama QF-algebra over K with the Kupisch series $e_n R, \ldots, e_1 R$. Suppose that R is $KNP(1 \to 1)$-type, put $Q = e_1 R e_1$ and let $Q = Qc$ and $d+1 = |Q|$. Then the following hold:*

(1) $Q = e_1 R e_1 \cong e_2 R e_2 \cong \cdots \cong e_n R e_n \cong K[x]/(x^{d+1})$.

(2)   *There exists a surjective algebra matrix homomorphism $\varphi = (\varphi_{ij})$:*

$$(Q)_{id_Q, c, n} = \begin{pmatrix} Q & \cdots & Q \\ & \cdots & \\ Q & \cdots & Q \end{pmatrix}_{id, c, n} \longrightarrow R = \begin{pmatrix} e_1 R e_1 & \cdots & e_1 R e_n \\ & \cdots & \\ e_n R e_1 & \cdots & e_n R e_n \end{pmatrix}$$

with

$$\mathrm{Ker}\, \varphi = \begin{pmatrix} 0 & & S(Q) \\ & \ddots & \\ S(Q) & & 0 \end{pmatrix}.$$

Thus R can be represented as

$$R \cong \begin{pmatrix} K[x]/(x^{d+1}) & \cdots & K[x]/(x^{d+1}) \\ & \cdots & \\ & \cdots & \\ K[x]/(x^{d+1}) & & K[x]/(x^{d+1}) \end{pmatrix}_{id,\overline{x},n} \Bigg/ \begin{pmatrix} 0 & & (x^d)/(x^{d+1}) \\ & \ddots & \\ (x^d)/(x^{d+1}) & & 0 \end{pmatrix}$$

$$\overset{put}{=} \begin{pmatrix} K[x]/(x^{d+1}) & K[x]/(x^d) & \cdots & K[x]/(x^d) \\ K[x]/(x^d) & \ddots & \ddots & \vdots \\ \vdots & \ddots & \ddots & K[x]/(x^d) \\ K[x]/(x^d) & \cdots & K[x]/(x^d) & K[x]/(x^{d+1}) \end{pmatrix}_{id,\overline{x},n}.$$

Proof. We represent R as

$$R = \begin{pmatrix} Q & A_{12} & \cdots & \cdots & A_{1n} \\ B_{21} & Q_2 & \ddots & & \cdots \\ \vdots & \ddots & \ddots & \ddots & \vdots \\ \vdots & & \ddots & \ddots & A_{n-1,n} \\ B_{n1} & \cdots & \cdots & B_{n,n-1} & Q_n \end{pmatrix},$$

where $Q_i = e_i R e_i$ for $i = 2, \ldots, n$ and $A_{ij} = e_i R e_j$ and $B_{ji} = e_j R e_i$ for any $i, j \in \{1, \ldots, n\}$ with $i < j$. Since R is $KNP(1 \rightarrow 1)$-type,

$$S(R) = \begin{pmatrix} S(Q) & & & & \\ & S(Q_2) & & \text{\Large 0} & \\ & & \ddots & & \\ & \text{\Large 0} & & \ddots & \\ & & & & S(Q_n) \end{pmatrix}.$$

By Proposition 7.3.1 and Theorem 7.3.3, $\overline{R} = R/S(R)$ is $KNP(1 \rightarrow n)$-type and there exists an algebra matrix isomorphism $\varphi = (\varphi_{ij})$ from $\overline{R}$ to the skew matrix algebra $(\overline{Q})_{id_{\overline{Q}},\overline{c},n}$:

$$(\overline{Q})_{id_{\overline{Q}},\overline{c},n} = \begin{pmatrix} \overline{Q} & \cdots & \overline{Q} \\ & \cdots & \\ \overline{Q} & \cdots & \overline{Q} \end{pmatrix}_{id_{\overline{Q}},\overline{c},n},$$

where $Qc = J(Q)$ and $\overline{c} = c + S(Q) \in \overline{Q} = Q/S(Q)$. Hence $K[x](x^d) \cong \overline{Q} = Q/S(Q) \cong \overline{Q_i} = Q_i/S(Q_i)$ for each $i = 2, \ldots, n$, and hence it follows that $K[x](x^{d+1}) \cong Q \cong Q_i$ for each $i = 1, \ldots, n$. By replacing Q_i by Q, we can represent R and $\overline{R}$ as follows:

$$R = \begin{pmatrix} Q & A_{12} & \cdots & & A_{1n} \\ B_{21} & Q & \ddots & & \vdots \\ \vdots & \ddots & \ddots & & A_{n-1,n} \\ B_{n1} & \cdots & B_{n,n-1} & & Q \end{pmatrix}$$

$$\overline{R} = \begin{pmatrix} \overline{Q} & A_{12} & \cdots & & A_{1n} \\ B_{21} & \overline{Q} & \ddots & & \vdots \\ \vdots & \ddots & \ddots & & A_{n-1,n} \\ B_{n1} & \cdots & B_{n,n-1} & & \overline{Q} \end{pmatrix} \cong \begin{pmatrix} \overline{Q} & \cdots & \overline{Q} \\ & \cdots & \\ \overline{Q} & \cdots & \overline{Q} \end{pmatrix}_{id_{\overline{Q}}, \overline{c}, n} = (\overline{Q})_{id_{\overline{Q}}, \overline{c}, n}$$

We take the skew matrix units $\{\, \alpha_{ij} \mid 1 \leq i < j \leq n \,\} \cup \{\, \beta_{ji} \mid 1 \leq i < j \leq n \,\}$ of $(\overline{Q})_{id_{\overline{Q}}, \overline{c}, n}$. Put $a_{ij} = \varphi_{ij}^{-1}(\alpha_{ij})$, $b_{ji} = \varphi_{ji}^{-1}(\beta_{ji})$ and $a_{ii} = \langle 1 \rangle_{ii}$ in R. Hence R is represented as

$$R = \begin{pmatrix} Q & Qa_{12} & \cdots & & Qa_{1n} \\ Qb_{21} & Q & \ddots & & \vdots \\ \vdots & \ddots & \ddots & & Qa_{n-1,n} \\ Qb_{n1} & \cdots & Qb_{n,n-1} & & Q \end{pmatrix}.$$

We put $t_i = b_{i1}a_{1i}$. Then $\overline{t}_i = \overline{c}$ in $\overline{Q}$ for each $i = 1, \ldots, n$. Consider the algebra automorphism γ_i of Q given by $k_0 + k_1 t_i + \cdots + k_d t_i^d \longmapsto k_0 + k_1 c + \cdots + k_d c^d$. Note that each γ_i induces the identity map of $\overline{Q}$. By these algebra automorphisms γ_i, we replace each Q_i by $Q_i^{\gamma_1}$ and represent R as follows:

$$R = \begin{pmatrix} Q^{\gamma_1} & Q \cdot a_{12} & \cdots & & Q \cdot a_{1n} \\ Q \cdot b_{21} & Q^{\gamma_2} & \ddots & & \vdots \\ \vdots & \ddots & \ddots & & Q \cdot a_{n-1,n} \\ Q \cdot b_{n1} & \cdots & Q \cdot b_{n,n-1} & & Q^{\gamma_n} \end{pmatrix}$$

where the multiplication "$\cdot$" means that $q \cdot a_{ij} = \gamma_i(q)a_{ij}$, $a_{ij} \cdot q = a_{ij}\gamma_j(q)$, $q \cdot b_{ji} = \gamma_j(q)b_{ji}$ and $b_{ji} \cdot q = b_{ji}\gamma_i(q)$ for any $q \in Q$ and $1 \leq i < j \leq n$. By this representation, we note that $a_{12}b_{21} = c$ and $b_{i1}a_{1i} = c$ for $i = 2, \ldots, n$. Moreover we show the following:

(a) (i) $a_{ij}b_{ji} = c$ for $i < j$ and
    (ii) $b_{ji}a_{ij} = c$ for $j < i$.

(b) (i) $q \cdot a_{ij} = qa_{ij}$, $a_{ij} \cdot q = a_{ij}q$ for any $q \in Q$ and
(ii) $q \cdot b_{ji} = qb_{ji}$, $b_{ji} \cdot q = b_{ji}q$ for any $q \in Q$.

We show (a) as follows: $a_{ij}b_{ji} = a_{ij}b_{j1}a_{1i} = b_{i1}a_{1i} = c$ and $b_{ji}a_{ij} = b_{j1}a_{1i}a_{ij} = b_{j1}a_{1j} = c$. To show (b), for $q \in Q$, $q \cdot a_{ij} = \gamma_i(q)a_{ij} = \gamma_i(q)a_{ij} = \bar{q}a_{ij} = qa_{ij}$, so $q \cdot a_{ij} = qa_{ij}$. Similarly the rest of (b) is shown.

By using (a) and (b), we can show the following relations:

$$\begin{cases} a_{ij}q = qa_{ij}, \\ q_{ij}a_{ij}q_{jk}b_{jk} = \begin{cases} q_{ij}q_{jk}ca_{ik} & \text{if } i < k, \\ q_{ij}q_{jk}c & \text{if } i = k, \\ q_{ij}q_{jk}b_{ij} & \text{if } i > k, \end{cases} \\ q_{ij}a_{ij}q_{jk}a_{jk} = q_{ij}q_{jk}a_{ik}, \end{cases} \qquad \begin{cases} b_{ij}q = qb_{ij}, \\ q_{ij}b_{ij}q_{jk}a_{jk} = \begin{cases} q_{ij}q_{jk}ca_{ik} & \text{if } i < k, \\ q_{ij}q_{jk}c & \text{if } i = k, \\ q_{ij}q_{jk}b_{ij} & \text{if } i > k, \end{cases} \\ q_{ij}b_{ij}q_{jk}a_{jk} = q_{ij}q_{jk}cb_{ik}. \end{cases}$$

Consequently the canonical map

$$\varphi : (Q)_{id,c,n} = \begin{pmatrix} Q & \cdots & Q \\ & \cdots & \\ Q & \cdots & Q \end{pmatrix}_{id,c,n} \longrightarrow R = \begin{pmatrix} Q & Qa_{12} & \cdots & Qa_{1n} \\ Qb_{21} & \ddots & & \vdots \\ \vdots & \ddots & \ddots & Qa_{n-1,n} \\ Qb_{n1} & \cdots & Qb_{n,n-1} & Q \end{pmatrix}$$

given by

$$\begin{pmatrix} q_{11} & \cdots & q_{1n} \\ & \cdots & \\ q_{n1} & \cdots & q_{nn} \end{pmatrix} \longmapsto \begin{pmatrix} q_{11} & q_{12}a_{12} & \cdots & q_{1n}a_{1n} \\ q_{21}b_{21} & \ddots & \ddots & \vdots \\ \vdots & \ddots & \ddots & q_{n-1,n}a_{n-1,n} \\ q_{n1}b_{n1} & \cdots & q_{n,n-1}b_{n,n-1} & q_{nn} \end{pmatrix}$$

is a surjective matrix algebra homomorphism and, comparing composition lengths of both (i,j)-positions, we get

$$\operatorname{Ker}\varphi = \begin{pmatrix} 0 & S(Q) & \cdots & S(Q) \\ S(Q) & \ddots & \ddots & \vdots \\ \vdots & \ddots & \ddots & S(Q) \\ S(Q) & \cdots & S(Q) & 0 \end{pmatrix}.$$

Thus we obtain

$$
R \cong
\begin{pmatrix}
Q & \overline{Q} & \cdots & \overline{Q} \\
\overline{Q} & \ddots & \ddots & \vdots \\
\vdots & \ddots & \ddots & \overline{Q} \\
\overline{Q} & \cdots & \overline{Q} & Q
\end{pmatrix}_{id,c,n}
$$

$$
=
\begin{pmatrix}
Q & Q & \cdots & Q \\
Q & \ddots & \ddots & \vdots \\
\vdots & \ddots & \ddots & Q \\
Q & \cdots & Q & Q
\end{pmatrix}_{id,c,n}
\Bigg/
\begin{pmatrix}
0 & S(Q) & \cdots & S(Q) \\
S(Q) & \ddots & \ddots & \vdots \\
\vdots & \ddots & \ddots & S(Q) \\
S(Q) & \cdots & S(Q) & 0
\end{pmatrix}.
$$

$\square$

Theorem 9.1.6. *Every basic indecomposable Nakayama algebra over an algebraically closed field K can be represented as a factor ring of a skew matrix algebra $(Q)_{id,\overline{x},n}$, where $Q = K[x]/(x^{d+1})$, $d+1 = |Q|$ and $\sharp Pi(R) = n$.*

Proof. Let R be a basic indecomposable Nakayama algebra over K with a Kupisch well-indexed set $Pi(R) = \{e_{ij}\}_{i=1,j=1}^{m, n(i)}$ and let $F(R)$ be its frame QF-subring of R. As noted in Remark 7.3.9, $F(R)$ is constructed from a certain subset $\{f_1 = e_{i_1 1}, f_2 = e_{i_2 1}, \ldots, f_y = e_{i_y 1}\} \subseteq \{e_{11}, e_{21}, \ldots, e_{m1}\}$ with $1 = i_1 < i_2 < \cdots < i_y \le m$ by setting $F(R) = (f_1 + \cdots + f_y)R(f_1 + \cdots + f_y)$. Then $F(R)$ is a basic indecomposble Nakayama QF-algebra and $Pi(F(R)) = \{f_i\}_{i=1}^{y}$ is a Kupisch well-indexed set. Further there exists $i \in \{1, \ldots, y\}$ for which

$$
\begin{pmatrix}
f_1 & f_2 & \cdots & f_{y-i+1} & f_{y-i+2} & \cdots & f_y \\
f_i & f_{i+1} & \cdots & f_y & f_1 & \cdots & f_{i-1}
\end{pmatrix}
$$

is the Nakayama permutation of $F(R)$. We put $Q = f_1 R f_1 = K[x]/(x^{d+1})$ and $T = (Q)_{id,\overline{x},y}$. Then, by Theorems 7.3.10 and 9.1.5, there exists a surjective matrix algebra homomorphism $\tau : T \to F(R)$ with its kernel $S_{y-i}(T)$. Now we represent R as an upper staircase factor ring of a suitable block extension $B(F(R)) = F(R)(k(1), \ldots, k(y))$ of $F(R)$ with $n = k(1) + \cdots + k(y)$. Consider $B(T) = F(T)(k(1), \ldots, k(y))$. As is easily seen, τ canonically extends a surjective matrix algebra homomorphism $\tau^* : B(T) \to B(F(R))$. On the other hand, by using Theorems 6.5.1 and 9.1.5, we can take a surjective matrix algebra homomorphism $\eta : (Q)_{id,\overline{x},n} \to B(F(R))$. Hence the composite map $\tau^*\eta : (Q)_{id,\overline{x},n} \to B(F(R))$ is a surjective matrix algebra homomorphism. Accordingly R can be expressed as a factor ring of $(Q)_{id,\overline{x},n}$. $\square$

Remark 9.1.7. By Theorem 9.1.5, if R is a basic indecomposable Nakayama QF-algebra over an algebraically closed field K, all its diagonal rings are isomorphic, i.e., $eRe \cong fRf$ for any $e, f \in Pi(R)$. However, in general for a Nakayama QF-ring with the identity Nakayama permutation, this fact need not be true as we saw in Section 7.4.

9.2 Nakayama Group Algebras

Nakayama group algebras are studied by many group and ring theorists, for example: Brauer [23], Osima [144], Morita [121], Higman [77], Dade [35], Murase [128], [129], Michler [114], Srinivasan [167], Janusz [86], Eisenbud-Griffith [43] and Isaacs [84]. For the history of these algebras, the reader is referred to Faith [47], Puninski [162] and Koshitani [102].

Now, in this section, we shall discuss Nakayama group algebras and their representations by skew matrix algebras.

Let G be a finite group. $|\, G \,|$ denotes the order of G, k means an arbitrary field, K always means an algebraically closed field and p denotes the characteristic of these fields.

The following are key facts for the study on Nakayama group algebras.

Theorem 9.2.1.

(1) (Maschkke [112]) *p is zero or prime to the group order $|\, G \,|$ if and only if the group algebra kG is semisimple.*

(2) (Higman [77], Kasch-Kneser-Kupisch [90]) *Assume that $p > 0$ and p divides $|\, G \,|$. Then the group algebra kG is of finite representation type if and only if G has a cyclic Sylow p-subgroup.*

Thus, if kG is a Nakayama algebra, then G must have a cyclic Sylow p-subgroup. More generally, a block B of kG, i.e., an indecomposable direct summand of kG as an ideal, is of finite representation type if and only if every "defect group" of B is cyclic. It is not so easy to define a defect group but it is a p-subgroup of G. Thus, if G has a cyclic Sylow p-subgroup, then so does the defect group. For details, see, for instance, Alperin [3] and Benson [22].

Now we first begin with summarizing the theory of Nakayama group algebras which was studied by Brauer [23]. Let G be a finite group with a cyclic Sylow p-subgroup. The structure of the group algebra KG is almost determined by a graph called the Brauer tree (cf. Janusz [86]). The Brauer tree T is graph with the following structure:

(1) T is a tree without loops.
(2) There exists an assignment of a positive integer, which is called the multiplicity, to a particular vertex of T, which is called the exceptional vertex.
(3) There exists a circular ordering of the edges emanating from each vertex.

The Brauer tree corresponds to the blocks of KG.

The following figures give examples of Brauer trees:

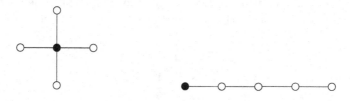

Fig. 9.1   Star and Open Polygon

We represent the exceptional vertex as the black (filled) circle and the other vertices as empty circles. Brauer trees of the two forms in Figure 9.1 are called a star and an open polygon, respectively.

The Brauer tree does not determine the structure of the algebra but it determine the structure of projective modules. Although the Brauer tree T does not determine the structure of the algebra, it exactly determines the Morita equivalence class of the algebra, and hence its projective modules. More specifically, an edge of T corresponds to a simple module S. Next let $P(S)$ be a projective cover of S. Then the heart $J(P(S)/S(P(S)))$ of $P(S)$ is a direct sum of uniserial modules U_1 and U_2. The structure of U_i is determined by the Brauer tree.

For example,

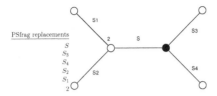

Fig. 9.2 Example

given the simple module S in Figure 9.2, we first consider the right end vertex of the edge labeled S. Rotating round this vertex from S to S, we get the simple components S_1 and S_2 of U_1. Next consider the left end vertex of the edge which is the exceptional vertex with multiplicity 2. Rotating around this vertex twice, we get the simple components S_3, S_4, S, S_3, and S_4 of U_2. Thus the structure of $P(S)$ is as given in Figure 9.3.

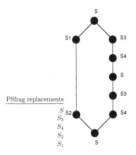

Fig. 9.3 Projective Module

The following gives us some informations on the structure of the Brauer tree (cf. Alperin [3]).

Theorem 9.2.2. *Let T be the Brauer tree of a block B of KG with a cyclic defect group of order p^n. If T has e edges (B has t simple modules), then t divides $p-1$ and the multiplicity of the exceptional vertex is $(p^n - 1)/t$.*

Now we consider the case where the block B is a Nakayama algebra. Here we can easily conclude that the Brauer tree is a star with the exceptional vertex in the center. Then the Brauer tree completely determines the Morita equivalence class of the algebra. Although no group theoretical criterion for B to be a Nakayama algebra is known, we can realize all basic algebras given by the possible forms of the Brauer trees as group algebras by Feit [50].

Fix p^n and t with $t \mid p - 1$. Put $m = (p^n - 1)/t$. Let P be a cyclic group of order p^n. Then the automorphism group of P is an abelian group of order $p^{n-1}(p - 1)$ and it has a unique subgroup H of order e. Define $G = P \rtimes H$ to be the semidirect product of P by H such that H acts on P faithfully. Then it is easy to see that the Brauer tree of KG is a star with exceptional vertex in the center and it has e edges and multiplicity m. Now we describe the algebra by a quiver and relations. The shape of the quiver is as follows: and the relation is $\alpha^{p^n} = 0$, where $\alpha = \sum_{i=1}^{e} \alpha_i$.

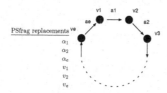

Fig. 9.4   Quiver

We summarize this fact as follows:

Theorem 9.2.3. *Let B be a block of KG. If B is a Nakayama algebra, then there exist n and e with $e \mid p - 1$ such that the following holds: The algebra B is Morita equivalent to the algebra defined by the quiver in Figure 9.4 and the relation $\alpha^{p^n} = 0$, where $\alpha = \sum_{i=1}^{e} \alpha_i$.*

Conversely, if a basic Nakayama algebra A is defined in this way, then it can be realized as a group algebra.

Again we put $G = P \rtimes H$, where we assume that P and H are cyclic groups of order p^n and e, respectively. In this case, KG is a basic indecomposable Nakayama algebra. Actually we shall represent KG as a skew matrix algebra. Let ζ be a primitive e-th root of unity. Put $P = \langle a \rangle$ and $H = \langle b \rangle$. The primitive idempotents of KH are

$$e_i = \frac{1}{|H|} \sum_{j=0}^{t-1} \zeta^{ij} b^j \quad (i = 1, 2, \ldots, t). \qquad (*)$$

Now $1 = \sum_{i=1}^{t} e_i$ is an orthogonal primitive idempotent decomposition in KG. Put $\alpha = a - 1$. Then α is a generator of the Jacobson radical $J(KG)$

of KG. For every i, there exists only one j such that $c_i J(KG)e_j \neq 0$, and this defines a cyclic ordering of the idempotents. Namely we have $\{e_1, e_2, \cdots, e_t\} = \{e_{(1)}, e_{(2)}, \cdots, e_{(t)}\}$ such that $e_{(i)} J(KG)e_{(i+1)} \neq 0$ for $i = 1, \ldots, t-1$ and $e_{(t)} J(KG)e_{(1)} \neq 0$.

Let $Q = K[x]/(x^{m+1})$, where $m = (p^n - 1)/t$. Put

$$
R = \begin{pmatrix} Q & \cdots & Q \\ & \cdots & \\ Q & \cdots & Q \end{pmatrix}_{id, \bar{x}, t} ,
$$

$$
I = \begin{pmatrix} 0 & (\overline{x^m}) & \cdots & \cdots & (\overline{x^m}) \\ (\overline{x^m}) & 0 & \ddots & & \vdots \\ \vdots & \ddots & \ddots & \ddots & \vdots \\ \vdots & & \ddots & 0 & (\overline{x^m}) \\ (\overline{x^m}) & \cdots & \cdots & (\overline{x^m}) & 0 \end{pmatrix}
$$

and $A = R/I$. Let E_i be the (i,i) matrix unit in A and let

$$
M = \begin{pmatrix} 0 & 1 & 0 & \cdots & \cdots & \cdots & 0 \\ \vdots & 0 & 1 & 0 & & & \vdots \\ \vdots & & \ddots & \ddots & \ddots & & \vdots \\ \vdots & & & \ddots & 1 & 0 \\ 0 & & & & & 0 & 1 \\ 1 & 0 & \cdots & \cdots & \cdots & & 0 \end{pmatrix} \in A.
$$

Then the map

$$
\alpha \longmapsto M
$$
$$
e_{(i)} \longmapsto E_i
$$

gives an algebra isomorphism from KG to A. Here we note that $\{e_{(1)}, e_{(2)}, \cdots, e_{(t)}, \alpha\}$ generates the group algebra KG.

Example 9.2.4. Let G be the symmetric group S_3 of permutations on $\{1, 2, 3\}$ and K a field of characteristic $p = 3$. Consider the group algebra $R = KG$. Let $H = \{1, b\}$ and $P = \{1, a, a^2\}$ be the cyclic subgroups of G generated by the permutations $a = (1, 2, 3)$ and $b = (1, 2)$, respectively. Then $G = P \rtimes H$, so, by the above and (*), we see that R is a basic Nakayama algebra with $Pi(R) = \{e_0 = 1 \cdot 1 + 2b, e_1 = 2 \cdot 2 + b\}$. Put $Q = K[x]/(x^2)$ and $\overline{Q} = Q/(x)$. Then

$$
R \cong A = \begin{pmatrix} Q & \overline{Q} \\ \overline{Q} & Q \end{pmatrix}_{id, \bar{x}, 2} .
$$

Put $\alpha = a - 1$ and further put
$$E_0 = \begin{pmatrix} 1 & 0 \\ 0 & 0 \end{pmatrix}, \quad E_1 = \begin{pmatrix} 0 & 0 \\ 0 & 1 \end{pmatrix}, \quad M = \begin{pmatrix} 0 & 1 \\ 1 & 0 \end{pmatrix}.$$

Then the map
$$\alpha \longmapsto M$$
$$e_i \longmapsto E_i$$
gives an algebra isomorphism from R to A.

Following results are featured in Puninski [162].

Fact 9.2.5. (Morita [121], Murase [128]) *Let k be a field of prime characteristic p and G a finite group such that p divides the order of G. If the p-Sylow subgroup is cyclic and normal in G, then kG is a Nakayama algebra.*

Fact 9.2.6. (Morita [121], Murase [129]) *Let k be a field of prime characteristic p. If a finite group G can be expressed as a semi-direct product of its cyclic p-Sylow subgroup and some normal subgroup, then kG is a Nakayama algebra.*

An indecomposable artinian ring R is called *primary* if $R/J(R)$ is a simple ring. Further, for an indecomposable artinian ring R, if there exists an integer $k > 0$ such that every simple R-module occurs exactly k times as a composition factor of $R/J(R)$, then R is called *quasi-primarily decomposable*.

Let G be a finite group and p a prime number. G is called p-nilpotent if G has a normal p'-subgroup H such that G/H is isomorphic to a Sylow p-subgroup P of G (and hence G is a semidirect product of H by P). G is called *p-solvable* if there exists a descending chain of normal subgroups: $G = H_0 \triangleright H_1 \triangleright H_2 \triangleright \cdots \triangleright H_k = \{e\}$ such that H_i/H_{i+1} is a p-group or p'-group (i.e., p does not divide $|H_i/H_{i+1}|$). In particular, G is called p-solvable of p-length one in the case that G has a normal p'-subgroup H such that, for every Sylow p-subgroup P of G, HP is a normal subgroup of G.

Let us record the following results (see Koshitani [102]):

Fact 9.2.7. (Osima [144], Morita [121]) *Let K be an algebraically closed field of prime characteristic p and G a finite group. Then the following are equivalent:*

(1) *For any block A of KG, A is primary.*

(2) *G is a p-nilpotent group.*

Fact 9.2.8. (Morita [121], cf. Michler [114]) *Let R be an indecomposable artinian ring. Then the following are equivalent:*

(1) $J(R) = Rx = yR$ *for some* $x, y \in J(R)$.

(2) *R is a quasi-primarily decomposable Nakayama ring.*

Let K be an algebraically closed field of prime characteristic p, G a finite group with p-Sylow subgroup P and H a largest normal subgroup of G with order prime to p. Then KG has the equivalent conditions above if and only if HP is a normal subgroup of G and P is cyclic.

Fact 9.2.9. (Morita [121], cf. Curtis-Reiner [34]) *Let K be an algebraically closed field of prime characteristic p and G a finite group. Then the following are equivalent:*

(1) *For any block A of KG, it follows that* $dim_K S = dim_K T$ *if S and T are simple KG-module in A.*

(2) *For any simple KG-module S in the principal block* $B_0(KG)$ *of KG, it follows that* $dim_K S = 0$.

(3) *G is p-solvable of p-length one (namely, G has a normal p'-subgroup H such that HP is a normal in G and G/HP is abelian).*

Fact 9.2.10. (Morita [121], cf. Curtis-Reiner (62.29) in [34]) *Let K be an algebraically closed field of prime charactristic p and G a finite group. Then the following are equivalent:*

(1) *KG satisfies the condition (1) in Fact 9.2.9. And moreover KG is a Nakayama algebra.*

(2) *There exist* $a, b \in KG$ *such that* $J(KG) = KGa = bKG$.

(3) *G is p-solvable p-length one and Sylow p-subgroups of G are cyclic.*

The following result is useful.

Fact 9.2.11. (Srinivasan [167]) *Let K be an algebraically closed field of prime characteristic p and G a finite p-solvable group. Then every block of KG with a cyclic defect group is a Nakayama algebra.*

By the results above, we know that Nakayama group algebras are well studied. However we cannot yet obtain a complete description for these algebras, that is, a criterion for KG to be Nakayama algebras are not obtained yet in terms of G and p. Can we find some conditions on the group G together with the condition "cyclic p-Sylow group" which characterize KG to be Nakayama algebras? This problem seems to be interesting.

In the rest of this section, for $n \leq 5$, we shall observe the conditions for which KS_n and KA_n are Nakayama algebras, where S_n and A_n are the symmetric group and alternating group of degree n, respectively.

We note that, if $p = 2$ and $n > 3$, then KS_n and KA_n are not Nakayama algebras since the Klein group is not cyclic and hence Sylow 2-group of these groups are not cyclic.

For $p = 3$, we see that S_3, A_3, S_4 and A_4 have cyclic 3-Sylow group. Therefore KS_3, KA_3, KS_5 and KA_4 are Nakayama algebras by Fact 9.2.11. Since A_5 is non-abelian simple, we note that A_5 and S_5 are not 3-solvable.

Now, in order to know detailed structures of KG for $G = S_n$ or $G = A_n$, we need to observe the following:

(1) The block decomposition $KG = B_1 \oplus \cdots \oplus B_t$.
(2) For each block B_i, all its simple modules.
(3) The K-dimension of each B_i and all K-dimensions of all simple modules over B_i.
(4) Related known facts.

However it seems not so easy to use this algorithm. For $p \geq 2$ and $2 \leq n \leq 5$, if KS_n or KA_n are Nakayama algebras, we shall record their representations by skew-matrix algebras.

[I] $p = 2$:

(1) KS_2 is a local Nakayama algebra with the block decomposition:

$$KS_2 \cong K[x]/(x^2).$$

(2) KS_3 is a Nakayama algebra with the block decomposition:

$$KS_3 \cong K[x]/(x^2) \oplus M_2(K).$$

(3) KA_3 is a Nakayama algebra with the block decomposition:

$$KA_3 \cong K \oplus K \oplus K.$$

[II] $p = 3$:

(1) KS_3 is a basic indecomposable Nakayama algebra with two simple modules with dimension 1, from which we get the block decomposition:

$$KS_3 \cong A = \begin{pmatrix} Q & \overline{Q} \\ \overline{Q} & Q \end{pmatrix}_{id,\overline{x},2},$$

where $Q = K[x]/(x^2)$ and $\overline{Q} = Q/((x)/(x^2))$.

(2) KA_3 is a local Nakayama algebra with the block decomposition:

$$KA_3 \cong K[x]/(x^3).$$

(3) KS_4 is a Nakayama algebra with the block decomposition:

$$KS_4 \cong A \oplus (K)_3 \oplus (K)_3,$$

where A is the algebra in [II](1).

(4) KA_4 is a Nakayama algebra with the block decomposition:

$$KA_4 \cong K[x]/(x^3) \oplus (K)_3.$$

(5) KS_5 is a Nakayama algebra with the block decomposition:

$$KS_5 = B_1 \oplus B_2 \oplus B_3$$

such that B_1 and B_2 are isomorphic blocks with dimension 42, and they have two simple modules with dimensions 1 and 4, respectively, and $B_3 \cong (K)_6$. We see from this situation and Remark 7.5.3 that

$$B_1 \cong B_2 \cong \begin{pmatrix} Q & Q & Q & Q & X \\ Q & Q & Q & Q & X \\ Q & Q & Q & Q & X \\ Q & Q & Q & Q & X \\ Y & Y & Y & Y & T \end{pmatrix},$$

where

$$\begin{pmatrix} Q & X \\ Y & T \end{pmatrix}$$

is A in [II] (1) with $Q = eAe$, $T = fAf$, $X = eAf$ and $Y = fAe$, where

$$e = \begin{pmatrix} 1 & 0 \\ 0 & 0 \end{pmatrix}, \quad f = \begin{pmatrix} 0 & 0 \\ 0 & 1 \end{pmatrix}.$$

(6) KA_5 is a Nakayama algebra with the block decomposition:

$$KA_5 = B_1 \oplus B_2 \oplus B_3,$$

where B_1 is an Nakayama algebra with dimension 42 and has two simple modules with dimensions 1 and 4, and $B_2 \cong B_3 \cong (K)_3$. B_1 is reprersented as

$$B_1 \cong \begin{pmatrix} Q & Q & Q & Q & X \\ Q & Q & Q & Q & X \\ Q & Q & Q & Q & X \\ Q & Q & Q & Q & X \\ Y & Y & Y & Y & T \end{pmatrix},$$

where

$$A = \begin{pmatrix} Q & X \\ Y & Q \end{pmatrix}$$

is A in [II] (1).

[III] $p = 5$:

(1) KS_5 is a not Nakayama algebra with the block decomposition:

$$KS_5 \cong B_1 \oplus (K)_5 \oplus (K)_5,$$

where B_1 is of finite representation type with dimension 70 and has 4 simple modules with dimensions $1, 1, 3, 3$.

(2) KA_5 is a not Nakayama algebra with the block decomposition:

$$KA_5 \cong B_1 \oplus (K)_5,$$

where B_1 is of finite representation type with dimension 35.

COMMENTS

The material in Sections 9.1 and 9.2 is taken from Hanaki-Koshitani-Oshiro [58] and Kositani [102]. For more information on Nakayama group algebras, the reader is referred to Faith [47], Puninski [162], Eisenbud-Griffith [43], and Koshitani [102].

Chapter 10

Local QF-rings

As we saw in the previous chapters, much of the content of this book is based on QF-rings, particularly, on local QF-rings and local Nakayama rings. Consequently, in this closing chapter, we discuss local QF-rings and give a construction of local QF-rings with Jacobson radical cubed zero. From our construction, we can see that there are many non-trivial local QF-rings which are not finite dimensional algebras over fields. Local QF-rings are important for the study of the Faith conjecture since the Faith conjecture is not solved even for local semiprimary rings with Jacobson radical cubed zero.

10.1 Local QF-rings

There are many open problems on QF-rings. Probably, two most famous longstanding unsolved problems are the Nakayama conjecture and the Faith conjecture. One may refer to Nicholson-Yousif [142] for the Faith conjecture as well as for several more recent questions on QF-rings.

The two conjectures are as follows.

The Nakayama conjecture (see Chang Chang Xi [26]). Let R be an artin algebra. If R has a minimal injective resolution $0 \to R_R \to I_1 \to I_2 \to I_2 \to \cdots$, where each I_i is projective, is R a QF-algebra? Nakayama conjectured "yes"in [135].

The Faith conjecture. Is a semiprimary right self-injective ring a QF-ring? Faith conjectured "no" in his book [47].

The following theorem is due to Osofsky.

Fact 10.1.1. ([159]) *A right perfect, right and left self-injective ring is a QF-ring.*

From this fact, we pose the following problem:

Problem 1. Is a one-sided or two-sided perfect one-sided self-injective ring a QF-ring?

In recent years, many people have tried to solve this conjecture. But, regrettably, little progress has been made.

However the following facts give some information on this question.

Fact 10.1.2. (Clark-Huynh [29]) *If R is a semiprimary right self-injective ring with $S_2(R_R)_R$ finitely generated, then R is a QF-ring.*

The following fact is an easy consequence of Baba-Oshiro [20, Theorem 2. B (d)].

Fact 10.1.3. (Baba-Oshiro [20]) *Let R be a semiprimary ring. If R_R is R_R-simple-injective, then R is a right self-injective ring. In particular, if R is a local semiprimary ring with Jacobson radical J cubed zero, then R is right self-injective if and only if both $S(R_R)_R$ and $_RS(_RR)$ are simple and, for any maximal right submodule M of J, there exists $a \in J - J^2$ satisfying $aM = 0$.*

The last part of this fact is shown as follows: We may show that R is right simple injective. In order to show this, let A be a right ideal of R with $A \subseteq J = J(R)$ and φ a non-zero homomorphism of A to $S = S(R_R)_R = {}_RS(_RR)$. Put $\overline{J} = J/S$. Take $0 \neq s \in S$, so $sR = Rs$. If φ is isomorphic, then $A = S$ and $\varphi(s) = us$ for some unit $u \in R$. Hence the left multiplication $(u)_L : R \to R$ is an extension of φ. Suppose that φ is not isomorphic. Put $K = \mathrm{Ker}\,\varphi$. Then K is a maximal submodule of A and $A = K + xR$ for some $x \in R$ such that $\overline{A} = \overline{K} \oplus \overline{xR}$ and $\overline{xR}$ is simple. Let $v \in R$ be a unit satisfying $\varphi(x) = vs$. We take a submodule B of J_R such that $\overline{J} = \overline{A} \oplus \overline{B}$. Then $K + B$ is a maximal submodule of J. Here using the assumption, we can take a in $J - S$ such that $a(K + B) = 0$ but $0 \neq ax \in Rs$. Let w be a unit such that $wax = vs$. Then the left multiplication $(wa)_L : R \to R$ is an extension of φ.

The Faith conjecture is not solved even for a local semiprimary ring with Jacobson radical cubed zero. Thus we record:

Problem 2. Is a semiprimary local right self-injective ring with Jacobson radical cubed zero a QF-ring?

We now provide a careful analysis of Problem 2 and translate this problem into a problem on two-sided vector space over a division ring.

Let R be a local semiprimary ring with $J^2 \neq 0$ and $J^3 = 0$. Let D denote the division ring R/J and put $S = J^2$ and $\overline{J} = J/S$. Then $\overline{J}$ and S are (D, D)-bispaces.

We now record some properties on R.

Fact 10.1.4.

(1) If $_RS$ and S_R are simple, then, by Remark 2.1.11, $r_Rl_R(A) = A$ and $l_Rr_R(B) = B$ for any finitely generated right submodule A and any finitely generated left submodule B of J, $_DS$ and S_D are one-dimensional spaces.

(2) If J_R is finitely generated, $_RS$ and S_R are simple, then R is QF. For this QF-ring R, we can make a new QF-ring T of a graded type as follows: Consider the (D, D)-bispace $T = D \times \overline{J} \times S$. In T, we define a multiplication by setting

$$t_1 t_2 = (\, d_1 d_2, \ d_1 \overline{a_2} + \overline{a_1} d_2, \ d_1 s_2 + s_1 d_2 + a_1 a_2 \,)$$

for $t_1 = (d_1, \overline{a}_1, s_1)$ and $t_2 = (d_2, \overline{a}_2, s_2) \in T$. Then T is a QF-ring with $J(T) = 0 \times \overline{J} \times S$, $J(T)^2 = 0 \times 0 \times S$ and $J(T)^3 = 0$. (In general, $R \not\cong T$.)

Fact 10.1.5. *Assume that R_R is (simple-)injective. Then the following hold:*

(1) $_RS$ and S_R are simple.

(2) For any maximal submodule M_R of J, $aM = 0$ for some $a \in J - S$.

(3) $r_Rl_R(A) = A$ for any "submodule A" of J_R and $l_Rr_R(B) = B$ for any "finitely generated submodule B" of $_RJ$.

(4) Put $J^* = \mathrm{Hom}_R(\overline{J}_R, S_R)$. Then, for any $a \in J$, the map $a \mapsto (a)_L$ (: left multiplication map) gives an (R, R)-isomorphism and a (D, D)-isomorphism: $_R\overline{J}_R \cong {}_RJ_R^*$ and $_D\overline{J}_D \cong {}_DJ_D^*$, respectively.

(5) Put $\alpha = \dim(\overline{J}_D)$. If α is finite, then R is QF, while, if α is infinite, then $\dim(_D\overline{J}) = (\#R)^\alpha = \#R > \alpha$. In particular, if $\alpha = \aleph_0$ and $\#R = \aleph$, then $\dim(_D\overline{J}) = \aleph$.

Most information on R emanates from Facts 10.1.4 and 10.1.5. In particular, (4) and (5) in Fact 10.1.5 are important for investigating Problem 2.

We now give a way of constructing local QF-rings. Let D be a division ring and $_DV_D$ a (D, D)-bispace. We put

$$T = D \times V \times (V \otimes_D V).$$

Then T is a (D, D)-bispace. In T, as above, we define a multiplication as follows:

$$t_1 t_2 = (\, d_1 d_2,\ d_1 v_2 + v_1 d_2,\ d_1 x_2 + x_1 d_2 + v_1 \otimes v_2 \,)$$

for $t_1 = (d_1, v_1, x_1)$ and $t_2 = (d_2, v_2, x_2)$. It is easy to see that T is a local semiprimary ring with Jacobson radical cubed zero and

$$J(T) = 0 \times V \times 0, \qquad J(T)^2 = 0 \times 0 \times V \otimes_D V.$$

We identify $(D \times 0 \times 0)$, $(0 \times V \times 0)$ and $(0 \times 0 \times V \otimes_D V)$ with D, V, and $V \otimes_D V$, respectively, and represent T as $T = \langle\, D, V, V \otimes_D V \,\rangle$.

We note the following result.

Proposition 10.1.6.

(1) *Assume that there exists a (D, D)-bisubspace I of $V \otimes_D V$ such that*

    (i) $\dim\left((V \otimes_D V)/I_D\right) = \dim\left((V \otimes_D V)/_D I\right) = 1$ *and*
    (ii) $vD \otimes_D V \not\subseteq I$ *and* $V \otimes_D Dv \not\subseteq I$ *for any* $0 \neq v \in V$.

    Then I is an ideal of T and $J(T/I)^2 = S(T/I)$ is simple as a left and as a right T/I-module.

(2) *Assume that $dim(V_D)$ is finite and such a (D, D)-bisubspace I satisfying (i) and (ii) exists. Then T/I is a local QF-ring with Jacobson radical cubed zero.*

Proof. (1) is easily seen and (2) follows from Fact 10.1.4.   □

Let pD be a one-dimensional right vector space and let $\rho \in Aut(D)$. Then pD becomes a one-dimensional left vector space by defining $dp = p\rho(d)$ for $d \in D$. We denote such a (D, D)-cyclic bispace by pD^ρ. We also put

$$V^* = \mathrm{Hom}_D(V_D, pD_D^\rho).$$

Then V^* is canonically a (D, D)-bispace. We assume the following:

Assumption A: There exists a (D, D)-isomorphism θ from V to V^*.

Since the map $(V, V) \mapsto pD^\rho$ given by $(v, w) \mapsto \theta(v)(w)$ is a bilinear (D, D)-onto map, the map

$$\lambda : \sum_i v_i \otimes w_i \mapsto \sum_i \theta(v_i)(w_i)$$

is a (D, D)-bispace onto homomorphism of $V \otimes_D V$ to pD^ρ. Put $I = \operatorname{Ker} \lambda$. Then I is an ideal of $T = D \times V \times (V \otimes_D V)$ satisfying (i) and (ii) in Proposition 10.1.9. We put

$$D\langle V, \theta, \rho, pD^\rho \rangle = \overline{T} = T/I.$$

Let $w \in \lambda^{-1}(p)$ be fixed and put $s = w + I \in \overline{V \otimes_D V}$. Then we can show the following result.

Theorem 10.1.7. *Put* $R = D\langle V, \theta, \rho, pD^\rho \rangle$. *Then the following hold:*

(1) $J = J(R) = \overline{V}$, $J^2 = \overline{V \otimes_D V} = Rs = sR$ *and* $J^3 = 0$.
(2) $S(R_R) = S(_RR) = J^2$ *and it is simple as a left and also a right ideal of* R.
(3) R *is a right self-injective ring.*
(4) R *is QF if and only if* $\dim(V_D)$ *is finite.*

In fact, (1), (2) and (4) follow from Proposition 10.1.6. To show (3), let M be a proper maximal submodule of J_R. By the Baba-Oshiro Fact 10.1.3, it suffices to show that there exists $a \in J - J^2$ satisfying $aM = 0$. Let X be a subspace of V_D with $X/I = M$. Then, since X is a proper subspace of V, we can take $v^* \neq 0 \in V^*$ such that $v^*(X) = 0$. Put $a = \overline{\theta^{-1}(v^*)}$. Then $a \in J - J^2$ and $aM = 0$.

By Theorem 10.1.7, we can translate Problem 2 into the following problem:

Problem 3. Do there exist a division ring D and a (D, D)-bispace V such that

$$\dim(V_D) = \infty \text{ and } {}_DV_D \cong {}_DV_D^* \ ((D, D)\text{-isomorphism})?$$

If such a bispace ${}_DV_D$ exists, Theorem 10.1.7 asserts that the Faith conjecture is true, that is, we can construct a semiprimary right self-injective ring which is not QF. However this problem is also extremely difficult. In fact, if we try to solve this, we immediately encounter pathologies. But, as a by-product of the study on Problem 3, we can obtain an important method of constructing local QF-rings. Let us state this construction.

The following is a key lemma.

Lemma 10.1.8. *Let V be a bispace over a division ring D with $n = \dim({}_D V) = \dim(V_D) < \infty$. Then we can take $x_1, \ldots, x_n \in V$ satisfying $V = Dx_1 \oplus \cdots \oplus Dx_n = x_1 D \oplus \cdots \oplus x_n D$.*

Proof. Let $x_1, \ldots, x_k, y, z \in V$ such that $x_1, \ldots, x_k, y$ and $x_1, \ldots, x_k, z$ are left and right independent over D, respectively. If $Dz \cap (\sum_{i=1}^{k} Dx_i) = 0$ or $yD \cap (\sum_{i=1}^{k} x_i D) = 0$, then $x_1, \ldots, x_k, z$ or $x_1, \ldots, x_k, y$ are left and right independent, respectively. If otherwise, i.e., $Dz \subset \sum_{1=1}^{k} Dx_i$ and $yD \subset \sum_{i=1}^{k} x_i D$, then $x_1, \ldots, x_k, y + z$ are left and right independent. By continuing this procedure, the statement is shown. ☐

Now let V_D be a finite dimensional vector space over a division ring D; say

$$V = x_1 D \oplus \cdots \oplus x_n D.$$

We consider a ring homomorphism $\sigma = (\sigma_{ij}) : D \to (D)_n$ defined by

$$d \mapsto \sigma(d) = \begin{pmatrix} \sigma_{11}(d) & \cdots & \sigma_{1n}(d) \\ & \cdots & \\ \sigma_{n1}(d) & \cdots & \sigma_{nn}(d) \end{pmatrix}.$$

By using σ, we define a left D-operation on V as follows: For $d \in D$,

$$dx_i = \sum_{j=1}^{n} x_j \sigma_{ji}(d),$$

namely,

$$d(x_1, \ldots, x_n) = (x_1, \ldots, x_n) \begin{pmatrix} \sigma_{11}(d) & \cdots & \sigma_{1n}(d) \\ & \cdots & \\ \sigma_{n1}(d) & \cdots & \sigma_{nn}(d) \end{pmatrix}.$$

Then V_D becomes a (D, D)-bispace. We denote this bispace by $V\langle x_1, \ldots, x_n; \sigma \rangle$ or simply V^{σ}. We note that pD^{ρ} mentioned above is $pD\langle p; \rho \rangle$.

The following result is a crucial information for making local QF-rings.

Proposition 10.1.9. *The following are equivalent:*

(1) $V^\sigma = Dx_1 \oplus \cdots \oplus Dx_n$, so $Dx_1 \oplus \cdots \oplus Dx_n = x_1 D \oplus \cdots \oplus x_n D$.

(2) *There exists a homomorphism $\xi = (\xi_{ij}) : D \to (D)_n$ such that, for $1 \le i, k \le n$, the following formulas hold:*

$$\textstyle\sum_{j=1}^n \sigma_{kj}(\xi_{ij}(d)) = \begin{cases} d & \text{if } k = i, \\ 0 & \text{if } k \neq i \end{cases}$$

and

$$\textstyle\sum_{j=1}^n \xi_{jk}(\sigma_{ji}(d)) = \begin{cases} d & \text{if } k = i, \\ 0 & \text{if } k \neq i. \end{cases}$$

When this is so,

$$\xi(d) \begin{pmatrix} x_1 \\ \vdots \\ x_n \end{pmatrix} = \begin{pmatrix} x_1 \\ \vdots \\ x_n \end{pmatrix} d$$

for any $d \in D$.

Proof. (1) $\Rightarrow$ (2). Let $\xi = (\xi_{ij}) : D \to (D)_n$ be the ring homomorphism

$$d \mapsto \xi(d) = \begin{pmatrix} \xi_{11}(d) & \cdots & \xi_{1n}(d) \\ & \cdots & \\ \xi_{n1}(d) & \cdots & \xi_{nn}(d) \end{pmatrix}$$

given by

$$\xi(d) \begin{pmatrix} x_1 \\ \vdots \\ x_n \end{pmatrix} = \begin{pmatrix} x_1 \\ \vdots \\ x_n \end{pmatrix} d.$$

Since $\sum_{j=1}^n \xi_{ij}(d)x_j = \sum_{j=1}^n (\sum_{k=1}^n x_k \sigma_{kj}(\xi_{ij}(d))) = x_i \sum_{j=1}^n \sigma_{ij}(\xi_{ij}(d)) = x_i d$, we see that

$$\textstyle\sum_{j=1}^n \sigma_{kj}(\xi_{ij}(d)) = \begin{cases} d & \text{if } k = i, \\ 0 & \text{if } k \neq i \end{cases}$$

and, similarly, the second formula is induced.

(2) $\Rightarrow$ (1). Since $\sum_{j=1}^n \xi_{ij}(d)x_j = \sum_{j=1}^n (\sum_{k=1}^n x_k \sigma_{kj}(\xi_{ij}(d))) = x_i \sum_{j=1}^n \sigma_{ij}(\xi_{ij}(d)) = x_i d$, we see that, for any $d \in D$,

$$\xi(d) \begin{pmatrix} x_1 \\ \vdots \\ x_n \end{pmatrix} = \begin{pmatrix} x_1 \\ \vdots \\ x_n \end{pmatrix} d$$

from which $V^\sigma = Dx_1 + \cdots + Dx_n$.

Next, to show that $\{Dx_1, \ldots, Dx_n\}$ is an independent set, suppose that $d_1 x_1 + \cdots + d_n x_n = 0$ for $d_1, \ldots, d_n \in D$. Since $\sum_{i=1}^{n}(\sum_{j=1}^{n} x_j \sigma_{ji}(d_i)) = 0$, we see that $\sum_{i=1}^{n} \sigma_{ji}(d_i) = 0$ for each $j = 1, \ldots, n$, and hence

$$\sigma_{11}(d_1) + \sigma_{12}(d_2) + \cdots + \sigma_{1n}(d_n) = 0,$$
$$\sigma_{21}(d_1) + \sigma_{22}(d_2) + \cdots + \sigma_{2n}(d_n) = 0,$$
$$\cdots$$
$$\sigma_{n1}(d_1) + \sigma_{n2}(d_2) + \cdots + \sigma_{nn}(d_n) = 0.$$

Hence it follows that

$$\xi_{1j}(\sigma_{11}(d_1)) + \cdots + \xi_{1j}(\sigma_{1j}(d_j)) + \cdots + \xi_{1j}(\sigma_{1n}(d_n))$$
$$+ \xi_{2j}(\sigma_{21}(d_1)) + \cdots + \xi_{2j}(\sigma_{2j}(d_j)) + \cdots + \xi_{2j}(\sigma_{2n}(d_n))$$
$$\cdots$$
$$+ \xi_{nj}(\sigma_{n1}(d_1)) + \cdots + \xi_{nj}(\sigma_{nj}(d_j)) + \cdots + \xi_{nj}(\sigma_{nn}(d_n))$$
$$= 0.$$

This implies that $0 = \sum_{i=1}^{n} \xi_{ij}(\sigma_{ij}(d_j)) = d_j$. Hence $d_j = 0$ for each $j = 1, \cdots, n$ as desired.   □

Under our observation above, we shall give a method of constructing local QF-rings with Jacobson radical cubed zero. We review our argument above. Let V be an n dimensional right vector space over a division ring D; say $V = x_1 D \oplus \cdots \oplus x_n D$. Let $\sigma = (\sigma_{ij})$ and $\xi = (\xi_{ij})$ be ring homomorphisms of D to $(D)_n$ satisfying formulas of Proposition 10.1.9. By defining $d(x_1, \cdots, x_n) = (x_1, \cdots, x_n)(\sigma_{ij}(d))$ for $d \in D$, V becomes a (D, D)-bispace such that $V = Dx_1 \oplus \cdots \oplus Dx_n = x_1 D \oplus \cdots \oplus x_n D$ and for $d \in D$,

$$\begin{pmatrix} \xi_{11}(d) & \cdots & \xi_{1n}(d) \\ & \cdots & \\ \xi_{n1}(d) & \cdots & \xi_{nn}(d) \end{pmatrix} \begin{pmatrix} x_1 \\ \vdots \\ x_n \end{pmatrix} = \begin{pmatrix} x_1 \\ \vdots \\ x_n \end{pmatrix} d.$$

In this case, we denote V by $V\langle x_1, \cdots, x_n; \sigma, \xi \rangle$. For this bispace, we construct T as mentioned above; $T = \langle D, V, V \otimes_D V \rangle$. Further considering a (D, D)-bispace $pD = pD^\rho$ with $\rho \in Aut(D)$, we make $V^* = \mathrm{Hom}_D(V_D, pD_D^\rho)$. Under this situation and Assumption A, i.e., there exists a (D, D)-isomorphism $\theta : {}_D V_D \cong {}_D V_D$, we can construct a local QF-ring $R = D\langle V, \theta, \rho, pD_D^\rho \rangle$ with Jacobson radical cubed zero. To indicate σ and ξ, we denote this ring by $R = D\langle V, \sigma, \xi, \theta, \rho, pD_D^\rho \rangle$. We review that $J = J(R) = \overline{V}$, $J^2 = \overline{V \otimes_D V} = Rs = sR$ and $J^3 = 0$.

Our argument above is reduced to the following theorem.

Theorem   10.1.10. *For a system $\langle V, \sigma, \xi, \rho, pD_D^\rho \rangle$ and a (D, D)-isomorphism $\theta : {}_D V_D \cong {}_D V_D$ (Assumption A), we can make $R = D\langle V, \sigma, \xi, \theta, \rho, pD_D^\rho \rangle$ which is a local QF-ring with Jacobson radical cubed zero.*

We proceed with our argument. Starting from a system $\langle V, \sigma, \xi, \rho, pD_D^\rho \rangle$, let us observe a construction of a (D, D)-bispace isomorphism $\theta : V \to V^*$ under some condition. For each x_i, we take $\alpha_i \in V^*$ satisfying

$$\alpha_i(x_j) = \begin{cases} p & \text{if } i = j, \\ 0 & \text{if } i \neq j. \end{cases}$$

Then ${}_D V_D^* = D\alpha_1 \oplus \cdots \oplus D\alpha_n$. Let $\tau = (\tau_{ij}) : D \to (D)_n$ be a homomorphism satisfying, for $d \in D$,

$$\tau(d) \begin{pmatrix} \alpha_1 \\ \vdots \\ \alpha_n \end{pmatrix} = \begin{pmatrix} \alpha_1 \\ \vdots \\ \alpha_n \end{pmatrix} d.$$

By calculating

$$\left(\begin{pmatrix} \alpha_1 \\ \vdots \\ \alpha_n \end{pmatrix} d \right) (x_1, \cdots, x_n) \quad \text{and} \quad \begin{pmatrix} \alpha_1 \\ \vdots \\ \alpha_n \end{pmatrix} (d (x_1, \cdots, x_n)),$$

we obtain

$$\tau(d) \begin{pmatrix} p & & 0 \\ & \ddots & \\ 0 & & p \end{pmatrix} = \begin{pmatrix} p & & 0 \\ & \ddots & \\ 0 & & p \end{pmatrix} \sigma(d).$$

This implies that $\tau_{ij}(d)p = p\sigma_{ij}(d)$ for any $i, j \in \{1, \ldots, n\}$, whence

$$(\rho\tau_{ij}) = \sigma \quad \cdots (*)$$

Since $\sum_{j=1}^n (\rho\tau_{ij}(\xi_{ij}(d))) = \sum_{j=1}^n \sigma_{ij}(\xi_{ij}(d)) = d$, ρ^{-1} is the map : $d \mapsto \sum_{j=1}^n \tau_{ij}(\xi_{ij}(d))$. By $(*)$, if ρ is id_D, then $\tau = \sigma$. Conversely, if $\tau = \sigma$, then $\rho^{-1}(d) = \sum_{j=1}^n \sigma_{ij}(\xi_{ij}(d)) = d$ for any $d \in D$, that is, $\rho = id_D$.

Hence we obtain the following theorem.

Theorem 10.1.11. *For the QF-ring $R = D\langle V, \sigma, \xi, \theta, \rho, pD^\rho \rangle$, the following hold:*

(1) ρ^{-1} *is determined by* ξ *and* τ *as the map* : $d \mapsto \sum_{j=1}^{n} \tau_{ij}(\xi_{ij}(d))$.
(2) $\tau = \sigma$ *if and only if* $\rho = id_D$.

We consider the case that $\sigma = \xi$, that is, for $1 \le i,\, k \le n$,

$$\sum_{j=1}^{n} \sigma_{kj}(\sigma_{ij}(d)) = \begin{cases} d & \text{if } k = i, \\ 0 & \text{if } k \ne i, \end{cases}$$

$$\sum_{j=1}^{n} \sigma_{jk}(\sigma_{ji}(d)) = \begin{cases} d & \text{if } k = i, \\ 0 & \text{if } k \ne i. \end{cases}$$

Then the map $\theta^* : {}_D V_D \to {}_D V_D^*$ given by $d_1 x_1 + \cdots + d_n x_n \mapsto d_1 \alpha_1 + \cdots + d_n \alpha_n$ is a (D, D)-isomorphism, and hence it follows that $\tau = \sigma = \xi$ and $\rho = id_D$. Therefore we obtain the following theorem.

Theorem 10.1.12. *Let* $V_D = x_1 D \oplus \cdots \oplus x_n D$ *be an* n-*dimensional vector space over a division ring* D *and* $\sigma = (\sigma_{ij}) : D \to (D)_n$ *a homomorphism satisfying the formulas: For* $1 \le i,\, k \le n$,

$$\sum_{j=1}^{n} \sigma_{kj}(\sigma_{ij}(d)) = \begin{cases} d & \text{if } k = i, \\ 0 & \text{if } k \ne i, \end{cases}$$

$$\sum_{j=1}^{n} \sigma_{jk}(\sigma_{ji}(d)) = \begin{cases} d & \text{if } k = i, \\ 0 & \text{if } k \ne i. \end{cases}$$

Then we can make a local QF-ring $D\langle V, \sigma, \sigma, \theta^*, id_D, 1D_D^{id_D} \rangle$.

10.2 Examples of Local *QF*-Rings with Radical Cubed Zero

Example 10.2.1. Let $V = x_1 D \oplus \cdots \oplus x_n D$ be an n-dimensional vector space over a division ring D and π a ring automorphism of D. Consider the ring homomorphism $\sigma : D \to (D)_n$ defined by

$$d \mapsto \begin{pmatrix} \pi(d) & & 0 \\ & \ddots & \\ 0 & & \pi(d) \end{pmatrix}$$

and make the (D, D)-bispace V^σ. Then

$$\begin{pmatrix} \pi^{-1}(d) & & 0 \\ & \ddots & \\ 0 & & \pi^{-1}(d) \end{pmatrix} \begin{pmatrix} x_1 \\ \vdots \\ x_n \end{pmatrix} = \begin{pmatrix} x_1 \\ \vdots \\ x_n \end{pmatrix} d.$$

Let $\xi = (\xi_{ij}) : D \to (D)_n$ be the map given by

$$d \mapsto \begin{pmatrix} \pi^{-1}(d) & & 0 \\ & \ddots & \\ 0 & & \pi^{-1}(d) \end{pmatrix}.$$

Then, by Theorem 10.1.12, we can construct a local QF-ring $D\langle V, \sigma, \xi, \theta^*, \pi^2, pD^{\pi^2} \rangle$.

Example 10.2.2. Let $\mathbb{C}$ be the field of complex numbers and $V = x_1\mathbb{C} \oplus x_2\mathbb{C}$ a 2-dimensional vector space over $\mathbb{C}$. We consider a ring homomorphism $\sigma : \mathbb{C} \to (\mathbb{C})_2$ defined by

$$a + bi \mapsto \begin{pmatrix} a & bi \\ bi & a \end{pmatrix}.$$

Then the map σ satisfies the formulas in Theorem 10.1.12. Hence we can make a local QF-ring $\mathbb{C}\langle V, \sigma, \sigma, \theta^*, id_\mathbb{C}, 1\mathbb{C}^{id_\mathbb{C}} \rangle$.

This example can be slightly generalized as the following:

Example 10.2.3. Let k be a commutative field and let $f(x) = x^n - a \in k[x]$ be irreducible with α a root, $D = k(\alpha)$ and $V = \sum_{i=1}^n \oplus x_i D$. We define a map $\sigma : D \to (D)_n$ by

$$\sum_{i=0}^{n-1} a_i \alpha^i \mapsto \begin{pmatrix} a_0 & a_1\alpha & a_2\alpha^2 & \cdots & a_{n-1}\alpha^{n-1} \\ a_{n-1}\alpha^{n-1} & a_0 & a_1\alpha & \cdots & a_{n-2}\alpha^{n-2} \\ \vdots & \ddots & \ddots & \ddots & \vdots \\ a_2\alpha^2 & \ddots & \ddots & \ddots & a_1\alpha \\ a_1\alpha & a_2\alpha^2 & \cdots & a_{n-1}\alpha^{n-1} & a_0 \end{pmatrix}.$$

Then σ is a ring homomorphism satisfying the formulas in Theorem 10.1.12. Hence, for a given n-dimensional vector space V over D, we can make a local QF-ring $D\langle V, \sigma, \sigma, \theta^*, id_D, 1D^{id_D} \rangle$.

Example 10.2.4. Let $\mathbb{H}$ be the quaternion algebra and let $V = x_1\mathbb{H} \oplus x_2\mathbb{H} \oplus x_3\mathbb{H} \oplus x_4\mathbb{H}$ be a 4-dimensional vector space over $\mathbb{H}$. We consider a ring homomorphism $\sigma : \mathbb{H} \to (\mathbb{H})_4$ defined by

$$a + bi + cj + dk \;\mapsto\; \begin{pmatrix} a & bi & cj & dk \\ bi & a & dk & cj \\ cj & dk & a & bi \\ dk & cj & bi & a \end{pmatrix}.$$

Then σ satisfies the formulas in Theorem 10.1.12. Hence we can make a local QF-ring $\mathbb{H}\langle V, \sigma, \sigma, \theta^*, id_\mathbb{H}, 1\mathbb{H}^{id_\mathbb{H}} \rangle$.

Further, using Theorem 10.1.12, we shall show two constructing ways of local QF-algebras R with radical cubed zero. One of which gives an example of a local QF-algebra which is not a finite dimensional algebra.

Example 10.2.5. Let E be a field and π an automorphism of E satisfying the following two conditions:

(1) $\pi^2 = id_E$.
(2) If $\alpha\pi(\alpha) + \beta\pi(\beta) = 0$ for $\alpha, \beta \in E$, then $\alpha = 0$ and $\beta = 0$.

We define a 2-dimensional vector space over E: Let $D = E \oplus Ei = \{\alpha + \beta i \mid \alpha, \beta \in E\}$ with the products $i^2 = -1$ and $i\alpha = \pi(\alpha)i$ for any $\alpha \in E$ (the addition, as well as the multiplication between elements of E being the natural ones). Then D is a division ring as it can be checked. See the product

$$(\alpha + \beta i)(\pi(\alpha) - \beta i) = \alpha\pi(\alpha) + \beta\pi(\beta)$$

and, if $\alpha + \beta i \neq 0$, then we have $\alpha\pi(\alpha) + \beta\pi(\beta) \neq 0$ by (3). Also the center of D is $K = \{a \in E \mid \pi(a) = a\}$.

Let $V = x_1 D \oplus x_2 D$ be a 2-dimensional vector space over D and consider a ring homomorphism $\sigma : D \to (D)_2$ defined by

$$\alpha + \beta i \;\mapsto\; \begin{pmatrix} \alpha & \beta i \\ \beta i & \alpha \end{pmatrix}.$$

Then σ satisfies the formulas in Theorem 10.1.12. Hence we can make a local QF-ring $R = D\langle V, \sigma, \sigma, \theta^*, id_D, 1D^{id_D} \rangle$.

We shall give some examples of fields E satisfying (1) and (2) above.

(i) Let $E = \mathbb{C}$ or an arbitrary imaginary quadratic field (e.g., $\mathbb{Q}(\sqrt{-3})$). And the map $\pi : E \to E$ defined by $\pi(\alpha) = \overline{\alpha}$, where $\overline{\alpha}$ denotes the conjugate of α.

(ii) Let K be a field and π an automorphism of K satisfying the conditions (1) and (2). Moreover, let $E = K(x)$ be the field of rational functions in x over K. For $f = a_n x^n + \cdots + a_1 x + a_0 \in K[x]$, we put $\overline{f} = \pi(a_n)x^n + \cdots + \pi(a_1)x + \pi(a_0)$. Then the map $\overline{\pi} : E \to E$ defined by $\overline{\pi}(f/g) = \overline{f}/\overline{g}$ is an automorphism of E. We see that the fixed field of $\overline{\pi}$ in E is $F(x)$, where F is the fixed field of π in K and E and $\overline{\pi}$ satisfies (1) and (2) again.

Example 10.2.6. Let E be a division ring such that E is infinite dimensional over its center K and $x^2 \neq -1$ holds for any element $x \in E$. We define a 2-dimensional vector space over E: Let $D = E \oplus Ei = \{\, \alpha + \beta i \mid \alpha, \beta \in E \,\}$. Define the products $i^2 = -1$ and $i\alpha = \alpha i$ for any $\alpha \in E$. Then D becomes a ring (the addition, as well as the multiplication between elements of E being the natural ones). Furthermore D is a division ring. Actually let $d = \alpha + \beta i$ be a non-zero element in D. If $\beta = 0$, then clearly $d^{-1} = \alpha^{-1}$. In case $\beta \neq 0$, it is easily checked that

$$(\alpha + \beta i) \cdot (\beta^{-1}\alpha - i)\beta^{-1}((\alpha\beta^{-1})^2 + 1)^{-1} = 1$$

and

$$((\beta^{-1}\alpha)^2 + 1)^{-1}(\beta^{-1}\alpha - i)\beta^{-1} \cdot (\alpha + \beta i) = 1.$$

This means that d is invertible.

Next let $V = x_1 D \oplus x_2 D$ be a 2-dimensional vector space over D and consider a ring homomorphism $\sigma : D \to (D)_2$ defined by

$$\alpha + \beta i \; \mapsto \; \begin{pmatrix} \alpha & \beta i \\ \beta i & \alpha \end{pmatrix}.$$

Then we see that σ satisfies the formulas in Theorem 10.1.12. Hence we can make a local QF-ring $R = D\langle V, \sigma, \sigma, \theta^*, id_D, 1D^{id_D} \rangle$ and we can see that R is an infinite dimensional algebra with K (its center).

Remark 10.2.7. We shall give an example of a division ring E in Example 10.2.6. Consider the functional field $L = \mathbb{R}(x)$ over the field $\mathbb{R}$ of real numbers and let σ be an into monomorphism of L defined by $\sigma(f(x)/g(x)) = f(x^2)/g(x^2)$. Let $L[y; \sigma]$ be a skew-polynomial ring associated with σ. Although $L[y; \sigma]$ is a non-commutative domain, it has a quotient ring E which is a division ring. As is easily seen, the center of E is $\mathbb{R}$, E is infinite dimensional over $\mathbb{R}$ and we have $a^2 \neq -1$ for any non-zero element $a \in E$.

Example 10.2.8. Let K be a field. We consider free algebra $k\langle x, y \rangle$. Let $q \in k$ and I an ideal of $k\langle x, y \rangle$ generated by x^2, y^2 and $xy - qyx$, i.e., $I = (x^2, y^2, xy - qyx)$. Then $k\langle x, y \rangle / I$ is a local QF-algebra with Jacobson radical cubed zero (cf. Example 1.3.4). This algebra can be obtained by using our method. Actually let $V = xk \oplus yk = kx \oplus ky$ and put $V^* = \mathrm{Hom}_k(V, k)$. For each $x, y \in V$, we take α, $\beta \in V^*$ satisfying $\alpha(x) = 0$, $\alpha(y) = 1$, $\beta(x) = q^{-1}$ and $\beta(y) = 0$. Then $V^* = \alpha K \oplus \beta k = k\alpha \oplus k\beta$. Let $\theta : V \to V^*$ be the isomorphism given by $xk_1 + yk_2 \mapsto \alpha k_1 + \beta k_2$. Let $\sigma = (\sigma_{ij}) : D \to (D)_2$ be a homomorphism given by

$$\sigma(d) = \begin{pmatrix} d & 0 \\ 0 & d \end{pmatrix}.$$

Then we can make a local QF-ring $\langle V, \sigma, \sigma, \theta, id_k, 1K^{id_k} \rangle$. This ring is isomorphic to $k\langle x, y \rangle / I$.

COMMENTS

Most of the material in this chapter is taken from Kikumasa-Oshiro-Yoshimura [94]. For the construction of local semiprimary right self-injective rings in Theorem 10.1.7, the reader is also referred to Yousif-Nicholson [142] in which the treatment is somewhat different. Faith-Huynh [48] provides a detailed description of the Faith conjecture. For more information about the Faith conjecture, the reader is referred to Ara-Park [6], Armendariz-Park [8], Koike [97], Clark-Huynh [28], [29], Nicholson-Yousif [140], [141] and Xue [185].

Open Questions

To complete this volume, we raise the following questions.

Question 1. If R is a perfect right R-simple injective ring, is R right self-injective?

Question 2. Let Q and T be local artinian rings. If there exist Morita dualities Q-$FMod \sim FMod$-T and T-$FMod \sim FMod$-Q, is then Q self-dual?

Remark. When we only assume that Q-$FMod \sim FMod$-T, then, in general, Q is not self-dual as is shown in Xue [183] and [184].

Question 3. Does every basic indecomposable QF-ring R with $\#Pi(R) = 2$ have a Nakayama automorphism? (cf. Example 5.3.2.)

Remark. If the answer of Problem 3 is affirmative, then so is the answer of Problem 2. Because, let $_QA_T$ and $_TB_Q$ be bimodules which induce the dualities Q-$FMod \sim FMod$-T and T-$FMod \sim FMod$-Q, respectively. Consider the ring

$$R = \begin{pmatrix} Q & A \\ B & T \end{pmatrix}$$

with the relations $AB = BA = 0$. Then R is a basic indecomposable QF-ring with a non-identity Nakayama permutation. Thus, if the answer of Problem 4 is affirmative, then R has a Nakayama automorphism, from which we see that $Q \cong T$. Thus Q is self-dual.

Question 4. Let Q be a local Nakayama ring and let τ be an automorphism of $Q/S(Q)$. Does there exist an automorphism σ of Q which induces τ?

Question 5. Let Q be a local Nakayama ring with $|Q| = n$. Does there exist a local Nakayama ring T such that $|T| = n + 1$ and $Q \cong T/S(T)$?

Question 6. Let G be a finite group and K be an algebraic closure of a field k. If kG is a Nakayama algebra, is KG a Nakayama algebra? And how is the converse?

Question 7. Let K be an algebraically closed field. Are there infinitely many KS_n or KA_n which are non-semisimple Nakayama algebras?

Bibliography

[1] H. Abe and M. Hoshino, Frobenius extensions and tilting complexes, Algebras and Representation Theory 11(3) (2008), 215–232.

[2] A. O. Al-attas and N. Vanaja, On $\sum$-extending modules, *Comm. Algebra* 25 (1997), 2365–2393.

[3] J. L. Alperin, "Local Representation Theory", Cambridge Univ. Press, Cambridge (1986).

[4] K. Amdal and F. Ringdal, Catégories unisérialles, *C. R. Acad. Sci. Paris, Série A-B* 267 (1968), A85–A87, B247–B249.

[5] F. W. Anderson and K. R. Fuller, "Rings and Categories of Modules (second edition)," Graduate Texts in Math. 13, Springer-Verlag, Berlin/Heidelberg/New York, 1991.

[6] P. M. Anh, Selfdualities and serial rings *Bull. London Math. Soc.* 38 (2006), 411–420.

[7] P. Ara and J. K. Park, On continuous semiprimary rings, *Comm. Algebra* 19 (1991), 1945–1957.

[8] E. P. Armendariz and J. K. Park, Self-injective rings with restricted chain conditions, *Arch. Math.* 58 (1992), 24–33.

[9] G. Azumaya, A duality theory for injective modules, *Amer. J. Math.* 81 (1959), 249–278.

[10] G. Azumaya, Corrections and supplements to my paper concerning Krull-Remak-Schmidt's theorem, *Nagoya Math. J.* 1 (1950), 117–124.

[11] Y. Baba, Injectivity of quasi-projective modules, projectivity of quasi-injective modules, and projective covers of injective modules, *J. Algebra* 155 (1993), 415–434.

[12] Y. Baba, Some classes of *QF-3* rings, *Comm. Algebra* 28 (2000), 2639–2669.

[13] Y. Baba, On Harada rings and quasi-Harada rings with left global dimension at most 2, *Comm. Algebra* 28 (2000), 2671–2684.

[14] Y. Baba, On self-duality of Auslander rings of local serial rings, *Comm. Algebra* 30 (2002), 2583–2592.

[15] Y. Baba, On quasi-projective modules and quasi-injective modules, *Scientiae Mathematicae Japonicae* 63 (2006), 589–596.

[16] Y. Baba, Self-duality of Harad ring of a component type, *J. of Algebra and its Applications* 7 (2008), 275–298.

[17] Y. Baba, Colocal pairs, Advances in Ring Theory, Proceedings of 5th China-Japan-Korea Symposium on Ring Theory, World Scientific (2008), 173–182.

[18] Y. Baba and K. Iwase, On quasi-Harada rings, *J. Algebra* 185 (1996), 415–434.

[19] Y. Baba and H. Miki, On strongly almost hereditary rings, *Tsukuba J. Math.* 26 (2002), 237–250.

[20] Y. Baba and K. Oshiro, On a Theorem of Fuller, *J. Algebra* 154 (1993), 86–94.

[21] H. Bass, Finitistic dimension and a homological generalization of semiprimary rings, *Trans. Amer. Math. Soc.* 95 (1960), 466–486.

[22] D. J. Benson, "Representations and Cohomology. I", Cambridge Studies in Advanced Mathematics, 30, Cambridge University Press, Cambridge (1991).

[23] R. Brauer, Investigations on group characters, *Ann. Math.* 42 (1941), 936–958.

[24] R. Brauer and C. Nesbitt, On regular representations of algebras, *Proc. Nat. Acad. Sci.* 23 (1937), 236–240.

[25] G. M. Brodskii and R. Wisbauer, On duality theory and AB5* Modules, *J. Pure Appl. Algebra* 121 (1997), 17–27.

[26] C. Xi, On the finitistic conjecture, Advances in Ring Theory, Proceedings of the 4th China-Japan-Korea International Conference on Ring Theory, World Scientific (2003), 282–294.

[27] J. Clark, On a question of Faith in commutative endomorphism rings, *Proc. Amer. Math. Soc.* 98 (1986), 196–198.

[28] J. Clark and D. V. Huynh, A note on perfect self-injective rings, *Quart. J. Math. (Oxford)* 45 (1994), 13–17.

[29] J. Clark and D. V. Huynh, When is a self-injective semiperfect ring quasi-Frobenius?, *J. Algebra* 165 (1994), 531–542.

[30] J. Clark and R. Wisbauer, $\sum$-extending modules, *J. Pure Appl. Algebra* 104 (1995), 19–32.

[31] J. Clark, C. Lomp, N. Vanaja and R. Wisbauer, "Lifting modules", Frontiers in Math., Birkhäuser Boston, Boston, 2006.

[32] R. R. Colby and E. A. Rutter, Jr., Generalizations of *QF-3* algebras, *Trans. Amer. Math. Soc.* 153 (1971), 371–386.

[33] P. Crawley and B. Jónsson, Refinements for infinite direct decompositions of algebraic systems, *Pacific J. Math.* 91(1980), 249–261.

[34] C. W. Curtis and I. Reiner, "Method of Representation Theory" Vol. II, John Wiley & Sons, New York, 1987.

[35] E. C. Dade, Blocks with cyclic defect groups, *Ann. Math.* 84 (1966), 20–48.

[36] P. Dan, Right perfect rings with the extending property on finitely generated free modules, *Osaka J. Math.* 26 (1989), 265–273.

[37] F. Dischinger and W. Müller, Einreihig zerlegbare artinsche Ringe sind

selbstdual, *Arch. Math.* 48 (1984), 132–136.

[38] F. Dischinger and W. Müller, Left PF is not right PF, *Comm. Algebra* 14 (1986), 1223–1227.

[39] Yu. A. Drozd and V. V. Kirichenko, "Finite Dimensional Algebras", Springer-Verlag, Berlin/Heidelberg/New York, 1993.

[40] N. V. Dung and P. F. Smith, Rings for which certain modules are CS, *J. Pure Appl. Algebra* 102 (1995), 273–287.

[41] N. V. Dung, D. V. Huynh, P. F. Smith, and R. Wisbauer, "Extending modules", Pitman Research Notes in Mathematics Series 313, London, 1994.

[42] B. Eckmann and A. Schopf, Über injective Moduln, *Arch Math.* 4 (1953), 75–78.

[43] D. Eisenbud and P. Griffith, Serial rings, *J. Algebra* 17 (1971), 389–400.

[44] G. Eisenbud and J. C. Robson, Hereditary noetherian prime rings, *J. Algebra* (1970), 86–104.

[45] C. Faith and E. A. Walker, Direct sum representations of injective modules, *J. Algebra* 59 (1967), 203–221.

[46] C. Faith, "Algebra I. Rings, Modules, and Categories.", Grundlehren Math. Wiss. 190, Springer-Verlag, Berlin/Heidelberg/New York, 1973.

[47] C. Faith, "Algebra II. Ring Theory", Grundlehren Math. Wiss. 192, Springer-Verlag, Berlin/Heidelberg/New York, 1976.

[48] C. Faith and D. V. Huynh, When self-injective rings are QF: A report on a problem, *J. Algebra Appl.* 1 (2002), 72–105.

[49] C. Faith, "Injective Modules and Injective Quotient Rings", Lecture Notes in Pure and Appl. Math. 72, Marcel Dekker, Basel/New York, 1982.

[50] W. Feit, Possible Brauer trees, *Ill. J. Math.* 28 (1984), 45–56.

[51] W. Von Frobenius, Theori der hyperkomplexen Großben, *Sitzung der phys.-math.* K1(1903), 504-538, 634-645.

[52] L. Fuchs, On quasi-injective modules, *Ann. Scuola Norm. Sup. Pisa* 23 (1969), 541–546.

[53] F. Fujita and Y. Sakai, Frobenius full matrix algebras and Gorenstein tiled orders, *Comm. Algebra* 34 (2006), 1181–1203.

[54] K. R. Fuller, On indecomposable injectives over artinian rings, *Pacific J. Math* 29 (1968), 115–135.

[55] K. R. Goodearl, "Ring Theory", Monographs and Textbooks in Pure and Appl. Math., 33, Dekker, New York, 1976.

[56] J. K. Haack, Self-duality and serial rings, *J. Algebra* 59 (1979), 345–363.

[57] K. Hanada, Y. Kuratomi and K. Oshiro, On direct sums of extending modules and internal exchange property, *J. Algebra* 250 (2002), 115–133.

[58] A. Hanaki, S. Koshitani and K. Oshiro, Nakayama algebras over algebraically closed fields, *preprint.*

[59] T. A. Hannula, The Morita context and the construction of QF rings, Proceedings of the Conference on Orders, Group Rings and Related Topics, Lect. Notes Math. 353, Springer-Verlag, Berlin/Heidelberg/New York (1973), 113–130.

[60] M. Harada, Non-small modules and non-cosmall modules, in "Ring The-

ory", Proceedings of 1978 Antwerp Conference (F. Van Oystaeyen, Ed.) Dekker, New York (1979), 669–690.

[61] M. Harada, On one-sided *QF-2* rings I, *Osaka J. Math.* 17 (1980), 421–431.

[62] M. Harada, On one-sided *QF-2* rings II, *Osaka J. Math.* 17 (1980), 433–438.

[63] M. Harada, On lifting property on direct sums of hollow modules, *Osaka J. Math.* 17 (1980), 783–791.

[64] M. Harada, "Factor Categories with Applications to Direct Decomposition of Modules," Lect. Notes Pure Appl. Math. 88, Dekker, New York, 1983.

[65] M. Harada, On almost relative injective on artinian modules, *Osaka J. Math.* 27 (1990), 963–971.

[66] M. Harada, A note on almost relative projectives and almost relative injectives, *Osaka J. Math.* 29 (1992), 435–446.

[67] M. Harada, Almost *QF* rings and almost $QF^{\#}$ rings, *Osaka J. Math.* 30 (1993), 887–892.

[68] M. Harada, Almost projective modules, *J. Algebra* 154 (1993), 150–159.

[69] M. Harada, On almost relative injective of finite length, *preprint*.

[70] M. Harada and T. Ishii, On perfect rings and the exchange property, *Osaka J. Math.* 12 (1975), 483–491.

[71] M. Harada and H. Kanbara, On categories of projective modules, *Osaka J. Math.* 8 (1971), 471–483.

[72] M. Harada and T. Mabuchi, On almost relative injectives on artinian modules, *Osaka J. Math.* 26 (1989), 837–848.

[73] M. Harada and K. Oshiro, On extending property on direct sums of uniform modules, *Osaka J. Math.* 18 (1981), 767–785.

[74] M. Harada and A. Tozaki, Almost *M*-projectives and Nakayama rings, *J. Algebra* 122 (1989), 447–474.

[75] A. Harmanci and P. F. Smith, Finite direct sums of CS-modules, *Houston J. Math* 19 (1993), 523–532.

[76] D. Herbera and A. Shamsuddin, On self-injective perfect rings, *Canad. Math. Bull.* 39 (1996), 55–58.

[77] D. G. Higman, Indecomposable representations at characteristic p, *Duke Math. J.* 21 (1954), 377–381.

[78] M. Hoshino, Strongly quasi-Frobenius rings, *Comm. Algebra* 28 (2000), 3585–3599.

[79] M. Hoshino and T. Sumioka, Injective pairs in perfect rings, *Osaka J. Math.* 35 (1998), 501–508.

[80] M. Hoshino and T. Sumioka, Colocal pairs in perfect rings, *Osaka J. Math.* 36 (1999), 587–603.

[81] D. V. Huynh and P. Dan, Some characterizations of right co-H-rings, *Math. J. Okayama Univ.* 34 (1992), 165–174.

[82] M. Ikeda, A characterization of quasi-Frobenius rings, *Osaka J. Math.* 4 (1952), 203–209.

[83] M. Ikeda and T. Nakayama, Some characteristic properties of quasi-Frobenius and regular rings, *Proc. Amer. Math. Soc.* 5 (1954), 15–19.

[84] I. M. Isaacs, Lifting Brauer characters of p-solvable groups, *Pacific J. Math.* 53 (1974), 171–188.

[85] K. Iwase, Remarks on QF-2 rings, QF-3 rings and Harada rings, Advances in Ring Theory, Proceedings of 5th China-Japan-Korea Ring Theory Symposium, World Scientific (2008), 210–224.

[86] G. J. Janusz, Indecomposable modules for finite groups, *Ann. Math.* 89 (1969), 209–241.

[87] L. Jeremy, Sur les modules et anneaux quasi-continus, *Canad. Math. Bull.* 17 (1974), 217–228.

[88] J. Kado, The maximal quotient rings of left H-rings, *Osaka J. Math.* 27 (1990), 247–252.

[89] J. Kado and K. Oshiro, Self-duality and Harada rings, *J. Algebra* 211 (1999), 384–408.

[90] F. Kasch, M. Kneser and H. Kupisch, Unzerlegbare modulare Darstellungen endlicher Gruppen mit zyklischer p-Sylow-Gruppen, *Arch. Math.* 8 (1957), 320–321.

[91] T. Kato, Self-injective rings, *Tohoku Math. J.* 19 (1967), 485–494.

[92] D. Keskin On lifting modules, *Comm. Algebra* 28 (2000), 3427–3440.

[93] D. Keskin, Discrete and quasi-discrete modules, *Comm. Algebra* 30 (2002), 5273–5282.

[94] I. Kikumasa, K. Oshiro and H. Yoshimura, A construction of local QF-rings with radical cube zero, Proceedings of the 39th Symposium on Ring Theory, Hiroshima (2007), 36–44.

[95] I. Kikumasa and H. Yoshimura, Commutative QF-algebras with radical cube zero, *Comm. Algebra* 31 (2003), 1837–1858.

[96] I. Kikumasa and H. Yoshimura, Isomorphism classes of QF-algebras with radical cube zero, Advances in Ring Theory, Proceedings of the 4th China-Japan-Korea International Symposium on Ring Theory, World Scientific (2003), 106–117.

[97] K. Koike, On self-injective semiprimary rings, *Comm. Algebra* 28 (2000), 4303–4319.

[98] K. Koike, Examples of QF rings without Nakayama automorphism and H-rings without self-duality, *J. Algebra* 241 (2001), 731–744.

[99] K. Koike, Almost self-duality and Harada rings, *J. Algebra* 254 (2002), 336–361.

[100] K. Koike, Self-duality of quasi-Harada rings and locally distributive rings, *J. Algebra* 302 (2006), 613–645.

[101] K. Koike, Morita duality and recent development, Proceedings of the 5th China-Japan-Korea International Symposium on Ring Theory, World Scientific (2008), 105–115.

[102] S. Koshitani, Some topics on modular group algebras of finite groups, *Proceeding 30th Symp. Ring Theory*, Shinsyu (1997), 81–90.

[103] J. Kraemer, "Characterizations of the existence of (quasi-)selfduality for complete tensor rings", Algebra Berichte 56, München, 1987.

[104] H. Kupisch, Über ein Klasse von Ringen mit Minimalbedingung II, *Arch. Math.* 26 (1975), 23–35.

[105] H. Kupisch, A characterization of Nakayama Rings, *Comm. Algebra* 23 (1995), 739–741.

[106] Y. Kuratomi, On direct sums of lifting modules and internal exxchange property, *Comm. Algebra* 33 (2005), 1795–1804.

[107] T. Y. Lam, "Lectures on Modules and Rings", Graduate Texts in Mathematics 189, Springer-Verlag, Berlin/Heidelberg/New York, 1998.

[108] J. Lambek, "Lectures on Rings and Modules," Blaisdell, Waltham/Toronto/London, 1966.

[109] W. T. H. Loggie, A simplified proof of Oshiro's theorem for co-H-rings, Rocky Mountain J. Math. 28 (1998), 643–648.

[110] T. Mano, The invariant system of serial rings and their applications to the theory of self-duality, *Proceedings of 16th Symp. Ring Theory*, Tokyo (1983), 48–53.

[111] T. Mano, Uniserial rings and skew polynomial rings, *Tokyo J. Math.* 7 (1984), 209–213.

[112] H. Maschke, Über den arithmetischen Character der Koeffizienten der Substitutionen endlicher linearer Substitutiongruppen, *Math. Ann.* 50 (1898), 482–498.

[113] E. Mares, Semiperfect modules, *Math. Z.* 82 (1963), 347–360.

[114] G. O. Michler, Green Correspondence between Blocks with Cyclic Defect groups, *J. Algebra* 39 (1976), 26–51.

[115] S. H. Mohamed and B. J. Müller, "Continuous Modules and Discrete Modules", London Math. Soc. Lect. Notes 147, Cambridge Univ. Press, Cambridge, 1990.

[116] S. H. Mohamed and B. J. Müller, Ojective modules, *Comm. Algebra*, 30 (2002), 1817–1829.

[117] S. H. Mohamed and B. J. Müller, Cojective modules, *Egyptian Math. Sci.* 12 (2004), 83–96.

[118] M. Morimoto and T. Sumioka, Generalizations of theorems of Fuller, *Osaka J. Math.* 34 (1997), 689–701.

[119] M. Morimoto and T. Sumioka, On dual pairs and simple-injective modules, *J. Algebra* 226 (2000), 191–201.

[120] M. Morimoto and T. Sumioka, Semicolocal pairs and finitely cogenerated injective modules, *Osaka J. Math.* 37 (2000), 801–810.

[121] K. Morita, On group rings over a modular field which possess radicals expressible as principal ideals, *Sci. Rep. Tokyo Bunrika Daigaku, Kyoiku Daigaku (Sec.A)* 4 (1951), 177–194.

[122] K. Morita, Duality for modules and its applications to the theory of rings with minimum condition, *Sci. Rep. Tokyo Kyoiku Daigaku* 6 (1958), 83–142.

[123] K. Morita and H. Tachikawa, Character modules, submodules of a free module, and quasi-Frobenius rings, *Math. Z.* 65 (1956), 414–428.

[124] B. J. Müller and S. T. Rizvi, Direct sums of indecomposable modules, *Osaka J. Math.* 21 (1984), 365–374.

[125] I. Murase, On the structure of generalized uniserial rings I, *Sci. Papers College Gen. Educ., Univ. Tokyo* 13 (1963), 1–22.

[126] I. Murase, On the structure of generalized uniserial rings II, *Sci. Papers College Gen. Educ., Univ. Tokyo* 13 (1963), 131–158.

[127] I. Murase, On the structure of generalized uniserial rings III, *Sci. Papers College Gen. Educ., Univ. Tokyo* 14 (1964), 11–25.

[128] I. Murase, Generalized uniserial group rings I, *Sci. Papers College Gen. Educ., Univ. Tokyo* 15 (1965), 15–28.

[129] I. Murase, Generalized uniserial group rings II, *Sci. Papers College Gen. Educ., Univ. Tokyo* 15 (1965), 111–128.

[130] H. Nagao and Y. Tsushima, "Representations of Finite Groups", Academic Press, Boston (1989).

[131] Y. Nagatomi, K. Oshiro, M. Uhara and K. Yamaura, Skew matrix rings and applications to QF-rings, Advances in Ring Theory, Proceedings of the 5th China-Japan-Korea International Symposium on Ring Theory, World Scientific (2008), 247–271.

[132] T. Nakayama, On Frobenius algebras I, *Ann. Math.* 40 (1939), 611–633.

[133] T. Nakayama, On Frobenius algebras II, *Ann. Math.* 42 (1941), 1–21.

[134] T. Nakayama, Note on uniserial and generalized uniserial rings, *Proc. Imp. Acad. Tokyo* 16 (1940), 285–289.

[135] T. Nakayama, On algebras with complete homology, *Abh. Math.Sem, Univ. Hanburg* 22 (1958), 300–307.

[136] T. Nakayama and C. Nesbitt, Note on symmetric algebras *Ann. Math.* 39 (1938), 659-668.

[137] T. Nakayama and G. Azumaya, "Daisuugaku II", *Iwanami*, 1954.

[138] C. J. Nesbitt, On the regular representations of algebras, *Ann. Math.* 39 (1938), 634–658.

[139] C. J. Nesbitt and R. M. Thrall, Some ring theorems with applications to modular representations, *Ann. Math.* 47 (1946), 551–567.

[140] W. K. Nicholson and M. F. Yousif, Mininjective rings, *J. Algebra* 187 (1997), 548–578.

[141] W. K. Nicholson and M. F. Yousif, On perfect simple-injective rings, *Proc. Amer. Math. Soc.* 125 (1997), 979–985.

[142] W. K. Nicholson and M. F. Yousif, "Quasi-Frobenius Rings", Cambridge Tracts in Mathematics 158, Cambridge University Press, Cambridge, 2003.

[143] K. Nonomura, On Nakayama rings, *Comm. Algebra* 32 (2004), 589–598.

[144] M. Osima, On primary decomposable of group rings, *Proc. Phys.-Math. Soc. Japan* 24 (1942),1–9.

[145] K. Oshiro, Continuous modules and quasi-continuous modules, *Osaka J. Math.* 20 (1983), 681–694.

[146] K. Oshiro, Semiperfect modules and quasi-semiperfect modules, *Osaka J. Math.* 20 (1983), 337–372.

[147] K. Oshiro, Lifting modules, extending modules and their applications to QF-rings, *Hokkaido Math. J.* 13 (1984), 310–338.

[148] K. Oshiro, lifting modules, extending modules and their applications to generalized uniserial rings, *Hokkaido Math. J.* 13 (1984), 339–346.

[149] K. Oshiro, Structure of Nakayama rings, Proceedings of 20th Symp. Ring

Theory, Okayama (1987), 109–133.

[150] K. Oshiro, On Harada rings I, *Math. J. Okayama Univ.* 31 (1989), 161–178.

[151] K. Oshiro, On Harada rings II, *Math. J. Okayama Univ.* 31 (1989), 179–188.

[152] K. Oshiro, On Harada rings III, *Math. J. Okayama Univ.* 32 (1990), 111–118.

[153] K. Oshiro, Theories of Harada in Artinian rings, Proceedings of the 3th International Symposium on Ring Theory, Birkhäuser Boston, Boston (2001), 279-328.

[154] K. Oshiro and S. Masumoto, The self-duality of H-rings and Nakayama automorphisms of *QF*-rings, Proceedings of the 18th Symp. Ring Theory, Yamaguchi (1985), 84–107.

[155] K. Oshiro and S. H. Rim, *QF*-rings with cyclic Nakayama permutations, *Osaka J. Math.* 34 (1997), 1–19.

[156] K. Oshiro and S. T. Rizvi, The exchange property of quasi-continuous modules with the finite exchange property, *Osaka J. Math.* 33 (1996), 217–234.

[157] K. Oshiro and K. Shigenaga, On *H*-rings with homogeneous socles, *Math. J. Okayama Univ.* 31 (1989), 189–196.

[158] K. Oshiro and R. Wisbauer, Modules with every subgenerated module lifting, *Osaka J. Math.* 32 (1995), 513–519.

[159] B. Osofsky, A generalization of quasi-Frobenius rings, *J. Algebra* 4 (1966), 373–387.

[160] D. Phan, Right perfect rings with the extending property on finitely generated free modules, *Osaka J. Math.* 26 (1989), 265–273.

[161] V. M. Purav and N. Vanaja, Characterizations of generalized uniserial rings, *Comm. Algebra* 20 (1992), 2253–2270.

[162] G. E. Puninski, "Serial Rings", Kluwer Academic Publishers, 2001.

[163] M. Rayar, "Small and Cosmall Modules," Ph.D. Dissertation, Indiana University, 1971.

[164] S. T. Rizvi, "Contributions to the theory of continuous modules", Ph.D. thesis, McMaster University, 1980.

[165] F. L. Sandomierski, Linearly compact modules and local Morita duality, *Proc. Conf. Ring Theory*, Utah 1971, Academic Press (1972), 333–346.

[166] D. Simson, Dualities and pure semisimple rings, in "Abelian groups, module theory, and topology (Padua, 1997)", 381–388. Dekker, New York (1998).

[167] B. Srinivasan, On the indecomposable representations of certain class of groups, Proc. Lond. Math. Soc. 10 (1960), 497–513.

[168] B. Stenström, "Rings of Quotients", Grundlehren der mathematischen Wisssenschaften, Vol.217. Springer-Verlag, Berlin/Heidelberg/New York, 1975.

[169] H. H. Storrer, Rings of quotients of perfect rings, *Math. Z.* 122 (1971), 151–165.

[170] T. Sumioka and S. Tozaki, On almost *QF*-rings, *Osaka J. Math.* 33 (1986),

649–661.

[171] H. Tachikawa, Duality theorem of character modules for rings with minimal condition, *Math. Z.* 68 (1958), 479–487.

[172] H. Tachikawa, "Quasi-Frobenius Rings and Generalizations", Lect. Notes Math. 351, Springer-Verlag, Berlin/Heidelberg/New York, 1973.

[173] A. Tarski, Sur la decomposition des ensembles en sous-ensembles presque disjoints, *Fund. Math.* 12 (1928), 188–205.

[174] R. M. Thrall, Some generalizations of quasi-Frobenius algebras, *Trans. Amer. Math. Soc.* 64 (1948), 173–183.

[175] S. Tozaki, Characterizations of almost QF rings, *Osaka J. Math.* 36 (1999), 195–201.

[176] Y. Utumi, On continuous regular rings, *Canad. Math. Bull.* 4 (1961), 63–69.

[177] N. Vanaja, Characterization of rings using extending and lifting modules, Ring theory, (Granville, OH), World Sci. Publ., River Edge (1992), 329–342.

[178] N. Vanaja and V. N. Purav, A note on generalized uniserial rings, *Comm. Algebra* 21 (1993), 1153–1159.

[179] R. B. Warfield, Jr., A Krull-Schmidt theorem for infinite sums of modules, *Proc. Amer. Math. Soc.* 22 (1969), 460–465.

[180] R. B. Warfield, Jr., Decompositions of injective modules, *Pacific J. Math.* 31 (1969), 263–276.

[181] J. Waschbüsch, Self-duality of serial rings, *Comm. Algebra* 14 (1986), 581–590.

[182] R. Wisbauer, "Foundations of Module and Ring Theory", Algebra, Logic and Applications Series 3, Gordon and Breach Science Publishers, 1991.

[183] W. Xue, Two examples of local artinian rings, *Proc. Amer. Math. Soc.* 107 (1989), 63–65.

[184] W. Xue, "Rings with Morita Duality", Lect. Notes Math. 1523, Springer-Verlag, Berlin/Heidelberg/New York, 1992.

[185] W. Xue, A note on perfect self-injective rings, *Comm. Algebra* 24 (1996), 749–755.

[186] W. Xue, On a theorem of Fuller, *J. Pure Appl. Algebra* 122 (1997), 159–168.

[187] W. Xue, Characterizations of Morita duality via idempotents for semiperfect rings, *Algebra Colloq.* 5 (1998), 99–110.

[188] K. Yamagata, On projective modules with exchange property, *Sci. Rep. Tokyo Kyoiku Daigaku, Sec. A* 12 (1974), 149–158.

[189] K. Yamagata, "Handbook of algebra", Vol.1, Elsevier Science B.V. (1996), 1841-1887.

[190] Y. Yukimoto, On decomposition of strongly quasi-Frobenius rings, *Comm. Algebra* 28 (2000), 1111–1113.

[191] B. Zimmerman-Huisgen and W. Zimmerman, Classes of modules with the exchange property, *J. Algebra* 88 (1984), 416–434.

Index